Honda
TRX300EX &
TRX400EX
ATV
Owners
Workshop
Manual

by Mike Stubblefield
and John H Haynes
Member of the Guild of Motoring Writers

Models covered:

TRX300EX, 1993 through 1999
TRX400EX, 1999

ABCDE
FGHIJ
KLMNO
PQRST

Haynes Publishing
Sparkford Nr Yeovil
Somerset BA22 7JJ England

Haynes North America, Inc
861 Lawrence Drive
Newbury Park
California 91320 USA

Acknowledgments

Our thanks to Craig Adams, owner of Cal Coast Motorsports, Oxnard, California, for providing the TRX300EX machine used in the production of this manual. Our thanks also to Honda of Milpitas, especially Service Manager Pete Sirett, for providing the TRX400EX machine used in some photographs.

© Haynes North America, Inc. 1999

With permission from J.H. Haynes & Co. Ltd.

A book in the Haynes Owners Workshop Manual Series

Printed in the U.S.A.

ISBN 1 56392 346 7

Library of Congress Catalog Card Number 99-63082

British Library Cataloguing in Publication Data
A catalogue record for this book is available from the British Library

Contents

1999 Honda TRX300EX

About this manual

Its purpose

The purpose of this manual is to help you get the best value from your motorcycle. It can do so in several ways. It can help you decide what work must be done, even if you choose to have it done by a dealer service department or a repair shop; it provides information and procedures for routine maintenance and servicing; and it offers diagnostic and repair procedures to follow when trouble occurs.

We hope you use the manual to tackle the work yourself. For many simpler jobs, doing it yourself may be quicker than arranging an appointment to get the vehicle into a shop and making the trips to leave it and pick it up. More importantly, a lot of money can be saved by avoiding the expense the shop must pass on to you to cover its labor and overhead costs. An added benefit is the sense of satisfaction and accomplishment that you feel after doing the job yourself.

Using the manual

The manual is divided into Chapters. Each Chapter is divided into numbered Sections, which are headed in bold type between horizontal lines. Each Section consists of consecutively numbered paragraphs or steps.

At the beginning of each numbered Section you will be referred to any illustrations which apply to the procedures in that Section. The reference numbers used in illustration captions pinpoint the pertinent Section and the Step within that Section. That is, illustration 3.2 means the illustration refers to Section 3 and Step (or paragraph) 2 within that Section.

Procedures, once described in the text, are not normally repeated. When it's necessary to refer to another Chapter, the reference will be given as Chapter and Section number. Cross references given without use of the word 'Chapter' apply to Sections and/or paragraphs in the same Chapter. For example, 'see Section 8' means in the same Chapter.

References to the left or right side of the vehicle assume you are sitting on the seat, facing forward.

Motorcycle manufacturers continually make changes to specifications and recommendations, and these, when notified, are incorporated into our manuals at the earliest opportunity.

Even though we have prepared this manual with extreme care, neither the publisher nor the authors can accept responsibility for any errors in, or omissions from, the information given.

NOTE

A **Note** provides information necessary to properly complete a procedure or information which will make the procedure easier to understand.

CAUTION

A **Caution** provides a special procedure or special steps which must be taken while completing the procedure where the Caution is found. Not heeding a Caution can result in damage to the assembly being worked on.

WARNING

A **Warning** provides a special procedure or special steps which must be taken while completing the procedure where the Warning is found. Not heeding a Warning can result in personal injury.

Introduction to the Honda TRX300EX and TRX400EX

The Honda TRX300EX and TRX400EX are highly successful and popular sport all-terrain vehicles.

The engine on both models is an air-cooled single-cylinder four-stroke with a single overhead camshaft. The TRX400EX employs Honda's Radial Four-Valve Combustion Chamber (RFVC) valve arrangement. Fuel is delivered to the cylinder by a single Keihin carburetor on all models.

The front suspension is a double wishbone design with a coil spring/shock absorber unit; the rear suspension consists of a swingarm, a shock absorber and, on the TRX400EX model, the links that connect the two.

The braking system consists of a disc brake at each front wheel and a single disc on the rear axle. All three brakes are hydraulically actuated.

All models are chain-driven.

Identification numbers

The frame serial number is stamped into the front of the frame **(see illustration)**. The vehicle identification number (VIN) is located on the front of the frame **(see illustration)**, behind and to the left of the frame serial number. The engine number is stamped into a pad, which is located on the right side of the crankcase **(see illustration)**. The carburetor identification number is stamped on the left side of the carburetor. The frame serial number and the engine serial number should be recorded and kept in a safe place so they can be furnished to law enforcement officials in the event of a theft.

The frame serial number, engine serial number and carburetor identification number should also be kept in a handy place (such as with your driver's license) so they are always available when purchasing or ordering parts for your machine.

The models covered by this manual are as follows:
Honda TRX300EX, 1993 through 1999
Honda TRX400EX, 1998 and 1999

Identifying model years

The procedures in this manual identify the vehicles by model year. The model year is included in a decal on the frame.

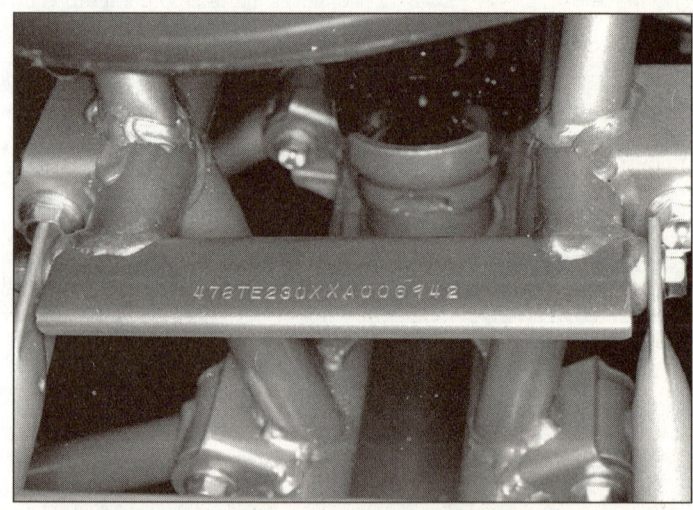

The frame serial number is located on the front of the frame, right behind the bumper

The VIN number is located on the frame, behind and to the left of the frame serial number

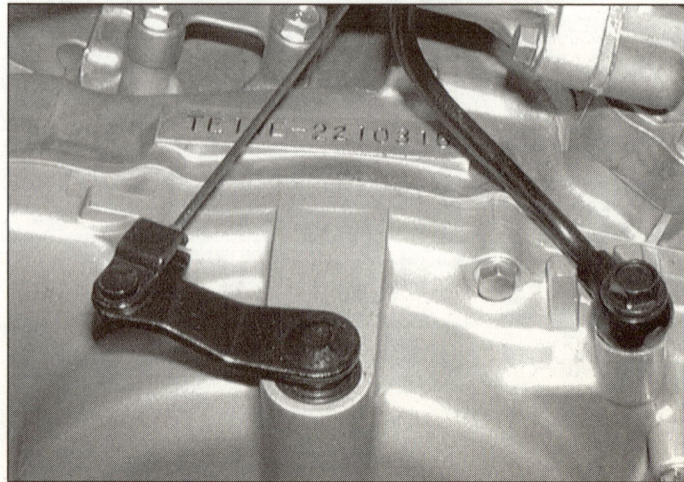

The engine serial number is located on the right side of the crankcase

Buying parts

Once you have found all the identification numbers, record them for reference when buying parts. Since the manufacturers change specifications, parts and vendors (companies that manufacture various components on the machine), providing the ID numbers is the only way to be reasonably sure that you are buying the correct parts.

Whenever possible, take the worn part to the dealer so direct comparison with the new component can be made. Along the trail from the manufacturer to the parts shelf, there are numerous places that the part can end up with the wrong number or be listed incorrectly.

The two places to purchase new parts for your motorcycle - the accessory store and the franchised dealer - differ in the type of parts they carry. While dealers can obtain virtually every part for your motor-cycle, the accessory dealer is usually limited to normal high wear items such as shock absorbers, tune-up parts, various engine gaskets, cables, chains, brake parts, etc. Rarely will an accessory outlet have major suspension components, cylinders, transmission gears, or cases.

Used parts can be obtained for roughly half the price of new ones, but you can't always be sure of what you're getting. Once again, take your worn part to the wrecking yard (breaker) for direct comparison.

Whether buying new, used or rebuilt parts, the best course is to deal directly with someone who specializes in parts for your particular make.

General specifications

Wheelbase
  TRX300EX.. 1150 mm (45.3 inches)
  TRX400EX.. 1230 mm (48.4 inches)
Overall length
  TRX300EX.. 1720 mm (67.7 inches)
  TRX400EX.. 1835 mm (72.2 inches)
Overall width
  TRX300EX.. 1105 mm (43.5 inches)
  TRX400EX.. 1150 mm (45.3 inches)
Overall height
  TRX300EX.. 1085 mm (42.7 inches)
  TRX400EX.. 1110 mm (43.7 inches)
Seat height
  TRX300EX.. 770 mm (30.3 inches)
  TRX400EX.. 810 mm (31.9 inches)
Ground clearance
  TRX300EX.. 125 mm (4.9 inches)
  TRX400EX.. 110 mm (4.3 inches)
Weight with oil and full fuel tank
  TRX300EX.. 177 kg (390 lbs)
  TRX400EX.. 178 kg (392 lbs)

Maintenance techniques, tools and working facilities

Basic maintenance techniques

There are a number of techniques involved in maintenance and repair that will be referred to throughout this manual. Application of these techniques will enable the amateur mechanic to be more efficient, better organized and capable of performing the various tasks properly, which will ensure that the repair job is thorough and complete.

Fastening systems

Fasteners, basically, are nuts, bolts and screws used to hold two or more parts together. There are a few things to keep in mind when working with fasteners. Almost all of them use a locking device of some type (either a lock washer, locknut, locking tab or thread adhesive). All threaded fasteners should be clean, straight, have undamaged threads and undamaged corners on the hex head where the wrench fits. Develop the habit of replacing all damaged nuts and bolts with new ones.

Rusted nuts and bolts should be treated with a penetrating oil to ease removal and prevent breakage. Some mechanics use turpentine in a spout type oil can, which works quite well. After applying the rust penetrant, let it -work for a few minutes before trying to loosen the nut or bolt. Badly rusted fasteners may have to be chiseled off or removed with a special nut breaker, available at tool stores.

If a bolt or stud breaks off in an assembly, it can be drilled out and removed with a special tool called an E-Z out (or screw extractor). Most dealer service departments and motorcycle repair shops can perform this task, as well as others (such as the repair of threaded holes that have been stripped out).

Flat washers and lock washers, when removed from an assembly, should always be replaced exactly as removed. Replace any damaged washers with new ones. Always use a flat washer between a lock washer and any soft metal surface (such as aluminum), thin sheet metal or plastic. Special locknuts can only be used once or twice before they lose their locking ability and must be replaced.

Tightening sequences and procedures

When threaded fasteners are tightened, they are often tightened to a specific torque value (torque is basically a twisting force). Over-tightening the fastener can weaken it and cause it to break, while under-tightening can cause it to eventually come loose. Each bolt, depending on the material it's made of, the diameter of its shank and the material it is threaded into, has a specific torque value, which is noted in the Specifications. Be sure to follow the torque recommendations closely.

Fasteners laid out in a pattern (i.e. cylinder head bolts, engine case bolts, etc.) must be loosened or tightened in a sequence to avoid warping the component. Initially, the bolts/nuts should go on finger tight only. Next, they should be tightened one full turn each, in a criss-cross or diagonal pattern. After each one has been tightened one full turn, return to the first one tightened and tighten them all one half turn, following the same pattern. Finally, tighten each of them one quarter turn at a time until each fastener has been tightened to the proper torque. To loosen and remove the fasteners the procedure would be reversed.

Disassembly sequence

Component disassembly should be done with care and purpose to help ensure that the parts go back together properly during reassembly. Always keep track of the sequence in which parts are removed. Take note of special characteristics or marks on parts that can be installed more than one way (such as a grooved thrust washer on a shaft). It's a good idea to lay the disassembled parts out on a clean surface in the order that they were removed. It may also be help-

ful to make sketches or take instant photos of components before removal.

When removing fasteners from a component, keep track of their locations. Sometimes threading a bolt back in a part, or putting the washers and nut back on a stud, can prevent mix-ups later. If nuts and bolts can't be returned to their original locations, they should be kept in a compartmented box or a series of small boxes. A cupcake or muffin tin is ideal for this purpose, since each cavity can hold the bolts and nuts from a particular area (i.e. engine case bolts, valve cover bolts, engine mount bolts, etc.). A pan of this type is especially helpful when working on assemblies with very small parts (such as the carburetors and the valve train). The cavities can be marked with paint or tape to identify the contents.

Whenever wiring looms, harnesses or connectors are separated, it's a good idea to identify the two halves with numbered pieces of masking tape so they can be easily reconnected.

Gasket sealing surfaces

Throughout any motorcycle, gaskets are used to seal the mating surfaces between components and keep lubricants, fluids, vacuum or pressure contained in an assembly.

Many times these gaskets are coated with a liquid or paste type gasket sealing compound before assembly. Age, heat and pressure can sometimes cause the two parts to stick together so tightly that they are very difficult to separate. In most cases, the part can be loosened by striking it with a soft-faced hammer near the mating surfaces. A regular hammer can be used if a block of wood is placed between the hammer and the part. Do not hammer on cast parts or parts that could be easily damaged. With any particularly stubborn part, always recheck to make sure that every fastener has been removed.

Avoid using a screwdriver or bar to pry apart components, as they can easily mar the gasket sealing surfaces of the parts (which must remain smooth). If prying is absolutely necessary, use a piece of wood, but keep in mind that extra clean-up will be necessary if the wood splinters.

After the parts are separated, the old gasket must be carefully scraped off and the gasket surfaces cleaned. Stubborn gasket material can be soaked with a gasket remover (available in aerosol cans) to soften it so it can be easily scraped off. A scraper can be fashioned from a piece of copper tubing by flattening and sharpening one end. Copper is recommended because it is usually softer than the surfaces to be scraped, which reduces the chance of gouging the part. Some gaskets can be removed with a wire brush, but regardless of the method used, the mating surfaces must be left clean and smooth. If for some reason the gasket surface is gouged, then a gasket sealer thick enough to fill scratches will have to be used during reassembly of the components. For most applications, a non-drying (or semi-drying) gasket sealer is best.

Hose removal tips

Hose removal precautions closely parallel gasket removal precautions. Avoid scratching or gouging the surface that the hose mates against or the connection may leak. Because of various chemical reactions, the rubber in hoses can bond itself to the metal spigot that the hose fits over. To remove a hose, first loosen the hose clamps that secure it to the spigot. Then, with slip joint pliers, grab the hose at the clamp and rotate it around the spigot. Work it back and forth until it is completely free, then pull it off (silicone or other lubricants will ease removal if they can be applied between the hose and the outside of the spigot). Apply the same lubricant to the inside of the hose and the outside of the spigot to simplify installation.

If a hose clamp is broken or damaged, do not reuse it. Also, do not reuse hoses that are cracked, split or torn.

Spark plug gap adjusting tool

Feeler gauge set

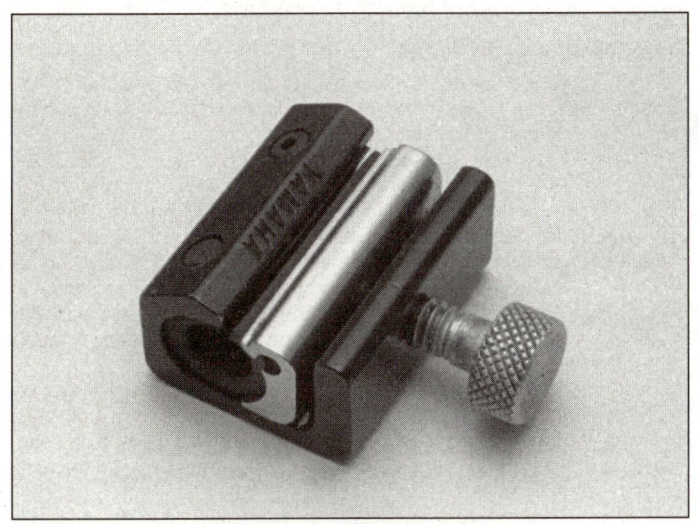

Control cable pressure luber

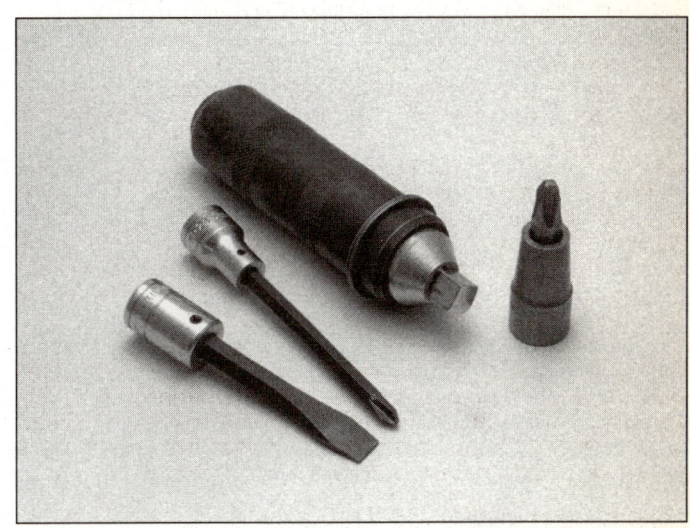

Hand impact screwdriver and bits

Tools

A selection of good tools is a basic requirement for anyone who plans to maintain and repair a motorcycle. For the owner who has few tools, if any, the initial investment might seem high, but when compared to the spiraling costs of routine maintenance and repair, it is a wise one.

To help the owner decide which tools are needed to perform the tasks detailed in this manual, the following tool lists are offered: *Maintenance and minor repair*, *Repair and overhaul* and *Special*. The newcomer to practical mechanics should start off with the *Maintenance and minor repair* tool kit, which is adequate for the simpler jobs. Then, as confidence and experience grow, the owner can tackle more difficult tasks, buying additional tools as they are needed. Eventually the basic kit will be built into the *Repair and overhaul* tool set. Over a period of time, the experienced do-it-yourselfer will assemble a tool set complete enough for most repair and overhaul procedures and will add tools from the *Special* category when it is felt that the expense is justified by the frequency of use.

Maintenance and minor repair tool kit

The tools in this list should be considered the minimum required for performance of routine maintenance, servicing and minor repair work. We recommend the purchase of combination wrenches (box end and open end combined in one wrench); while more expensive than

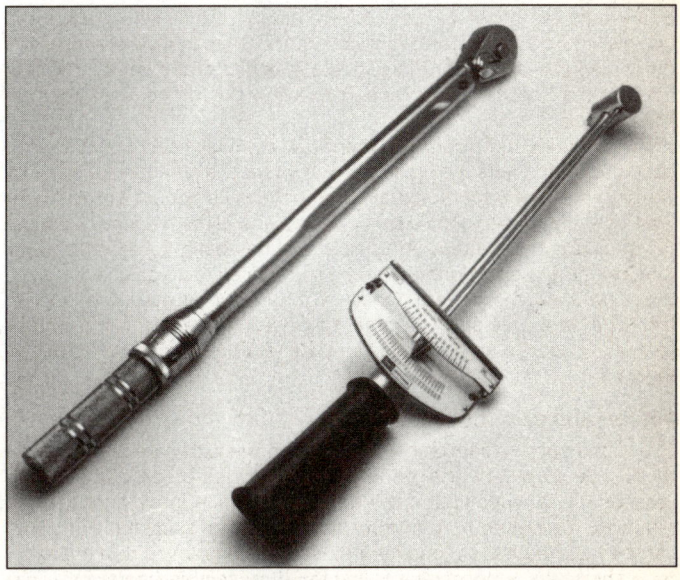

Torque wrenches (left - click; right - beam type)

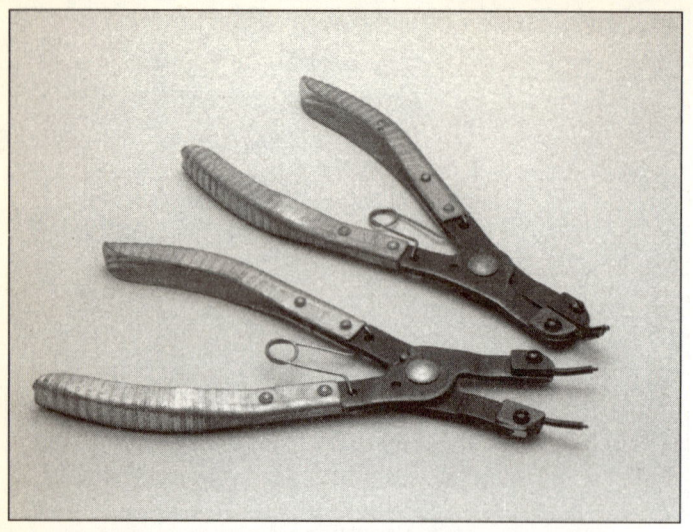

Snap-ring pliers (top - external; bottom - internal)

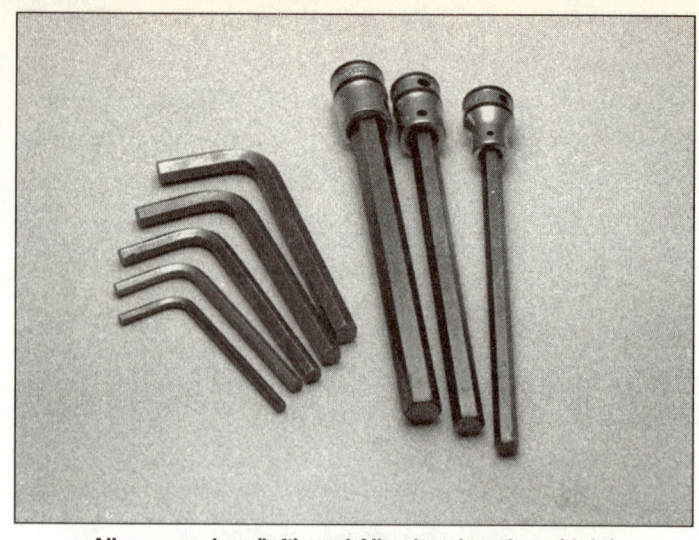

Allen wrenches (left), and Allen head sockets (right)

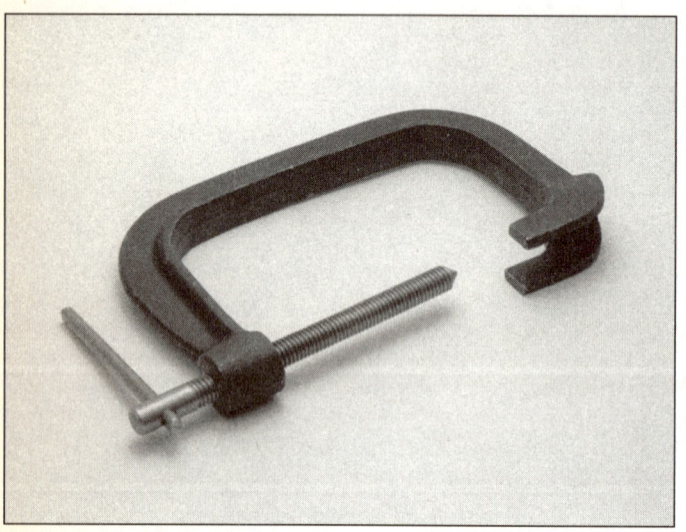

Valve spring compressor

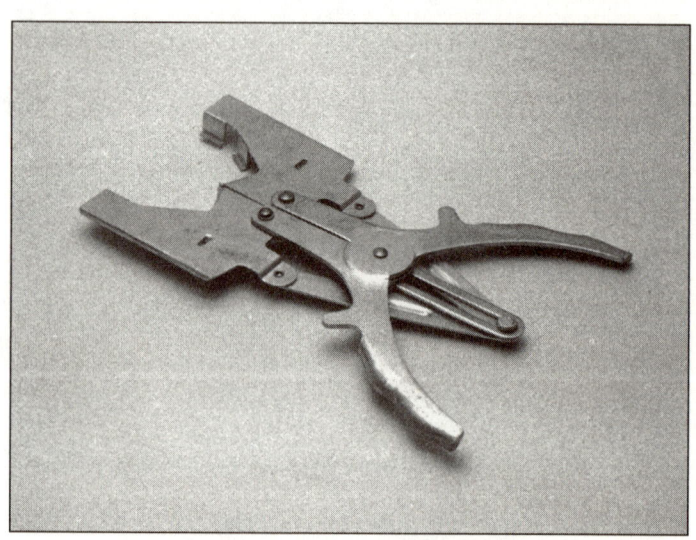

Piston ring removal/installation tool

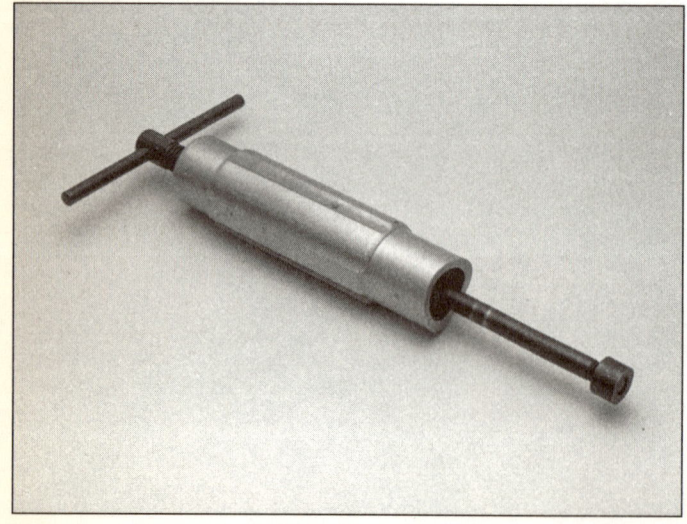

Piston pin puller

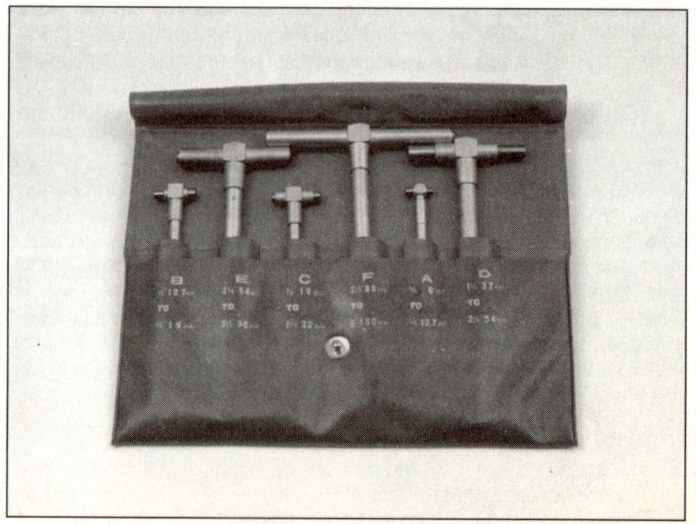

Telescoping gauges

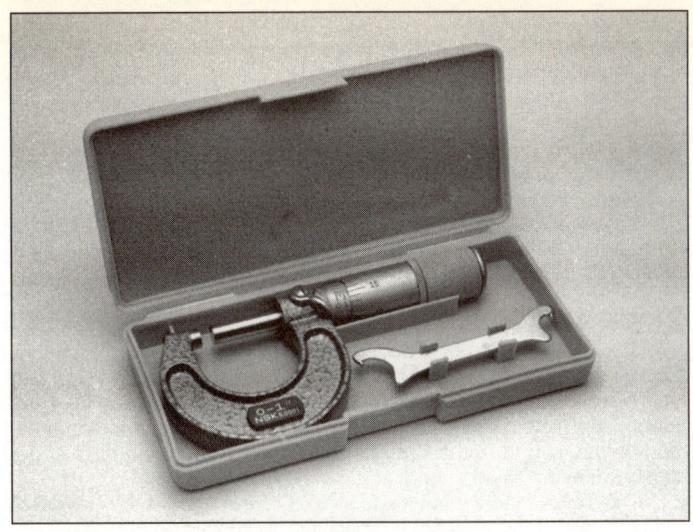

0-to-1 inch micrometer

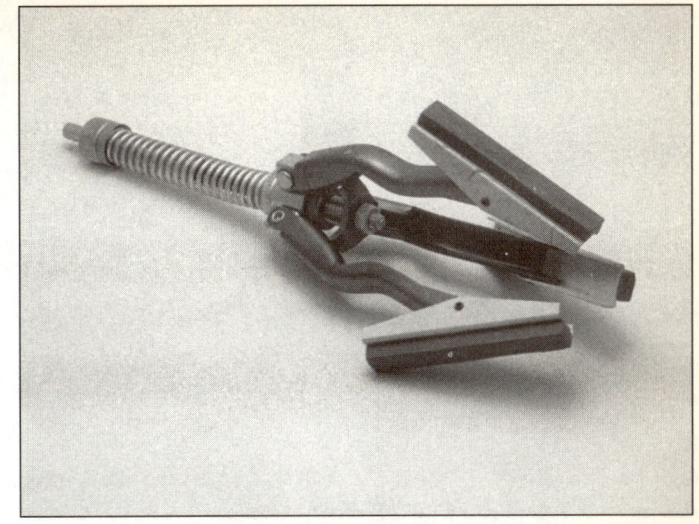

Cylinder surfacing hone

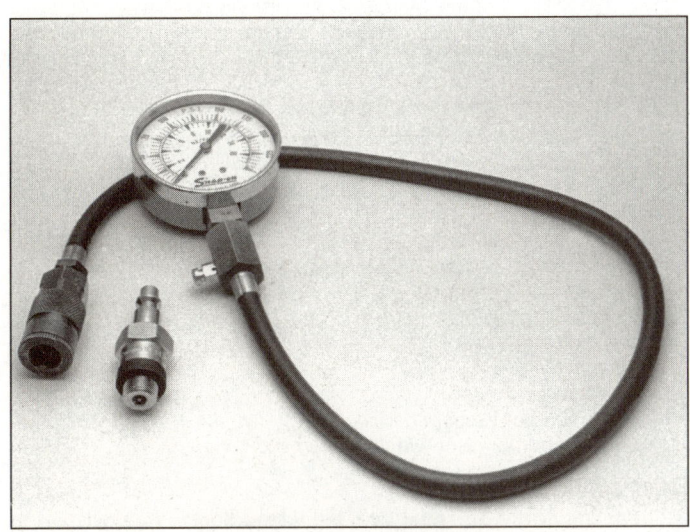

Cylinder compression gauge

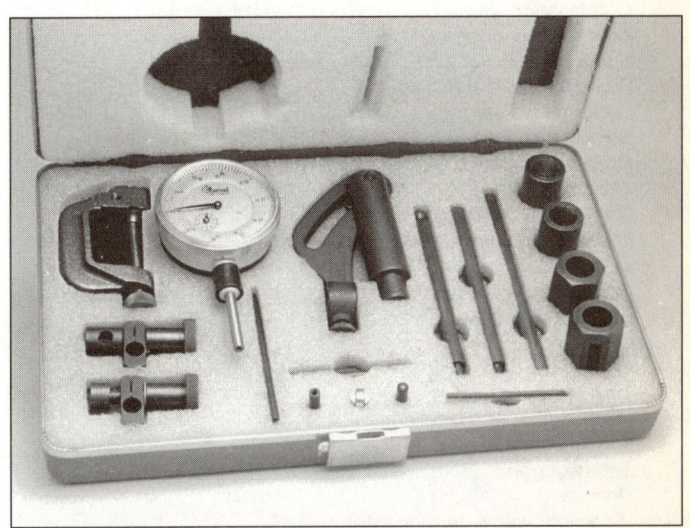

Dial indicator set

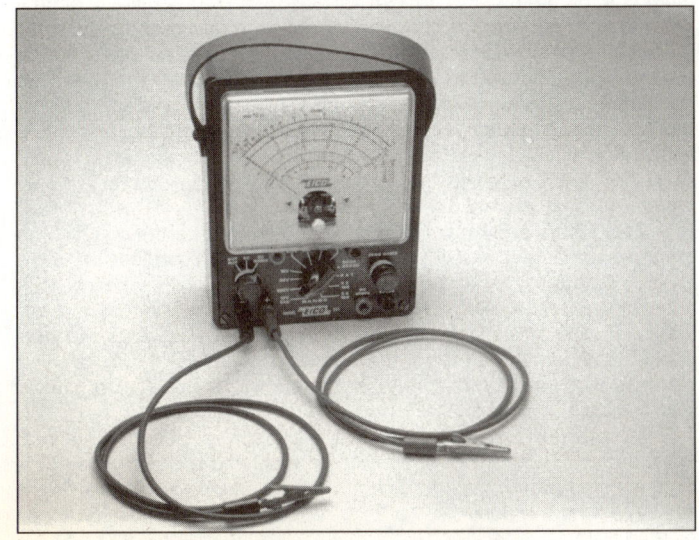

Multimeter (volt/ohm/ammeter)

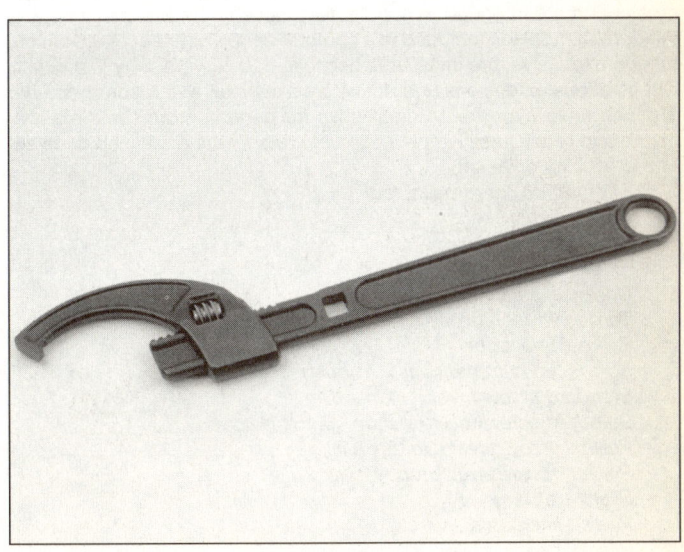

Adjustable spanner

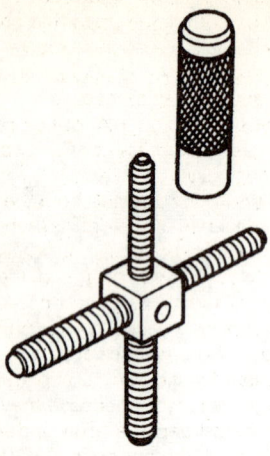

Alternator rotor puller

open-ended ones, they offer the advantages of both types of wrench.

   Combination wrench set (6 mm to 22 mm)
   Adjustable wrench - 8 in
   Spark plug socket (with rubber insert)
   Spark plug gap adjusting tool
   Feeler gauge set
   Standard screwdriver (5/16 in x 6 in)
   Phillips screwdriver (No. 2 x 6 in)
   Allen (hex) wrench set (4 mm to 12 mm)
   Combination (slip-joint) pliers - 6 in
   Hacksaw and assortment of blades
   Tire pressure gauge
   Control cable pressure luber
   Grease gun
   Oil can
   Fine emery cloth
   Wire brush
   Hand impact screwdriver and bits
   Funnel (medium size)
   Safety goggles
   Drain pan
   Work light with extension cord

Repair and overhaul tool set

These tools are essential for anyone who plans to perform major repairs and are intended to supplement those in the Maintenance and minor repair tool kit. Included is a comprehensive set of sockets which, though expensive, are invaluable because of their versatility (especially when various extensions and drives are available). We recommend the 3/8 inch drive over the 1/2 inch drive for general motorcycle maintenance and repair (ideally, the mechanic would have a 3/8 inch drive set and a 1/2 inch drive set).

   Alternator rotor removal tool
   Socket set(s)
   Reversible ratchet
   Extension - 6 in
   Universal joint
   Torque wrench (same size drive as sockets)
   Ball pein hammer - 8 oz
   Soft-faced hammer (plastic/rubber)
   Standard screwdriver (1/4 in x 6 in)
   Standard screwdriver (stubby - 5/16 in)
   Phillips screwdriver (No. 3 x 8 in)
   Phillips screwdriver (stubby - No. 2)
   Pliers - locking
   Pliers - lineman's
   Pliers - needle nose
   Pliers - snap-ring (internal and external)

   Cold chisel - 1/2 in
   Scriber
   Scraper (made from flattened copper tubing)
   Center punch
   Pin punches (1/16, 1/8, 3/16 in)
   Steel rule/straightedge - 12 in
   Pin-type spanner wrench
   A selection of files
   Wire brush (large)

Note: *Another tool which is often useful is an electric drill with a chuck capacity of 3/8 inch (and a set of good quality drill bits).*

Special tools

The tools in this list include those which are not used regularly, are expensive to buy, or which need to be used in accordance with their manufacturer's instructions. Unless these tools will be used frequently, it is not very economical to purchase many of them. A consideration would be to split the cost and use between yourself and a friend or friends (i.e. members of a motorcycle club).

This list primarily contains tools and instruments widely available to the public, as well as some special tools produced by the vehicle manufacturer for distribution to dealer service departments. As a result, references to the manufacturer's special tools are occasionally included in the text of this manual. Generally, an alternative method of doing the job without the special tool is offered. However, sometimes there is no alternative to their use. Where this is the case, and the tool can't be purchased or borrowed, the work should be turned over to the dealer service department or a motorcycle repair shop.

   Paddock stand (for models not fitted with a centerstand)
   Valve spring compressor
   Piston ring removal and installation tool
   Piston pin puller
   Telescoping gauges
   Micrometer(s) and/or dial/Vernier calipers
   Cylinder surfacing hone
   Cylinder compression gauge
   Dial indicator set
   Multimeter
   Adjustable spanner
   Manometer or vacuum gauge set
   Small air compressor with blow gun and tire chuck

Buying tools

For the do-it-yourselfer who is just starting to get involved in motorcycle maintenance and repair, there are a number of options available when purchasing tools. If maintenance and minor repair is the extent of the work to be done, the purchase of individual tools is satisfactory. If, on the other hand, extensive work is planned, it would be a good idea to purchase a modest tool set from one of the large retail chain stores. A set can usually be bought at a substantial savings over the individual tool prices (and they often come with a tool box). As additional tools are needed, add-on sets, individual tools and a larger tool box can be purchased to expand the tool selection. Building a tool set gradually allows the cost of the tools to be spread over a longer period of time and gives the mechanic the freedom to choose only those tools that will actually be used.

Tool stores and motorcycle dealers will often be the only source of some of the special tools that are needed, but regardless of where tools are bought, try to avoid cheap ones (especially when buying screwdrivers and sockets) because they won't last very long. There are plenty of tools around at reasonable prices, but always aim to purchase items which meet the relevant national safety standards. The expense involved in replacing cheap tools will eventually be greater than the initial cost of quality tools.

It is obviously not possible to cover the subject of tools fully here. For those who wish to learn more about tools and their use, there is a book entitled *Motorcycle Workshop Practice Manual* (Book no. 1454) available from the publishers of this manual. It also provides an introduction to basic workshop practice which will be of interest to a home mechanic working on any type of motorcycle.

Care and maintenance of tools

Good tools are expensive, so it makes sense to treat them with respect. Keep them clean and in usable condition and store them properly when not in use. Always wipe off any dirt, grease or metal chips before putting them away. Never leave tools lying around in the work area.

Some tools, such as screwdrivers, pliers, wrenches and sockets, can be hung on a panel mounted on the garage or workshop wall, while others should be kept in a tool box or tray. Measuring instruments, gauges, meters, etc. must be carefully stored where they can't be damaged by weather or impact from other tools.

When tools are used with care and stored properly, they will last a very long time. Even with the best of care, tools will wear out if used frequently. When a tool is damaged or worn out, replace it; subsequent jobs will be safer and more enjoyable if you do.

Working facilities

Not to be overlooked when discussing tools is the workshop. If anything more than routine maintenance is to be carried out, some sort of suitable work area is essential.

It is understood, and appreciated, that many home mechanics do not have a good workshop or garage available and end up removing an engine or doing major repairs outside (it is recommended, however, that the overhaul or repair be completed under the cover of a roof).

A clean, flat workbench or table of comfortable working height is an absolute necessity. The workbench should be equipped with a vise that has a jaw opening of at least four inches.

As mentioned previously, some clean, dry storage space is also required for tools, as well as the lubricants, fluids, cleaning solvents, etc. which soon become necessary.

Sometimes waste oil and fluids, drained from the engine or cooling system during normal maintenance or repairs, present a disposal problem. To avoid pouring them on the ground or into a sewage system, simply pour the used fluids into large containers, seal them with caps and take them to an authorized disposal site or service station. Plastic jugs (such as old antifreeze containers) are ideal for this purpose.

Always keep a supply of old newspapers and clean rags available. Old towels are excellent for mopping up spills. Many mechanics use rolls of paper towels for most work because they are readily available and disposable. To help keep the area under the motorcycle clean, a large cardboard box can be cut open and flattened to protect the garage or shop floor.

Whenever working over a painted surface (such as the fuel tank) cover it with an old blanket or bedspread to protect the finish.

Safety first!

Professional mechanics are trained in safe working procedures. However enthusiastic you may be about getting on with the job at hand, take the time to ensure that your safety is not put at risk. A moment's lack of attention can result in an accident, as can failure to observe simple precautions.

There will always be new ways of having accidents, and the following is not a comprehensive list of all dangers; it is intended rather to make you aware of the risks and to encourage a safe approach to all work you carry out on your bike.

Essential DOs and DON'Ts

DON'T start the engine without first ascertaining that the transmission is in neutral.

DON'T suddenly remove the pressure cap from a hot cooling system - cover it with a cloth and release the pressure gradually first, or you may get scalded by escaping coolant.

DON'T attempt to drain oil until you are sure it has cooled sufficiently to avoid scalding you.

DON'T grasp any part of the engine or exhaust system without first ascertaining that it is cool enough not to burn you.

DON'T allow brake fluid or antifreeze to contact the machine's paint work or plastic components.

DON'T siphon toxic liquids such as fuel, hydraulic fluid or antifreeze by mouth, or allow them to remain on your skin.

DON'T inhale dust - it may be injurious to health (see *Asbestos* heading).

DON'T allow any spilled oil or grease to remain on the floor - wipe it up right away, before someone slips on it.

DON'T use ill fitting wrenches or other tools which may slip and cause injury.

DON'T attempt to lift a heavy component which may be beyond your capability - get assistance.

DON'T rush to finish a job or take unverified short cuts.

DON'T allow children or animals in or around an unattended vehicle.

DON'T inflate a tire to a pressure above the recommended maximum. Apart from over stressing the carcase and wheel rim, in extreme cases the tire may blow off forcibly.

DO ensure that the machine is supported securely at all times. This is especially important when the machine is blocked up to aid wheel or fork removal.

DO take care when attempting to loosen a stubborn nut or bolt. It is generally better to pull on a wrench, rather than push, so that if you slip, you fall away from the machine rather than onto it.

DO wear eye protection when using power tools such as drill, sander, bench grinder etc.

DO use a barrier cream on your hands prior to undertaking dirty jobs - it will protect your skin from infection as well as making the dirt easier to remove afterwards; but make sure your hands aren't left slippery. Note that long-term contact with used engine oil can be a health hazard.

DO keep loose clothing (cuffs, ties etc. and long hair) well out of the way of moving mechanical parts.

DO remove rings, wristwatch etc., before working on the vehicle - especially the electrical system.

DO keep your work area tidy - it is only too easy to fall over articles left lying around.

DO exercise caution when compressing springs for removal or installation. Ensure that the tension is applied and released in a controlled manner, using suitable tools which preclude the possibility of the spring escaping violently.

DO ensure that any lifting tackle used has a safe working load rating adequate for the job.

DO get someone to check periodically that all is well, when working alone on the vehicle.

DO carry out work in a logical sequence and check that everything is correctly assembled and tightened afterwards.

DO remember that your vehicle's safety affects that of yourself and others. If in doubt on any point, get professional advice.

IF, in spite of following these precautions, you are unfortunate enough to injure yourself, seek medical attention as soon as possible.

Asbestos

Certain friction, insulating, sealing and other products - such as brake pads, clutch linings, gaskets, etc. - contain asbestos. *Extreme care must be taken to avoid inhalation of dust from such products since it is hazardous to health.* If in doubt, assume that they *do* contain asbestos.

Fire

Remember at all times that gasoline (petrol) is highly flammable. Never smoke or have any kind of naked flame around, when working on the vehicle. But the risk does not end there - a spark caused by an electrical short-circuit, by two metal surfaces contacting each other, by careless use of tools, or even by static electricity built up in your body under certain conditions, can ignite gasoline (petrol) vapor, which in a confined space is highly explosive. Never use gasoline (petrol) as a cleaning solvent. Use an approved safety solvent.

Always disconnect the battery ground (earth) terminal before working on any part of the fuel or electrical system, and never risk spilling fuel on to a hot engine or exhaust.

It is recommended that a fire extinguisher of a type suitable for fuel and electrical fires is kept handy in the garage or workplace at all times. Never try to extinguish a fuel or electrical fire with water.

Fumes

Certain fumes are highly toxic and can quickly cause unconsciousness and even death if inhaled to any extent. Gasoline (petrol) vapor comes into this category, as do the vapors from certain solvents such as trichloroethylene. Any draining or pouring of such volatile fluids should be done in a well ventilated area.

When using cleaning fluids and solvents, read the instructions carefully. Never use materials from unmarked containers - they may give off poisonous vapors.

Never run the engine of a motor vehicle in an enclosed space such as a garage. Exhaust fumes contain carbon monoxide which is extremely poisonous; if you need to run the engine, always do so in the open air or at least have the rear of the vehicle outside the workplace.

The battery

Never cause a spark, or allow a naked light near the vehicle's battery. It will normally be giving off a certain amount of hydrogen gas, which is highly explosive.

Always disconnect the battery ground (earth) terminal before working on the fuel or electrical systems (except where noted).

If possible, loosen the filler plugs or cover when charging the battery from an external source. Do not charge at an excessive rate or the battery may burst.

Take care when topping up, cleaning or carrying the battery. The acid electrolyte, even when diluted, is very corrosive and should not be allowed to contact the eyes or skin. Always wear rubber gloves and goggles or a face shield. If you ever need to prepare electrolyte yourself, always add the acid slowly to the water; never add the water to the acid.

Electricity

When using an electric power tool, inspection light etc., always ensure that the appliance is correctly connected to its plug and that, where necessary, it is properly grounded (earthed). Do not use such appliances in damp conditions and, again, beware of creating a spark or applying excessive heat in the vicinity of fuel or fuel vapor. Also ensure that the appliances meet national safety standards.

A severe electric shock can result from touching certain parts of the electrical system, such as the spark plug wires (HT leads), when the engine is running or being cranked, particularly if components are damp or the insulation is defective. Where an electronic ignition system is used, the secondary (HT) voltage is much higher and could prove fatal.

ATV chemicals and lubricants

A number of chemicals and lubricants are available for use in vehicle maintenance and repair. They include a wide variety of products ranging from cleaning solvents and degreasers to lubricants and protective sprays for rubber, plastic and vinyl.

Contact point/spark plug cleaner is a solvent used to clean oily film and dirt from points, grime from electrical connectors and oil deposits from spark plugs. It is oil free and leaves no residue. It can also be used to remove gum and varnish from carburetor jets and other orifices.

Carburetor cleaner is similar to contact point/spark plug cleaner but it usually has a stronger solvent and may leave a slight oily residue. It is not recommended for cleaning electrical components or connections.

Brake system cleaner is used to remove grease or brake fluid from brake system components (where clean surfaces are absolutely necessary and petroleum-based solvents cannot be used); it also leaves no residue.

Silicone-based lubricants are used to protect rubber parts such as hoses and grommets, and are used as lubricants for hinges and locks.

Multi-purpose grease is an all purpose lubricant used wherever grease is more practical than a liquid lubricant such as oil. Some multi-purpose grease is colored white and specially formulated to be more resistant to water than ordinary grease.

Gear oil (sometimes called gear lube) is a specially designed oil used in transmissions and final drive units, as well as other areas where high friction, high temperature lubrication is required. It is available in a number of viscosities (weights) for various applications.

Motor oil, of course, is the lubricant specially formulated for use in the engine. It normally contains a wide variety of additives to prevent corrosion and reduce foaming and wear. Motor oil comes in various weights (viscosity ratings) of from 5 to 80. The recommended weight of the oil depends on the seasonal temperature and the demands on the engine. Light oil is used in cold climates and under light load conditions; heavy oil is used in hot climates and where high loads are encountered. Multi-viscosity oils are designed to have characteristics of both light and heavy oils and are available in a number of weights from 5W-20 to 20W-50. On these machines, the engine and transmission share the same oil supply.

Gas additives perform several functions, depending on their chemical makeup. They usually contain solvents that help dissolve gum and varnish that build up on carburetor and intake parts. They also serve to break down carbon deposits that form on the inside surfaces of the combustion chambers. Some additives contain upper cylinder lubricants for valves and piston rings.

Brake fluid is a specially formulated hydraulic fluid that can withstand the heat and pressure encountered in brake systems. Care must be taken that this fluid does not come in contact with painted surfaces or plastics. An opened container should always be resealed to prevent contamination by water or dirt.

Chain lubricants are formulated especially for use on the final drive chain. A good chain lube should adhere well and have good penetrating qualities to be effective as a lubricant inside the chain and on the side plates, pins and rollers. Most chain lubes are either the foaming type or quick drying type and are usually marketed as sprays.

Degreasers are heavy duty solvents used to remove grease and grime that may accumulate on engine and frame components. They can be sprayed or brushed on and, depending on the type, are rinsed with either water or solvent.

Solvents are used alone or in combination with degreasers to clean parts and assemblies during repair and overhaul. The home mechanic should use only solvents that are non-flammable and that do not produce irritating fumes.

Gasket sealing compounds may be used in conjunction with gaskets, to improve their sealing capabilities, or alone, to seal metal-to-metal joints. Many gasket sealers can withstand extreme heat, some are impervious to gasoline and lubricants, while others are capable of filling and sealing large cavities. Depending on the intended use, gasket sealers either dry hard or stay relatively soft and pliable. They are usually applied by hand, with a brush, or are sprayed on the gasket sealing surfaces.

Thread cement is an adhesive locking compound that prevents threaded fasteners from loosening because of vibration. It is available in a variety of types for different applications.

Moisture dispersants are usually sprays that can be used to dry out electrical components such as the fuse block and wiring connectors. Some types can also be used as treatment for rubber and as a lubricant for hinges, cables and locks.

Waxes and polishes are used to help protect painted and plated surfaces from the weather. Different types of paint may require the use of different types of wax polish. Some polishes utilize a chemical or abrasive cleaner to help remove the top layer of oxidized (dull) paint on older vehicles. In recent years, many non-wax polishes (that contain a wide variety of chemicals such as polymers and silicones) have been introduced. These non-wax polishes are usually easier to apply and last longer than conventional waxes and polishes.

Troubleshooting

Contents

Engine doesn't start or is difficult to start

1   Starter motor does not rotate

1   Engine kill switch Off.
2   Fuse blown. Check fuse (Chapter 8).
3   Battery voltage low. Check and recharge battery (Chapter 8).
4   Starter motor defective. Make sure the wiring to the starter is secure. Test starter relay (Chapter 8). If the relay is good, then the fault is in the wiring or motor.
5   Starter relay faulty. Check it according to the procedure in Chapter 8.
6   Starter switch not contacting. The contacts could be wet, corroded or dirty. Disassemble and clean the switch (Chapter 8).
7   Wiring open or shorted. Check all wiring connections and harnesses to make sure that they are dry, tight and not corroded. Also check for broken or frayed wires that can cause a short to ground (see wiring diagrams at the end of the manual).
8   Ignition (main) switch defective. Check the switch according to the procedure in Chapter 8. Replace the switch with a new one if it is defective.
9   Engine kill switch defective. Check for wet, dirty or corroded contacts. Clean or replace the switch as necessary (Chapter 8).
10   Starting circuit cut-off relay, neutral switch, reverse switch or front brake switch defective. Check the relay and switches according to the procedure in Chapter 8. Replace the switch with a new one if it is defective.

2   Starter motor rotates but engine does not turn over

1   Starter motor clutch defective. Inspect and repair or replace (Chapter 8).
2   Damaged starter idle or wheel gears. Inspect and replace the damaged parts (Chapter 8).

3   Starter works but engine won't turn over (seized)

Seized engine caused by one or more internally damaged components. Failure due to wear, abuse or lack of lubrication. Damage can include seized valves, rocker arms, camshaft, piston, crankshaft, connecting rod bearings, or transmission gears or bearings. Refer to Chapter 2 for engine disassembly.

4   No fuel flow

1   No fuel in tank.
2   Tank cap air vent obstructed. Usually caused by dirt or water. Remove it and clean the cap vent hole.
3   Clogged strainer in fuel tap. Remove and clean the strainer (Chapter 1).
4   Fuel line clogged. Pull the fuel line loose and carefully blow through it.
5   Inlet needle valve clogged. A very bad batch of fuel with an unusual additive may have been used, or some other foreign material has entered the tank. Many times after a machine has been stored for many months without running, the fuel turns to a varnish-like liquid and forms deposits on the inlet needle valve and jets. The carburetor should be removed and overhauled if draining the float chamber does not solve the problem.

5   Engine flooded

1   Float level too high. Check as described in Chapter 3 and replace the float if necessary.

2   Needle valve worn or stuck open. A piece of dirt, rust or other debris can cause the float valve to seat improperly, causing excess fuel to be admitted to the float bowl. In this case, the float chamber should be cleaned and the float valve and seat inspected. If the valve and seat are worn, then the leaking will persist and the parts should be replaced (Chapter 3).
3   Starting technique incorrect. Under normal circumstances (i.e., if all the carburetor functions are sound) the machine should start with little or no throttle. When the engine is cold, the choke should be operated and the engine started without opening the throttle. When the engine is at operating temperature, only a very slight amount of throttle should be necessary. If the engine is flooded, turn the fuel tap off and hold the throttle open while cranking the engine. This will allow additional air to reach the cylinder. Remember to turn the fuel tap back on after the engine starts.

6   No spark or weak spark

1   Ignition switch Off.
2   Engine kill switch turned to the Off position.
3   Spark plug dirty, defective or worn out. Locate reason for fouled plug using spark plug condition chart and follow the plug maintenance procedures in Chapter 1.
4   Spark plug cap or secondary wiring faulty. Check condition. Replace either or both components if cracks or deterioration are evident (Chapter 4).
5   Spark plug cap not making good contact. Make sure that the plug cap fits snugly over the plug end.
6   Alternator charging coil (TRX300EX) or exciter coil (TRX400EX) defective. Check the unit, referring to Chapter 8 or Chapter 4 for details.
7   CDI unit defective. Check the unit, referring to Chapter 4 for details.
8   Ignition coil defective. Check the coil, referring to Chapter 4.
9   Ignition or kill switch shorted. This is usually caused by water, corrosion, damage or excessive wear. The kill switch can be disassembled and cleaned with electrical contact cleaner. If cleaning does not help, replace the switches (Chapter 8).
10   Wiring shorted or broken between:
  a) *Ignition switch and engine kill switch (or blown fuse)*
  b) *CDI unit and engine kill switch*
  c) *CDI unit and ignition coil*
  d) *Ignition coil and plug*
  e) *CDI unit and alternator*
    Make sure that all wiring connections are clean, dry and tight. Look for chafed and broken wires (Chapters 4 and 8).
11   Reverse switch defective (TRX300EX). Check and replace if necessary (Chapter 8).
12   Ignition pulse generator defective (Chapter 4).

7   Compression low

1   Spark plug loose. Remove the plug and inspect the threads. Reinstall and tighten to the specified torque (Chapter 1).
2   Cylinder head not sufficiently tightened down. If the cylinder head is suspected of being loose, then there's a chance that the gasket or head is damaged if the problem has persisted for any length of time. The head nuts and bolts should be tightened to the proper torque in the correct sequence (Chapter 2).
3   Incorrect valve clearance. This means that the valve is not closing completely and compression pressure is leaking past the valve. Check and adjust the valve clearances (Chapter 1).
4   Cylinder and/or piston worn. Excessive wear will cause compression pressure to leak past the rings. This is usually accompanied by worn rings as well. A top end overhaul is necessary (Chapter 2).
5   Piston rings worn, weak, broken, or sticking. Broken or sticking piston rings usually indicate a lubrication or carburetion problem that

causes excess carbon deposits or seizures to form on the pistons and rings. Top end overhaul is necessary (Chapter 2).

6　Piston ring-to-groove clearance excessive. This is caused by excessive wear of the piston ring lands. Piston replacement is necessary (Chapter 2).

7　Cylinder head gasket damaged. If the head is allowed to become loose, or if excessive carbon build-up on a piston crown and combustion chamber causes extremely high compression, the head gasket may leak. Retorquing the head is not always sufficient to restore the seal, so gasket replacement is necessary (Chapter 2).

8　Cylinder head warped. This is caused by overheating or incorrectly tightened head nuts and bolts. Machine shop resurfacing or head replacement is necessary (Chapter 2).

9　Valve spring broken or weak. Caused by component failure or wear; the spring(s) must be replaced (Chapter 2).

10　Valve not seating correctly. This is caused by a bent valve (from over-revving or incorrect valve adjustment), burned valve or seat (incorrect carburetion) or an accumulation of carbon deposits on the seat (from carburetion or lubrication problems). The valves must be cleaned and/or replaced and the seats serviced if possible (Chapter 2).

8　Stalls after starting

1　Incorrect choke action. Make sure the choke knob or lever is getting a full stroke and staying in the out position.

2　Ignition malfunction (Chapter 4).

3　Carburetor malfunction (Chapter 3).

4　Fuel contaminated. The fuel can be contaminated with either dirt or water, or can change chemically if the machine is allowed to sit for several months or more. Drain the tank and float bowl and refill with fresh fuel (Chapter 3).

5　Intake air leak. Check for loose carburetor-to-intake joint connections or loose carburetor top (Chapter 3).

6　Engine idle speed incorrect. Turn throttle stop screw until the engine idles at the specified rpm (Chapter 1).

9　Rough idle

1　Ignition malfunction (Chapter 4).

2　Idle speed incorrect (Chapter 1).

3　Carburetor malfunction (Chapter 3).

4　Idle fuel/air mixture incorrect (Chapter 3).

5　Fuel contaminated. The fuel can be contaminated with either dirt or water, or can change chemically if the machine is allowed to sit for several months or more. Drain the tank and float bowl (Chapter 3).

6　Intake air leak. Check for loose carburetor-to-intake joint connections, loose or missing vacuum gauge access port cap or hose, or loose carburetor top (Chapter 3).

7　Air cleaner clogged. Service or replace air cleaner element (Chapter 1).

Poor running at low speed

10　Spark weak

1　Battery voltage low. Check and recharge battery (Chapter 8).

2　Spark plug fouled, defective or worn out. Refer to Chapter 1 for spark plug maintenance.

3　Spark plug cap or secondary wiring defective. Refer to Chapters 1 and 4 for details on the ignition system.

4　Spark plug cap not making contact.

5　Incorrect spark plug. Wrong type, heat range or cap configuration. Check and install correct plug listed in Chapter 1. A cold plug or one with a recessed firing electrode will not operate at low speeds without fouling.

6　CDI unit defective (Chapter 4).

7　Alternator charging coil (TRX300EX) or exciter coil (TRX400EX) defective. Check the unit, referring to Chapter 8 or Chapter 4 for details.

8　Ignition coil defective (Chapter 4).

9　Ignition pulse generator defective (Chapter 4).

11　Air/fuel mixture incorrect

1　Pilot screw out of adjustment (Chapter 3).

2　Pilot jet or air passage clogged. Remove and overhaul the carburetor (Chapter 3).

3　Air bleed holes clogged. Remove carburetor and blow out all passages (Chapter 3).

4　Air cleaner clogged, poorly sealed or missing.

5　Air cleaner-to-carburetor air intake tube poorly sealed. Look for cracks, holes or loose clamps and replace or repair defective parts.

6　Float level too high or too low. Check and replace the float if necessary (Chapter 3).

7　Fuel tank air vent obstructed. Make sure that the air vent passage in the filler cap is open.

8　Carburetor intake joint loose. Check for cracks, breaks, tears or loose clamps or bolts. Repair or replace the rubber boot and its O-ring.

12　Compression low

1　Spark plug loose. Remove the plug and inspect the threads. Reinstall and tighten to the specified torque (Chapter 1).

2　Cylinder head not sufficiently tightened down. If the cylinder head is suspected of being loose, then there's a chance that the gasket and head are damaged if the problem has persisted for any length of time. The head nuts and bolts should be tightened to the proper torque in the correct sequence (Chapter 2).

3　Incorrect valve clearance. This means that the valve is not closing completely and compression pressure is leaking past the valve. Check and adjust the valve clearances (Chapter 1).

4　Cylinder and/or piston worn. Excessive wear will cause compression pressure to leak past the rings. This is usually accompanied by worn rings as well. A top end overhaul is necessary (Chapter 2).

5　Piston rings worn, weak, broken, or sticking. Broken or sticking piston rings usually indicate a lubrication or carburetion problem that causes excess carbon deposits or seizures to form on the pistons and rings. Top end overhaul is necessary (Chapter 2).

6　Piston ring-to-groove clearance excessive. This is caused by excessive wear of the piston ring lands. Piston replacement is necessary (Chapter 2).

7　Cylinder head gasket damaged. If the head is allowed to become loose, or if excessive carbon build-up on the piston crown and combustion chamber causes extremely high compression, the head gasket may leak. Retorquing the head is not always sufficient to restore the seal, so gasket replacement is necessary (Chapter 2).

8　Cylinder head warped. This is caused by overheating or improperly tightened head nuts and bolts. Machine shop resurfacing or head replacement is necessary (Chapter 2).

9　Valve spring broken or weak. Caused by component failure or wear; the spring(s) must be replaced (Chapter 2).

10　Valve not seating properly. This is caused by a bent valve (from over-revving or improper valve adjustment), burned valve or seat (improper carburetion) or an accumulation of carbon deposits on the seat (from carburetion, lubrication problems). The valves must be cleaned and/or replaced and the seats serviced if possible (Chapter 2).

13　Poor acceleration

1　Carburetor leaking or dirty. Overhaul the carburetor (Chapter 3).

2　Timing not advancing. The CDI unit may be defective. If so, it must be replaced with a new one, as it can't be repaired.

3   Engine oil viscosity too high. Using a heavier oil than that recommended in Chapter 1 can damage the oil pump or lubrication system and cause drag on the engine.

4   Brakes dragging. Usually caused by a sticking brake caliper piston, warped disc or bent axle. Repair as necessary (Chapter 6).

Poor running or no power at high speed

14   Firing incorrect

1   Air cleaner restricted. Clean or replace element (Chapter 1).

2   Spark plug fouled, defective or worn out. See Chapter 1 for spark plug maintenance.

3   Spark plug cap or secondary wiring defective. See Chapters 1 and 4 for details of the ignition system.

4   Spark plug cap not in good contact (Chapter 4).

5   Incorrect spark plug. Wrong type, heat range or cap configuration. Check and install correct plugs listed in Chapter 1. A cold plug or one with a recessed firing electrode will not operate at low speeds without fouling.

6   Ignition coil defective (Chapter 4).

7   Ignition pulse generator defective (Chapter 4).

15   Fuel/air mixture incorrect

1   Pilot screw out of adjustment. See Chapter 3 for adjustment procedures.

2   Main jet clogged. Dirt, water or other contaminants can clog the main jets. Clean the fuel tap strainer and in-tank strainer, the float bowl area, and the jets and carburetor orifices (Chapter 3).

3   Main jet wrong size (Chapter 3).

4   Throttle shaft-to-carburetor body clearance excessive. Refer to Chapter 4 for inspection and part replacement procedures.

5   Air bleed holes clogged. Remove and overhaul carburetor (Chapter 3).

6   Air cleaner clogged, poorly sealed, or missing.

7   Air cleaner-to-carburetor air intake tube poorly sealed. Look for cracks, holes or loose clamps, and replace or repair defective parts.

8   Float level too high or too low. Check float level and replace the float if necessary (Chapter 3).

9   Fuel tank air vent obstructed. Make sure the air vent passage in the filler cap is open.

10   Carburetor intake joint loose. Check for cracks, breaks, tears or loose clamps or bolts. Repair or replace the rubber boots (Chapter 3).

11   Fuel tap clogged. Remove the tap and clean it (Chapter 1).

12   Fuel line clogged. Pull the fuel line loose and carefully blow through it.

16   Compression low

1   Spark plug loose. Remove the plug and inspect the threads. Reinstall and tighten to the specified torque (Chapter 1).

2   Cylinder head not sufficiently tightened down. If the cylinder head is suspected of being loose, then there's a chance that the gasket and head are damaged if the problem has persisted for any length of time. The head nuts and bolts should be tightened to the proper torque in the correct sequence (Chapter 2).

3   Incorrect valve clearance. This means that the valve is not closing completely and compression pressure is leaking past the valve. Check and adjust the valve clearances (Chapter 1).

4   Cylinder and/or piston worn. Excessive wear will cause compression pressure to leak past the rings. This is usually accompanied by worn rings as well. A top end overhaul is necessary (Chapter 2).

5   Piston rings worn, weak, broken, or sticking. Broken or sticking piston rings usually indicate a lubrication or carburetion problem that causes excess carbon deposits or seizures to form on the pistons and rings. Top end overhaul is necessary (Chapter 2).

6   Piston ring-to-groove clearance excessive. This is caused by excessive wear of the piston ring lands. Piston replacement is necessary (Chapter 2).

7   Cylinder head gasket damaged. If a head is allowed to become loose, or if excessive carbon build-up on the piston crown and combustion chamber causes extremely high compression, the head gasket may leak. Retorquing the head is not always sufficient to restore the seal, so gasket replacement is necessary (Chapter 2).

8   Cylinder head warped. This is caused by overheating or improperly tightened head nuts and bolts. Machine shop resurfacing or head replacement is necessary (Chapter 2).

9   Valve spring broken or weak. Caused by component failure or wear; the spring(s) must be replaced (Chapter 2).

10   Valve not seating properly. This is caused by a bent valve (from over-revving or improper valve adjustment), burned valve or seat (improper carburetion) or an accumulation of carbon deposits on the seat (from carburetion or lubrication problems). The valves must be cleaned and/or replaced and the seats serviced if possible (Chapter 2).

17   Knocking or pinging

1   Carbon build-up in combustion chamber. Use of a fuel additive that will dissolve the adhesive bonding the carbon particles to the crown and chamber is the easiest way to remove the build-up. Otherwise, the cylinder head will have to be removed and decarbonized (Chapter 2).

2   Incorrect or poor quality fuel. Old or improper grades of fuel can cause detonation. This causes the piston to rattle, thus the knocking or pinging sound. Drain old fuel and always use the recommended fuel grade.

3   Spark plug heat range incorrect. Uncontrolled detonation indicates the plug heat range is too hot. The plug in effect becomes a glow plug, raising cylinder temperatures. Install the proper heat range plug (Chapter 1).

4   Incorrect air/fuel mixture. This will cause the cylinder to run hot, which leads to detonation. Clogged jets or an air leak can cause this imbalance (Chapter 3).

18   Miscellaneous causes

1   Throttle valve doesn't open fully. Adjust the cable slack (Chapter 1).

2   Clutch slipping. May be caused by improper adjustment or loose or worn clutch components. Refer to Chapter 1 for adjustment or Chapter 2 for clutch overhaul procedures.

3   Timing not advancing.

4   Engine oil viscosity too high. Using heavier oil than the one recommended in Chapter 1 can damage the oil pump or lubrication system and cause drag on the engine.

5   Brakes dragging. Usually caused by a sticking brake caliper piston, warped disc or bent axle. Repair as necessary.

Overheating

19   Engine overheats

1   Engine oil level low. Check and add oil (Chapter 1).

2   Wrong type of oil. If you're not sure what type of oil is in the engine, drain it and fill it with the correct type (Chapter 1).

3   Air leak at carburetor air intake tube. Check and tighten or replace

as necessary (Chapter 3).

4   Fuel level in float chamber low. Check and adjust if necessary (Chapter 3).

5   Worn oil pump or clogged oil passages. Replace pump or clean passages as necessary.

6   Clogged external oil line. Remove and check for foreign material (see Chapter 2).

7   Carbon build-up in combustion chambers. Use of a fuel additive that will dissolve the adhesive bonding the carbon particles to the piston crown and chambers is the easiest way to remove the build-up. Otherwise, the cylinder head will have to be removed and decarbonized (Chapter 2).

8   Operation in high ambient temperatures.

20   Firing incorrect

1   Spark plug fouled, defective or worn out. See Chapter 1 for spark plug maintenance.

2   Incorrect spark plug (Chapter 1).

3   Faulty ignition coil(s) (Chapter 4).

21   Air/fuel mixture incorrect

1   Pilot screw out of adjustment (Chapter 3).

2   Main jet clogged. Dirt, water and other contaminants can clog the main jet. Clean the fuel tap strainer, the float bowl area and the jets and carburetor orifices (Chapter 3).

3   Main jet wrong size. The standard jetting is for sea level atmospheric pressure and oxygen content.

4   Air cleaner poorly sealed or missing.

5   Air cleaner-to-carburetor air intake tube poorly sealed. Look for cracks, holes or loose clamps and replace or repair.

6   Fuel level in float chamber too low. Check fuel level and float level and adjust or replace the float if necessary (Chapter 3).

7   Fuel tank air vent obstructed. Make sure that the air vent passage in the filler cap is open.

8   Carburetor intake manifold loose. Check for cracks or loose clamps or bolts. Check the carburetor-to-manifold gasket and the manifold-to-cylinder head O-ring (Chapter 3).

22   Compression too high

1   Carbon build-up in combustion chamber. Use of a fuel additive that will dissolve the adhesive bonding the carbon particles to the piston crown and chamber is the easiest way to remove the build-up. Otherwise, the cylinder head will have to be removed and decarbonized (Chapter 2).

2   Improperly machined head surface or installation of incorrect gasket during engine assembly.

23   Engine load excessive

1   Clutch slipping. Can be caused by damaged, loose or worn clutch components. Refer to Chapter 2 for overhaul procedures.

2   Engine oil level too high. The addition of too much oil will cause pressurization of the crankcase and inefficient engine operation. Check Specifications and drain to proper level (Chapter 1).

3   Engine oil viscosity too high. Using heavier oil than the one recommended in Chapter 1 could damage the oil pump or lubrication system as well as cause drag on the engine.

4   Brakes dragging. Usually caused by a sticking brake caliper piston, warped disc or bent axle. Repair as necessary (Chapter 6).

24   Lubrication inadequate

1   Engine oil level too low. Friction caused by intermittent lack of lubrication or from oil that is overworked can cause overheating. The oil provides a definite cooling function in the engine. Check the oil level (Chapter 1).

2   Poor quality engine oil or incorrect viscosity or type. Oil is rated not only according to viscosity but also according to type. Some oils are not rated high enough for use in this engine. Check the Specifications and change to the correct oil (Chapter 1).

3   Camshaft or journals worn. Excessive wear causing drop in oil pressure. Replace cam or cylinder head. Abnormal wear could be caused by oil starvation at high rpm from low oil level, incorrect oil viscosity or incorrect type of oil (Chapter 1).

4   Crankshaft and/or bearings worn. Same problems as Paragraph 3. Check and replace crankshaft if necessary (Chapter 2).

Clutch problems

25   Clutch slipping

1   Clutch friction plates worn or warped. Overhaul the clutch (Chapter 2).

2   Clutch metal plates worn or warped (Chapter 2).

3   Clutch spring(s) broken or weak. Old or heat-damaged spring(s) (from slipping clutch) should be replaced with new ones (Chapter 2).

4   Clutch release mechanism defective. Replace any defective parts (Chapter 2).

5   Clutch housing unevenly worn. This causes improper engagement of the plates. Replace the damaged or worn parts (Chapter 2).

26   Clutch not disengaging completely

1   Clutch cable incorrectly adjusted (see Chapter 1).

2   Clutch plates warped or damaged. This will cause clutch drag, which in turn will cause the machine to creep. Overhaul the clutch (Chapter 2).

3   Sagged or broken clutch spring(s). Check and replace the spring(s) (Chapter 2).

4   Engine oil deteriorated. Old, thin, worn out oil will not provide proper lubrication for the discs, causing the clutch to drag. Replace the oil and filter (Chapter 1).

5   Engine oil viscosity too high. Using a thicker oil than recommended in Chapter 1 can cause the clutch plates to stick together, putting a drag on the engine. Change to the correct viscosity oil (Chapter 1).

6   Clutch housing seized on shaft. Lack of lubrication, severe wear or damage can cause the housing to seize on the shaft. Overhaul of the clutch, and perhaps transmission, may be necessary to repair the damage (Chapter 2).

7   Clutch release mechanism defective. Worn or damaged release mechanism parts can stick and fail to apply force to the pressure plate. Overhaul the release mechanism (Chapter 2).

8   Loose clutch center nut. Causes housing and center misalignment putting a drag on the engine. Engagement adjustment continually varies. Overhaul the clutch (Chapter 2).

Gear shifting problems

27   Doesn't go into gear or lever doesn't return

1   Clutch not disengaging. See Section 26.

2   Shift fork(s) bent or seized. May be caused by lack of lubrication.

Overhaul the transmission (Chapter 2).

3   Gear(s) stuck on shaft. Most often caused by a lack of lubrication or excessive wear in transmission bearings and bushings. Overhaul the transmission (Chapter 2).

4   Shift drum binding. Caused by lubrication failure or excessive wear. Replace the drum and bearing (Chapter 2).

5   Shift shaft return spring weak or broken (Chapter 2).

6   Shift shaft broken. Splines stripped out of pedal or shaft, caused by allowing the pedal to get loose. Replace necessary parts (Chapter 2).

7   Shift mechanism broken or worn. Full engagement and rotary movement of shift drum results. Replace shift mechanism (Chapter 2).

28   Jumps out of gear

1   Shift fork(s) worn. Overhaul the transmission (Chapter 2).

2   Gear groove(s) worn. Overhaul the transmission (Chapter 2).

3   Gear dogs or dog slots worn or damaged. The gears should be inspected and replaced. No attempt should be made to service the worn parts.

29   Overshifts

1   Pawl spring weak or broken (Chapter 2).

2   Shift drum stopper lever not functioning (Chapter 2).

Abnormal engine noise

30   Knocking or pinging

1   Carbon build-up in combustion chamber. Use of a fuel additive that will dissolve the adhesive bonding the carbon particles to the piston crown and chamber is the easiest way to remove the build-up. Otherwise, the cylinder head will have to be removed and decarbonized (Chapter 2).

2   Incorrect or poor quality fuel. Old or improper fuel can cause detonation. This causes the pistons to rattle, thus the knocking or pinging sound. Drain the old fuel (Chapter 3) and always use the recommended grade fuel (Chapter 1).

3   Spark plug heat range incorrect. Uncontrolled detonation indicates that the plug heat range is too hot. The plug in effect becomes a glow plug, raising cylinder temperatures. Install the proper heat range plug (Chapter 1).

4   Incorrect air/fuel mixture. This will cause the cylinder to run hot and lead to detonation. Clogged jets or an air leak can cause this imbalance. See Chapter 3.

31   Piston slap or rattling

1   Cylinder-to-piston clearance excessive. Caused by improper assembly. Inspect and overhaul top end parts (Chapter 2).

2   Connecting rod bent. Caused by over-revving, trying to start a badly flooded engine or from ingesting a foreign object into the combustion chamber. Replace the damaged parts (Chapter 2).

3   Piston pin or piston pin bore worn or seized from wear or lack of lubrication. Replace damaged parts (Chapter 2).

4   Piston ring(s) worn, broken or sticking. Overhaul the top end (Chapter 2).

5   Piston seizure damage. Usually from lack of lubrication or overheating. Replace the pistons and bore the cylinder, as necessary (Chapter 2).

6   Connecting rod upper or lower end clearance excessive. Caused by excessive wear or lack of lubrication. Replace worn parts.

32   Valve noise

1   Incorrect valve clearance. Adjust the clearance (Chapter 1).

2   Valve spring broken or weak. Check and replace weak valve springs (Chapter 2).

3   Camshaft or cylinder head worn or damaged. Lack of lubrication at high rpm is usually the cause of damage. Insufficient oil or failure to change the oil at the recommended intervals is the chief cause.

33   Other noise

1   Cylinder head gasket leaking. Tighten cylinder head bolts or replace head gasket (Chapter 2).

2   Exhaust pipe leaking at cylinder head connection. Caused by incorrect fit of pipe, damaged gasket(s) or loose exhaust pipe flange(s). All exhaust fasteners should be tightened evenly and carefully (Chapter 3). Failure to do this will lead to a leak.

3   Crankshaft runout excessive. Caused by a bent crankshaft (from over-revving). Check and, if necessary, replace crankshaft (Chapter 2).

4   Engine mounting bolts or nuts loose. Tighten all engine mounting bolts and nuts to the specified torque (Chapter 2).

5   Crankshaft bearings worn (Chapter 2).

6   Camshaft chain tensioner defective. Replace according to the procedure in Chapter 2.

7   Camshaft chain, sprockets or guides worn (Chapter 2).

Abnormal driveline noise

34   Clutch noise

1   Clutch housing/friction plate clearance excessive (Chapter 2).

2   Loose or damaged clutch pressure plate and/or bolts (Chapter 2).

35   Transmission noise

1   Bearings worn. Also includes the possibility that the shafts are worn. Overhaul the transmission (Chapter 2).

2   Gears worn or chipped (Chapter 2).

3   Metal chips jammed in gear teeth. Probably pieces from a broken gear or shift mechanism that were picked up by the gears. This will cause early bearing failure (Chapter 2).

4   Engine oil level too low. Causes a howl from transmission. Also affects engine power and clutch operation (Chapter 1).

36   Final drive noise

1   Dry or dirty chain. Inspect, clean and lubricate (see Chapter 1).

2   Chain out of adjustment. Adjust chain slack (see Chapter 1).

3   Chain and sprockets damaged or worn. Inspect the chain and sprockets and replace them as necessary (Chapter 5).

4   Sprockets loose (Chapter 5).

Abnormal chassis noise

37   Suspension noise

1   Spring weak or broken. Makes a clicking or scraping sound.

2   Steering shaft bearings worn or damaged. Clicks when braking. Check and replace as necessary (Chapter 5).

3   Shock absorber fluid level incorrect. Indicates a leak caused by defective seal. Shock will be covered with oil. Replace shock (Chapter 5).

4   Defective shock absorber with internal damage. This is in the body of the shock and can't be remedied. The shock must be replaced with a new one (Chapter 5).

5   Bent or damaged shock body. Replace the shock with a new one (Chapter 5).

38   Brake noise

1   Squeal caused by dust on brake pads. Usually found in combination with glazed pads. Clean using brake cleaning solvent (Chapter 6).

2   Contamination of brake pads. Grease, water or dirt causing pads to chatter or squeal. Clean or replace pads (Chapter 6).

3   Pads glazed. Caused by excessive heat from prolonged use or from contamination. Do not use sandpaper, emery cloth, carborundum cloth or any other abrasives to roughen pad surface; abrasives will stay in the pad material and damage the disc. A very fine flat file can be used, but pad replacement is suggested as a cure (Chapter 6).

4   Disc warped. Can cause chattering, clicking or intermittent squeal. Usually accompanied by a pulsating lever and uneven braking. Replace the disc (Chapter 6).

5   Loose or worn wheel bearings. Check and replace as necessary (Chapter 6).

Excessive exhaust smoke

39   White smoke

1   Piston oil ring worn. The ring may be broken or damaged, causing oil from the crankcase to be pulled past the piston into the combustion chamber. Replace the rings with new ones (Chapter 2).

2   Cylinder worn, cracked, or scored. Caused by overheating or oil starvation. If worn or scored, the cylinder will have to be rebored and a new piston installed. If cracked, the cylinder will have to be replaced (see Chapter 2).

3   Valve oil seal damaged or worn. Replace oil seals with new ones (Chapter 2).

4   Valve guide worn. Perform a complete valve job (Chapter 2).

5   Engine oil level too high, which causes the oil to be forced past the rings. Drain oil to the proper level (Chapter 1).

6   Head gasket broken between oil return passage and cylinder. Causes oil to be pulled into the combustion chamber. Replace the head gasket and check the head for warpage (Chapter 2).

7   Excessive crankcase pressurization, which forces oil past the rings. Clogged breather or hoses usually the cause (Chapter 2).

40   Black smoke

1   Air cleaner clogged. Clean or replace the element (Chapter 1).

2   Main jet too large or loose. Compare the jet size to the Specifications (Chapter 3).

3   Choke lever stuck (Chapter 3).

4   Fuel level too high. Check the fuel level and float level (Chapter 3).

5   Float valve held off seat. Clean the float chamber and fuel line and replace the float valve if necessary (Chapter 3).

41   Brown smoke

1   Main jet too small or clogged. Lean condition caused by wrong size main jet or by a restricted orifice. Clean float chamber and jets and compare jet size to Specifications (Chapter 3).

2   Fuel flow insufficient. Fuel inlet needle valve stuck closed due to chemical reaction with old fuel. Float level incorrect; check and replace float if necessary. Restricted fuel line. Clean line and float chamber (Chapter 3).

3   Carburetor air intake tube loose (Chapter 3).

4   Air cleaner poorly sealed or not installed (Chapter 1).

Poor handling or stability

42   Handlebar hard to turn

1   Steering shaft nut too tight (Chapter 5).

2   Lower bearing or upper bushing damaged. Roughness can be felt as the bars are turned from side-to-side. Replace bearing and bushing (Chapter 5).

3   Steering shaft bearing lubrication inadequate. Causes are grease getting hard from age or being washed out by high pressure car washes. Remove steering shaft and replace bearing (Chapter 5).

4   Steering shaft bent. Caused by a collision, hitting a pothole or by rolling the machine. Replace damaged part. Don't try to straighten the steering shaft (Chapter 5).

5   Front tire air pressure too low (Chapter 1).

43   Handlebar shakes or vibrates excessively

1   Tires worn or out of balance (Chapter 1 or 6).

2   Swingarm bearings worn. Replace worn bearings by referring to Chapter 6.

3   Wheel rim(s) warped or damaged. Inspect wheels (Chapter 6).

4   Wheel bearings worn. Worn front or rear wheel bearings can cause poor tracking. Worn front bearings will cause wobble (Chapter 6).

5   Wheel hubs installed incorrectly (Chapter 6).

6   Handlebar clamp bolts or bracket nuts loose (Chapter 5).

7   Steering shaft bearing holder bolts loose. Tighten them to the specified torque (Chapter 5).

8   Motor mount bolts loose. Will cause excessive vibration with increased engine rpm (Chapter 2).

44   Handlebar pulls to one side

1   Uneven tire pressures (Chapter 1).

2   Frame bent. Definitely suspect this if the machine has been rolled. May or may not be accompanied by cracking near the bend. Replace the frame.

3   Wheel out of alignment. Caused by incorrect toe-in adjustment (Chapter 1) or bent tie-rod (Chapter 5).

4   Swingarm bent or twisted. Caused by age (metal fatigue) or impact damage. Replace the swingarm (Chapter 5).

5   Steering stem bent. Caused by impact damage or by rolling the vehicle. Replace the steering stem (Chapter 5).

45   Poor shock absorbing qualities

1   Too hard:
a) *Shock internal damage.*
b) *Tire pressure too high (Chapters 1 and 6).*
2   Too soft:
a) *Shock oil insufficient and/or leaking (Chapter 5).*
d) *Fork springs weak or broken (Chapter 5).*

Braking problems

46   Brakes are spongy, don't hold

1   Disc brake pads worn (Chapters 1 and 6).
2   Disc brake pads contaminated by oil, grease, etc. Clean or replace pads (Chapter 6).
3   Disc warped. Replace disc (Chapter 6).

47   Brake lever or pedal pulsates

1   Disc warped. Replace disc (Chapter 6).
2   Axle bent. Replace axle (Chapter 6).
3   Brake caliper bolts loose (see Chapter 6).
4   Brake caliper slide pin and/or caliper bracket pin damaged or sticking, causing caliper to bind. Lube the pins or replace them if they're corroded or bent (Chapter 6).
5   Wheel warped or otherwise damaged (Chapter 6).
6   Wheel hub or axle bearings damaged or worn (Chapter 6).

48   Brakes drag

1   Lever or pedal balky or stuck. Check pivot and lubricate (Chapter 6).
2   Brake caliper binds. Caused by inadequate lubrication or damage to caliper shafts (Chapter 6).
3   Brake caliper piston seized in bore. Caused by wear or ingestion of dirt getting past deteriorated seal (Chapter 6).
4   Brake pad(s) damaged. Pad material separated from backing plate. Usually caused by faulty manufacturing process or contact with chemicals. Replace pads (Chapter 6).
5   Pads incorrectly installed (Chapter 6).
6   Brake pedal or lever freeplay insufficient (Chapter 1).

Chapter 1
Tune-up and routine maintenance

Contents

Specifications

Engine

Spark plugs
  Type
    TRX300EX
      Standard .... NGK DR8ES-L or ND X24ESR-U
      Cold climate (below 41 degrees F) .... NGK DR7ES or ND X22ESR-U
      Extended high speed riding .... NGK DR8ES or ND X27ESR-U
    TRX400EX
      Standard .... NGK DPR8Z or ND X24GPR-U
      Extended high speed riding .... NGK DPR9Z or ND X27GPR-U
  Gap .... 0.6 to 0.7 mm (0.024 to 0.028 inch)
Clutch lever freeplay .... 10 to 20 mm (3/8 to 3/4 inch)
Valve clearance
  TRX300EX (Intake and exhaust) .... 0.08 to 0.12 mm (0.003 to 0.005
  TRX400EX
    Intake .... 0.10 mm (0.004 inch)
    Exhaust .... 0.12 mm (0.005 inch)

Engine (continued)

Engine idle speed .... 1300 to 1500 rpm
Pilot screw initial opening.. 2-1/4 turns out
Compression
    TRX300EX .... 11.9 kg/cm² to 14.0 kg/cm² (170.8 to 199.2 psi)
    TRX400EX .... 7.0 to 9.0 kg/cm² (100 to 128 psi)

Chassis

Brake pad thickness limit (front and rear).. Replace pads when pad wear indicator is aligned
    with index mark on caliper
Rear brake pedal maximum travel.. About 80 mm (3.1 inches)
Rear brake pedal height above footrest .... Adjust to desired height*
Throttle lever freeplay .... 3 to 8 mm (1/8 to 5/16 inch)
Clutch lever freeplay .... 25 to 30 mm (1 to 1-1/8 inch)
Reverse selector cable freeplay... 2 to 4 mm (1/16 to 1/8 inch)
Drive chain slack
    TRX300EX .... 35 to 45 mm (1-3/8 to 1-3/4 inches)
    TRX400EX .... 30 to 40 mm (1-1/4 to 1-5/8 inches)
Minimum tire tread depth .... 4 mm (0.16 inch)
Tire pressures (cold)
    TRX300EX
        Front.. 30 kPa (0.30 kg/cm², 4.4 psi)
        Rear... 20 kPa (0.20 kg/cm², 2.9 psi)
    TRX400EX (front and rear)
        Standard.. 27 kPa (0.275 kg/cm2, 4.0 psi)
        Minimum .... 23 kPa (o.235 kg/cm2, 3.4 psi)
        Maximum .... 31 kPa (0.315 kg/cm2, 4.6 psi)
*Pushrod-to-brake pedal clearance must be at least 1 mm (0.04 inch)

Torque specifications

Axle bearing holder pinch bolts
    TRX300EX .... 22 Nm (16 ft-lbs)
    TRX400EX .... 21 Nm (15 ft-lbs)
Oil filter cover bolts
    TRX300EX .... 10 Nm (84 in-lbs)
    TRX400EX .... 12 Nm (108 in-lbs)
Oil drain plug .... 25 Nm (18 ft-lbs)
Spark plug .... 18 Nm (156 in-lbs)
Timing hole plug .... 10 Nm (84 in-lbs)
Crankshaft hole plug
    TRX300EX .... 14 Nm (132 in-lbs)
    TRX400EX .... 8 Nm (70 in-lbs)
Valve adjuster cover bolts
    TRX300EX .... 18 Nm (156 in-lbs)
    TRX400EX .... 15 Nm (132 in-lbs)
*Apply non-permanent thread locking agent to the locknut threads.

Recommended lubricants and fluids

Engine oil
    TRX300EX
        Type .... Honda 4-stroke oil or equivalent 10W40
        Capacity
            At oil change.. 1.6 liters (1.7 quarts)
            After engine overhaul.. 2.0 liters (2.1 quarts)
    TRX400EX
        Type ....
        Capacity
            At oil change..
            After engine overhaul..
Miscellaneous
    Wheel bearings... Medium weight, lithium-based multi-purpose grease
    Swingarm pivot bearings.. Medium weight, lithium-based multi-purpose grease
    Steering shaft bushings.. Medium weight, lithium-based multi-purpose grease
    Cables and lever pivots.. Medium weight, lithium-based multi-purpose grease
    Brake pedal/shift lever/throttle lever pivots..................................... Medium weight, lithium-based multi-purpose grease

1   Introduction to tune-up and routine maintenance

Refer to illustration 1.1

This Chapter covers in detail the checks and procedures necessary for the tune-up and routine maintenance of your vehicle. Section 1 includes the routine maintenance schedule, which is designed to keep the machine in proper running condition and prevent possible problems. The remaining Sections contain detailed procedures for carrying out the items listed on the maintenance schedule, as well as additional maintenance information designed to increase reliability. Maintenance information is also printed on decals, which are mounted in various locations on the vehicle **(see illustration)**. Where information on the decals differs from that presented in this Chapter, use the decal information.

Since routine maintenance plays such an important role in the safe and efficient operation of your vehicle, it is presented here as a comprehensive checklist. These lists outline the procedures and checks that should be done on a routine basis.

Deciding where to start or plug into the routine maintenance schedule depends on several factors. If you have a vehicle whose warranty has recently expired, and it has been maintained according to the warranty standards, you may want to pick up routine maintenance as it coincides with the next mileage or calendar interval. If you have owned the machine for some time but have never performed any maintenance on it, then you may want to start at the nearest interval and include some additional procedures to ensure that nothing important is overlooked. If you have just had a major engine overhaul, then you may want to start the maintenance routine from the beginning. If you have a used machine and have no knowledge of its history or maintenance record, you may desire to combine all the checks into one large service initially and then settle into the maintenance schedule prescribed.

The Sections which actually outline the inspection and mainte-

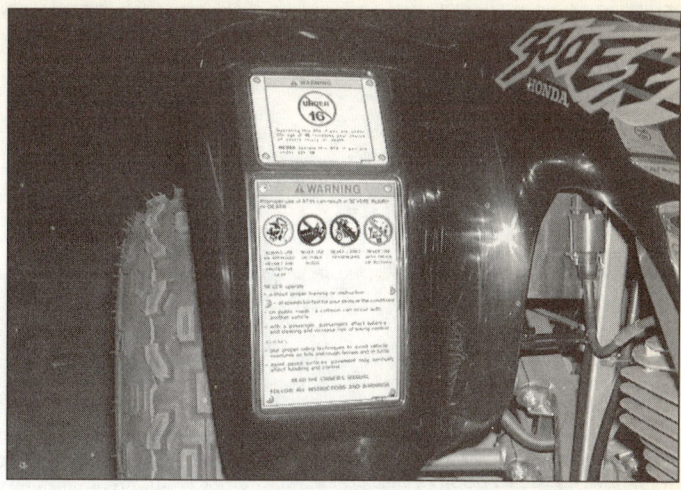

1.1 Decals on the vehicle include maintenance and safety information

nance procedures are written as step-by-step comprehensive guides to the actual performance of the work. They explain in detail each of the routine inspections and maintenance procedures on the checklist. References to additional information in applicable Chapters are also included and should not be overlooked.

Before beginning any actual maintenance or repair, the machine should be cleaned thoroughly, especially around the oil filler plug, spark plug, engine covers, carburetor, etc. Cleaning will help ensure that dirt does not contaminate the engine and will allow you to detect wear and damage that could otherwise easily go unnoticed.

2 Honda TRX300EX and TRX400EX Routine maintenance intervals

Note: *The pre-ride inspection outlined in the owner's manual covers checks and maintenance that should be carried out on a daily basis. It's condensed and included here to remind you of its importance. Always perform the pre-ride inspection at every maintenance interval (in addition to the procedures listed). The intervals listed below are the shortest intervals recommended by the manufacturer for each particular operation during the model years covered in this manual. Your owner's manual may have different intervals for your model.*

Daily or before riding

Check the operation of both brakes - check the brake lever and
 pedal for correct freeplay
Check the throttle for smooth operation and correct freeplay
Make sure the engine kill switch works correctly
Check the tires for damage, the presence of foreign objects and
 correct air pressure
Check the engine oil
Check the fuel level and inspect for leaks
Check the air cleaner drain tube and clean it if necessary
If the drain tube is clogged, clean the air filter element
Inspect the drive chain
Make sure the steering operates smoothly
Verify that the headlight and taillight are operating
 satisfactorily
Check all fasteners, including wheel nuts and axle nuts,
 for tightness
Check the underbody for mud or debris that could start a fire or
 interfere with vehicle operation

Every 20 to 40 hours

Clean the air filter element and replace it if necessary*
Clean the spark arrester

Every six months

Change the engine oil
Clean and gap and, if necessary, replace the spark plug
Inspect the fuel tap and the fuel line
Check the throttle for smooth operation and correct freeplay
Check choke operation
Check idle speed and adjust it if necessary
Inspect the front and rear brake discs
Check brake operation and brake lever and pedal freeplay
Check the clutch for smooth operation and correct
 lever freeplay
Lubricate the steering shaft and front suspension
Lubricate and inspect the drive chain, sprockets and rollers
Check steering system operation and freeplay
Inspect the wheels and tires
Check the wheel bearings for looseness or damage
Inspect the front and rear suspension
Check all chassis fasteners for tightness
Check the skid plates for looseness or damage
Check the exhaust system for leaks and check
 fastener tightness
*More often in dusty or wet conditions.

3.2a Make sure that the brake fluid level in the front brake master cylinder reservoir is above the LOWER mark

3.2b Make sure that the brake fluid level in the rear brake master cylinder reservoir is above the LOWER mark; this is the TRX300EX reservoir . . .

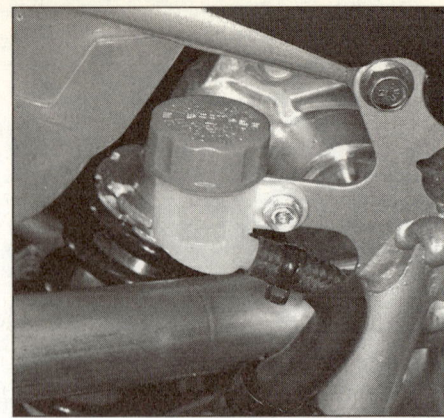

3.2c . . . and this is the TRX400EX reservoir

3 Fluid levels - check

Brake fluid

Refer to illustrations 3.2a, 3.2b, 3.2c, 3.4a, 3.4b and 3.4c

1   With the vehicle in a level position, turn the handlebars until the top of the front brake master cylinder is as level as possible.

2   The fluid level is visible through a window in the master cylinder reservoir or through the reservoir itself **(see illustrations)**. Make sure that the fluid level is above the Lower mark on the reservoir.

3   If the brake fluid level is low, the fluid must be replenished. Before removing the master cylinder cover, place rags beneath the reservoir (to protect the paint from brake fluid spills) and remove all dust and dirt from the area around the cap.

4   To add fluid to the front brake master cylinder, remove the cover screws, then lift off the cover, diaphragm plate and diaphragm **(see illustrations)**. **Caution:** *Don't operate the brake lever with the cover removed.* To add fluid to the rear brake master cylinder, remove the seat and rear fenders if necessary for access (see Chapter 7). If you're working on a TRX300EX, remove the cover screws, lift off the cover and remove the diaphragm plate and diaphragm **(see illustration 3.2b and the accompanying illustration)**. If you're working on a TRX400EX, unscrew the reservoir cap **(see illustration 3.2c)**. **Caution:** *Don't operate the brake pedal with the cover or cap removed.*

5   Add new, clean brake fluid of the recommended type to bring the level above the Lower mark. Don't mix different brands of brake fluid in the reservoir, as they may not be compatible. Also, don't mix different

specifications (DOT 3 with DOT 4).

6   Reinstall the diaphragm, diaphragm plate and cover (TRX300EX) or screw on the cap (TRX400EX). Tighten the cover screws securely (don't overtighten them, or you'll strip the threads).

7   Wipe any spilled fluid off the reservoir body.

8   If the brake fluid level was low, inspect the front brake system for leaks (see Section 4).

Engine oil

Refer to illustrations 3.11a and 3.11b

9   Support the vehicle in a level position, then start the engine and allow it to reach normal operating temperature. **Warning:** *Do not run the engine in an enclosed space such as a garage or shop.*

10   Stop the engine and allow the machine to sit undisturbed in a level position for about five minutes.

11   To check the engine oil on TRX300EX models, unscrew the dipstick from the right side of the crankcase **(see illustration)**. To check the oil on TRX400EX models, unscrew the dipstick from the left side of the oil tank located in front of the engine **(see illustration)**. Pull out the dipstick, wipe it off with a clean rag, and reinsert it (let the dipstick rest on the threads; don't screw it back in). Pull the dipstick out and check the oil level on the dipstick scale **(see illustration 3.11b)**. The oil level should be between the Maximum and Minimum level marks on the scale.

12   If the level is below the Minimum mark, add oil through the dipstick hole. Add enough oil of the recommended grade and type to bring the level up to the Maximum mark. Do not overfill.

3.4a To add fluid to the front brake master cylinder, remove the cover screws and the cover . . .

3.4b . . . then lift out the diaphragm plate and diaphragm (arrow)

3.4c The TRX300EX rear master cylinder also has a diaphragm plate and diaphragm

3.11a On TRX300EX models, unscrew the engine oil filler cap from the right front corner of the engine and pull out the dipstick tube

3.11b The oil filler cap/dipstick tube on TRX400EX models is at the upper left corner of the oil tank (in front of the engine)

4.4a Replace the brake pads when the wear indicator reaches the limit line; this is a TRX300EX front caliper . . .

4.4b . . . this is a TRX300EX rear caliper . . .

4.4c . . . this is a TRX400EX front caliper . . .

1

4   Brake system - general check

Refer to illustrations 4.4a, 4.4b, 4.4c, 4.4d, 4.5 and 4.6

1   Always inspect the brakes before riding! A routine pre-ride general check will ensure that problems are discovered and remedied before they become dangerous.

2   Inspect the brake lever and pedal for loose pivots, excessive play, bending, cracking and other damage. Replace any damaged parts (see Chapter 6).

3   Make sure all brake fasteners are tight. Check the brake pads for wear as follows.

4   Inspect the front and rear brake pads for wear. Look at the wear indicator on each caliper **(see illustrations)**. If it's aligned with the reference line on the caliper, the pads are worn out. It's time for new pads (see Chapter 6).

5   Apply the rear brake pedal and measure the distance it travels from its fully released position to its fully applied position **(see illustration)**. It should travel no more than the distance listed in this Chapter's Specifications. If pedal travel is excessive, or if the pedal feels spongy when

4.4d . . . and this is a TRX400EX rear caliper

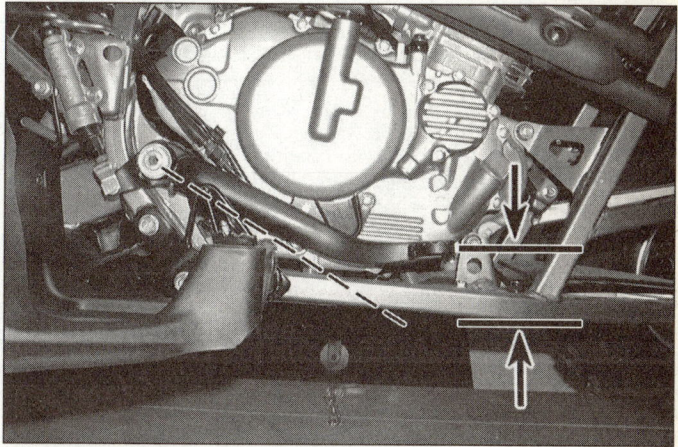

4.5 If the rear brake pedal travels more than the distance listed in this Chapter's Specifications, there's air in the system

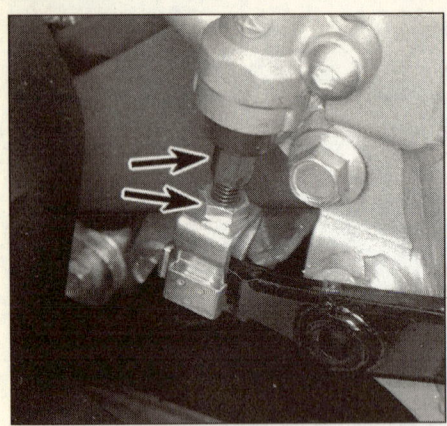

4.6 Loosen the locknut (lower arrow), adjust pedal height with the pushrod adjuster (upper arrow), then tighten the locknut securely

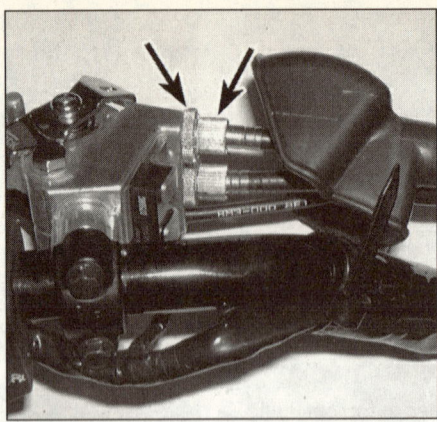

5.3 To put slack in the parking brake cable, loosen the locknut (left arrow) and turn the adjuster (right arrow) all the way in

5.4 Parking brake adjuster bolt (left) and locknut (right)

applied, there is probably air in the system. Inspect the rear brake master cylinder, brake hose and caliper for a leak (see Chapter 6).

6   There is no specified brake pedal height (the height of the brake pedal above the footrest). If the pedal feels too low or too high, adjust it by loosening the locknut on the pushrod for the rear brake master cylinder (**see illustration**) and turning the pushrod. When the pedal height is satisfactory, tighten the locknut securely. If you're lowering the pedal height, make sure that the clearance between the pushrod and the pedal is at least 1 mm (0.04 inch).

5   Parking brake - adjustment

Refer to illustrations 5.3 and 5.4

1   Verify that the parking brake system will hold the vehicle on an incline. If it won't, adjust the parking brake as follows.

2   Remove the handlebar cover (see Chapter 7).

3   Temporarily loosen the clutch lever (see Section 18) so that it has more than 1-1/8 inch freeplay, then loosen the locknut on the parking brake cable and screw in the adjuster all the way (**see illustration**).

4   Using a backup wrench on the adjuster bolt, loosen the locknut on the parking brake lever at the rear caliper, turn the adjusting bolt 1/8 turn counterclockwise (**see illustration**), then tighten the locknut securely.

5   Back out the adjuster at the handlebar to take the slack out of the cable and tighten the locknut.

6   Test the parking brake again and verify that it now holds the vehicle on an incline.

6   Steering system - inspection and toe-in adjustment

Inspection

1   This vehicle is equipped with bearings at the upper and lower ends of the steering shaft. These can become dented, rough or loose during normal use of the machine. In extreme cases, worn or loose parts can cause steering wobble that is potentially dangerous.

2   To check, block the rear wheels so the vehicle can't roll, jack up the front end and support it securely on jackstands.

3   Point the wheel straight ahead and slowly move the handlebar from side-to-side. Dents or roughness in the bearing will be felt and the bars will not move smoothly. **Note:** *Make sure any hesitation in movement is not being caused by the cables and wiring harnesses that run to the handlebar.*

4   If the handlebar doesn't move smoothly, or if it has excessive lateral play, remove and inspect the steering shaft bushings (see Chapter 5).

5   Look at the tie-rod ends (inner and outer) while slowly turning the handlebar from side-to-side. If there's any vertical movement in the tie-rod balljoints, replace them (see Chapter 5).

Toe-in adjustment

Refer to illustrations 6.8 and 6.11

6   Roll the vehicle forward onto a level surface and stop it with the front wheels pointing straight ahead.

7   Make a mark at the front and center of each tire, even with the centerline of the front hub.

8   Measure the distance between the marks with a toe-in gauge or steel tape measure (**see illustration**).

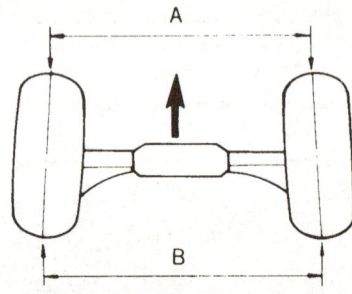

6.8 Toe-in measurement (B minus A = toe-in)

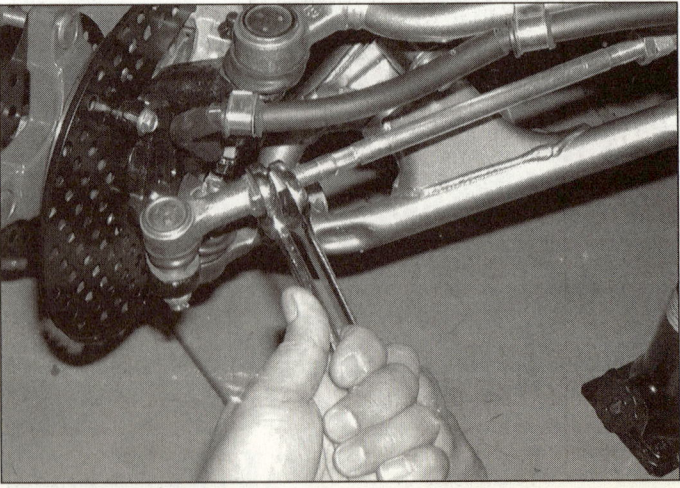

6.11 Hold the locknut with one wrench and turn the tie-rod with a second wrench on the flat near the outer end

9   Have an assistant push the vehicle backward while you watch the marks on the tires. Stop pushing when the tires have rotated exactly one-half turn, so the marks are at the backs of the tires.

10   Again, measure the distance between the marks. Subtract the front measurement from the rear measurement **(see illustration 6.8)** to get toe-in.

11   If toe-in is not as specified in this Chapter's Specifications, hold each tie-rod with a wrench on the flats and loosen the locknuts **(see illustration)**. Turn the tie-rods an equal amount to change toe-in. When toe-in is set correctly, tighten the locknuts securely.

7   Suspension - check

Refer to illustration 7.3

1   The suspension components must be maintained in top operating condition to ensure rider safety. Loose, worn or damaged suspension parts decrease the vehicle's stability and control.

2   Lock the front brake and push on the handlebars to compress the front shock absorbers several times. See if they move up-and-down smoothly without binding. If binding is felt, the shocks should be inspected (see Chapter 5).

3   Check the tightness of all front suspension nuts and bolts **(see illustration)** to be sure none have worked loose.

4   Inspect the rear shock absorber for fluid leakage and tightness of the mounting nuts and bolts. If leakage is found, the shock should be replaced.

5   Raise the rear of the vehicle and support it securely on jack-stands. Grab the swingarm on each side, just ahead of the axle. Rock the swingarm from side to side - there should be no discernible movement at the rear. If there's a little movement or a slight clicking can be heard, make sure the swingarm pivot shaft is tight. If the pivot shaft is tight but movement is still noticeable, the swingarm will have to be removed and the bearings replaced (see Chapter 5).

6   Inspect the tightness of the rear suspension nuts and bolts.

8   Drive chain and sprockets - inspection, adjustment and lubrication

Inspection

Refer to illustrations 8.3, 8.5 and 8.6

1   A neglected drive chain won't last long and can quickly damage the sprockets. Routine chain adjustment isn't difficult and will ensure maximum chain and sprocket life.

2   To check the chain, support the vehicle securely on jackstands with the rear wheels off the ground. Place the transmission in neutral.

7.3 It's a good idea to inspect all front suspension fasteners (arrows) before riding

3   Push down and pull up on the lower run of the chain and measure the slack midway between the two sprockets **(see illustration)**, then compare the measurements to the value listed in this Chapter's Specifications. As wear occurs, the chain will actually stretch, which means adjustment is necessary to remove some slack from the chain. In some cases where lubrication has been neglected, corrosion and galling may cause the links to bind and kink, which effectively shortens the chain's length. If the chain is tight between the sprockets, rusty or kinked, it's time to replace it with a new one. **Note:** *Repeat the chain slack measurement along the length of the chain - ideally, every inch or so. If you find a tight area, mark it with felt pen or paint and repeat the measurement after the machine has been ridden. If the chain's still tight in the same areas, it may be damaged or worn. Because a tight or kinked chain can damage the transmission countershaft bearing, it's a good idea to replace it.*

4   Inspect the entire length of the chain for damaged rollers or O-rings, loose links and loose pins.

5   Look through the slots in the engine sprocket cover and inspect the engine sprocket **(see illustration)**. Check the teeth on the engine sprocket and the rear sprocket for wear (see Chapter 5). Refer to Chapter 7 for the sprocket replacement procedure if the sprockets appear to be worn excessively. **Note:** *Never install a new chain on old sprockets and never use the old chain if you install new sprockets - replace the chain and sprockets as a set.*

6   Inspect the chain roller and slider **(see illustration)**. If a roller or slider is worn, replace it (see Chapter 5).

8.3 Inspect the drive chain at a point on the lower run about halfway between the sprockets

8.5 Inspect the front sprocket teeth for excessive wear; make sure there's no play in the sprocket

8.6 Inspect the drive chain slider (upper arrow) and the chain roller (lower arrow) for wear, damage or rough roller movement (TRX300EX model shown; TRX400EX similar)

8.8a To adjust the drive chain on a TRX300EX, loosen these two axle bearing holder pinch bolts (arrows) . . .

8.8b . . . to adjust the chain on a TRX400EX, loosen these four axle bearing holder pinch bolts (arrows) . . .

Adjustment

Refer to illustrations 8.8a, 8.8b and 8.8c

7   Rotate the rear wheels until the chain is positioned with the least amount of slack present.

8   Loosen the pinch bolts on the axle bearing holder **(see illustrations)**. Turn the adjuster **(see illustration)** until the chain tension is correct. If the specified chain tension cannot be achieved with the adjuster, the chain is excessively worn and should be replaced (see Chapter 5).

Lubrication

Note: *If the chain is dirty, it should be removed and cleaned before it's lubricated (see Chapter 5).*

9   The best time to lubricate the chain is after the vehicle has been ridden. When the chain is warm, the lubricant will penetrate the joints between the side plates to provide lubrication. Honda specifies SAE 30 to 50 engine oil only; do not use chain lube, which may contain solvents that can damage the chain's rubber O-rings. Apply the oil to the area where the side plates overlap - not to the middle of the rollers.

10   Apply the lubricant along the top of the lower chain run, so that when the machine is ridden centrifugal force will move the lubricant into the chain, rather than throwing it off.

11   After applying the lubricant, let it soak in a few minutes before wiping off any excess.

9   Tires/wheels - general check

Refer to illustration 9.4

1   Routine tire and wheel checks should be made with the realization that your safety depends to a great extent on their condition.

2   Check the tires carefully for cuts, tears, embedded nails or other sharp objects and excessive wear. Operation of the vehicle with excessively worn tires is extremely hazardous, as traction and handling are directly affected. Measure the tread depth at the center of the tire and replace worn tires with new ones when the tread depth is less than that listed in this Chapter's Specifications.

3   Repair or replace punctured tires as soon as damage is noted. Do not try to patch a torn tire, as wheel balance and tire reliability may be impaired.

4   Check the tire pressures when the tires are cold and keep them properly inflated **(see illustration)**. Proper air pressure will increase tire life and provide maximum stability and ride comfort. Keep in mind that low tire pressures may cause the tire to slip on the rim or come off, while high tire pressures will cause abnormal tread wear and unsafe handling. **Caution:** *ATV tires operate at very low pressures. Overinflation may rupture them.*

5   Some ATV tires are directional; that is, they are designed to rotate in only one forward direction. The direction of forward rotation on these

8.8c . . . and turn the adjuster with a prybar until the chain tension is correct (TRX300EX model shown, TRX400EX similar)

9.4 Check tire pressure with a gauge that will read accurately at the low pressures used in ATV tires

10.3 Rock the wheel and tire from side-to-side to check wheel bearing play

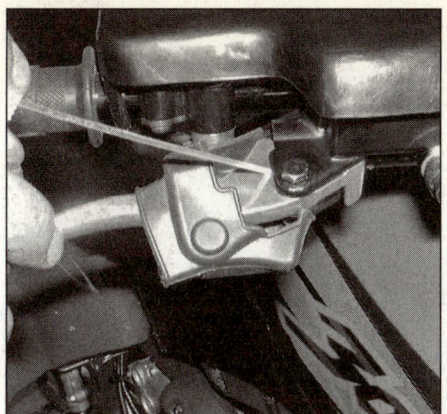

11.2a Lubricate the brake lever pivot bolt . . .

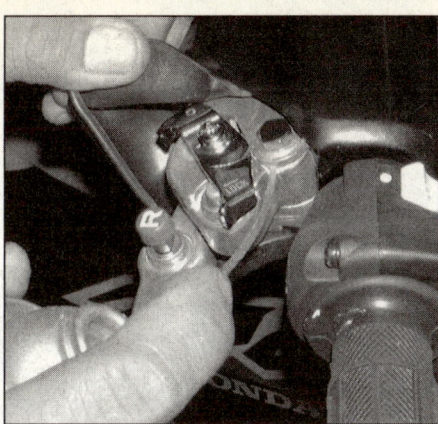

11.2b . . . and the clutch lever pivot bolt regularly

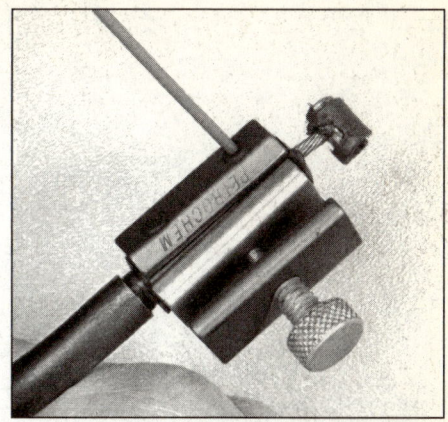

11.3 Lubricating a cable with a pressure lube adapter (make sure the tool seats around the inner cable)

tires is indicated by an arrow molded into the tire sidewall. If your vehicle is equipped with directional tires, make sure the tires are installed on the correct side of the vehicle.

6   The wheels used on this machine are virtually maintenance free, but they should be kept clean and checked periodically for cracks, bending and rust. Never attempt to repair damaged wheels; they must be replaced with new ones.

7   Check the valve stem locknuts to make sure they're tight. Also, make sure the valve stem cap is in place and tight. If it is missing, install a new one made of metal or hard plastic.

10   Front wheel bearings - check

Refer to illustration 10.3

1   Raise the front of the vehicle and support it securely on jackstands.

2   Spin the front wheels by hand. Listen for noise, which indicates dry or worn wheel bearings.

3   Grasp the top and bottom of the tire and try to rock it back-and-forth **(see illustration)**. If there's more than a very small amount of play, the wheel bearings are in need of adjustment or replacement. Refer to Chapter 6 for service procedures.

11   Lubrication - general

Refer to illustration 11.2a, 11.2b and 11.3

1   Since the controls, cables and various other components of an ATV are exposed to the elements, they should be lubricated periodically to ensure safe and trouble-free operation.

2   The throttle lever, brake levers and brake pedal should be lubricated frequently. Use motor oil inside the cables and for the pivot points of levers **(see illustrations)** and pedals. Multi-purpose lithium grease is recommended for cable ends and other lubrication points, such as suspension and steering bushings. In order for the lubricant to be applied where it will do the most good, the component should be disassembled. However, if chain and cable lubricant is being used, it can be applied to the pivot joint gaps and will usually work its way into the areas where friction occurs. If motor oil or light grease is being used, apply it sparingly as it may attract dirt (which could cause the controls to bind or wear at an accelerated rate). **Note:** *One of the best lubricants for the control lever pivots is a dry-film lubricant (available from many sources by different names).*

3   The throttle and clutch cables should be removed and treated with a commercially available cable lubricant which is specially formu-lated for use on ATV control cables. Small adapters for pressure lubricating the cables with spray can lubricants are available and ensure that the cable is lubricated along its entire length **(see illustration)**. When attaching a cable to its handlebar lever, be sure to lubricate the barrel-shaped fitting at the end with multi-purpose grease.

4   To lubricate the cables, disconnect one end, then lubricate the cable with a pressure lube adapter **(see illustration 11.3)**. (For the clutch cable, see Chapter 2; for the throttle cable, see Chapter 3.)

5   Refer to Chapter 5 for the following lubrication procedures:

   a)   *Upper steering shaft bushing*
   b)   *Swingarm bearings and dust seals*

6   Refer to Chapter 6 for the following lubrication procedures:

   a)   *Rear brake pedal pivot*
   b)   *Front wheel bearings*

12   Fasteners - check

1   Since vibration of the machine tends to loosen fasteners, all nuts, bolts, screws, etc. should be periodically checked for proper tightness. Also make sure all cotter pins or other safety fasteners are correctly installed.

2   Pay particular attention to the following:

   Spark plug
   Transmission oil drain plug
   Gearshift pedal
   Brake pedal
   Footpegs
   Engine mount bolts
   Shock absorber mount bolts
   Front axle nuts
   Rear axle nuts
   Skid plate bolts

3   If a torque wrench is available, use it along with the torque specifications at the beginning of this, or other, Chapters.

13   Skid plates - check

1   Check the skid plates under the vehicle for damage (see Chapter 7). The front skid plate also includes the front bumper. Have damaged plates repaired, or else replace them.

2   Make sure the skid plate fasteners are all in position and tightly secured.

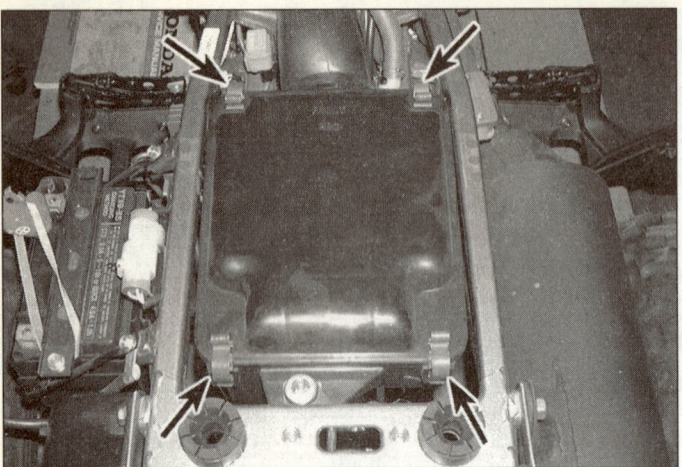

14.2a To remove the air cleaner housing cover on a TRX300EX, disengage these four spring clips (arrows)

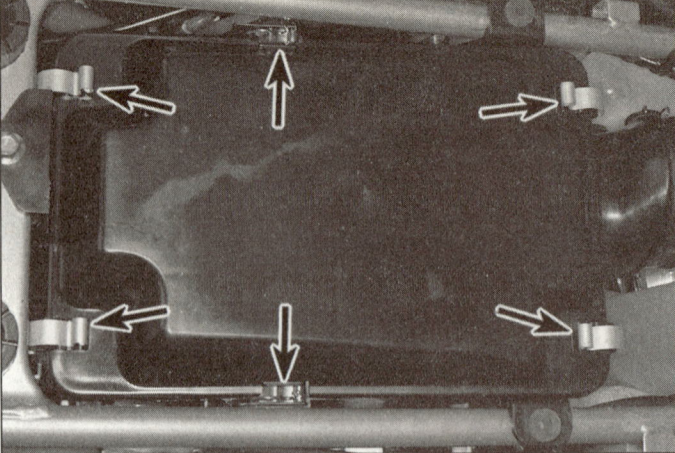

14.2b The air cleaner housing cover on TRX400EX models also has four spring clips (arrows) and two wire retainer clips (arrows)

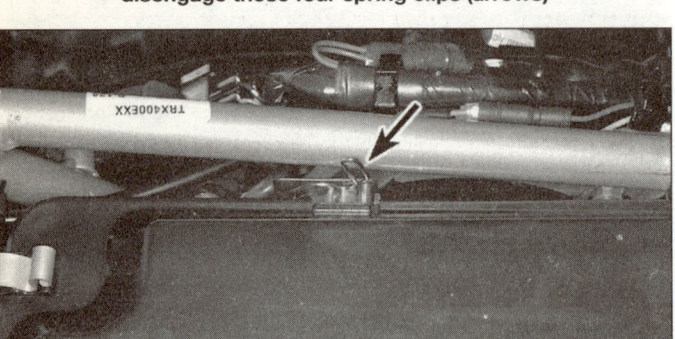

14.2c Wire retainer clip in closed position (TRX400EX)

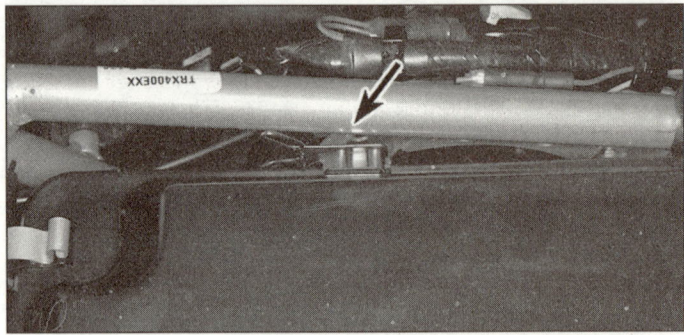

14.2d Wire retainer clip in open position (TRX400EX)

14   Air filter element and drain tube - cleaning

Element cleaning

Refer to illustrations 14.2a, 14.2b, 14.2c, 14.2d, 14.3a, 14.3b and 14.7

1   Remove the seat and rear fenders (see Chapter 7).

2   Remove the cover from the air cleaner housing. On TRX300EX models, it's secured by four spring clips **(see illustration)**; on TRX400EX models, the cover has four spring clips and a pair of wire clips **(see illustrations)**.

3   Remove the air filter element **(see illustration)** and separate the element from the frame **(see illustration)**.

4   Clean the element and frame in a high flash point solvent,

squeeze the solvent out of the foam and let the guide and element dry completely.

5   Soak the foam element in the foam filter oil listed in this Chapter's Specifications, then squeeze it firmly to remove the excess oil. Don't wring it out or the foam may be damaged. The element should be thoroughly oil-soaked, but not dripping.

6   Reassemble the element and frame **(see illustration 14.3b)**.

7   Before installing the air cleaner housing cover, inspect the condition of the cover gasket. If it's damaged or worn, replace it **(see illustration)**.

8   When installing the air cleaner housing cover, make sure that the cover is placed on the housing with the FRONT marking **(see illustration)** at the forward end of the housing.

9   Installation is otherwise the reverse of removal.

14.3a Loosen this hose clamp and lift the air element and frame out of the housing

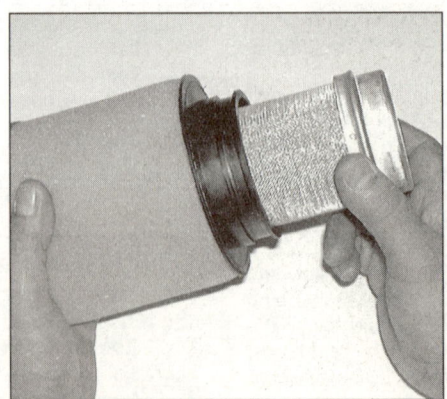

14.3b To detach the air filter element from the frame, simply pull it off

14.7 Before installing the air cleaner housing cover, inspect and, if necessary, replace the cover gasket

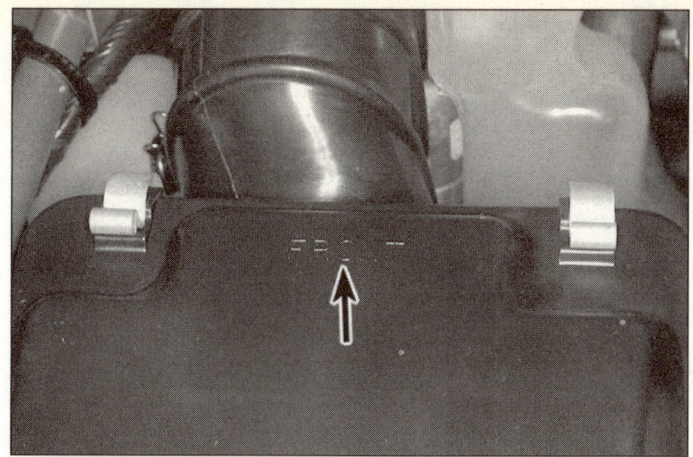

14.8 Be sure to install the air cleaner housing cover with the FRONT marking facing forward

Drain tube cleaning

Refer to illustration 14.10

10   Check the drain tube **(see illustration)** for accumulated water and oil. If oil or water has built up in the tube, squeeze its clamp, remove the tube from the air cleaner housing and clean it out. Install the drain tube on the housing and secure it with the clamp. **Note:** *A drain tube that's full indicates the need to clean the filter element and the inside of the air cleaner housing.*

15   Fuel system - inspection

Refer to illustrations 15.1a, 15.1b, 15.5a, 15.5b and 15.6

Warning: *Gasoline is extremely flammable, so take extra precautions when you work on any part of the fuel system. Don't smoke or allow open flames or bare light bulbs near the work area, and don't work in a garage where a natural gas-type appliance (such as a water heater or clothes dryer) is present. Since gasoline is carcinogenic, wear latex gloves when there's a possibility of being exposed to fuel, and if you spill any fuel on your skin, rinse it off immediately with soap and water. Mop up any fuel spills immediately and do not store fuel-soaked rags where they could ignite. When you perform any kind of work on the fuel system, wear safety glasses and have a fire extinguisher suitable for class B type fires (flammable liquids) on hand.*

1   Inspect the fuel tank, the fuel tap, the fuel line and the carburetor for leaks and evidence of damage **(see illustrations)**.
2   If carburetor gaskets are leaking, the carburetor should be disassembled and rebuilt (see Chapter 3).

14.10 Inspect the drain tube (arrow) (TRX400EX shown, TRX300EX similar)

15.1a Inspect the fuel tap and the fuel line (arrow) for leaks (fuel tank raised slightly for clarity)

3   If the fuel tap is leaking, tightening the screws may help. If leakage persists, the tap should be disassembled and repaired or replaced with a new one.
4   If the fuel line is cracked or otherwise deteriorated, replace it with a new one.
5   Place the fuel tap lever in the Off position. Remove the fuel tank (see Chapter 3) and drain it into a suitable container. Remove the strainer housing cover bolt and remove the strainer **(see illustrations)**. Then unscrew the fuel tap nut and remove the fuel tap from the tank.

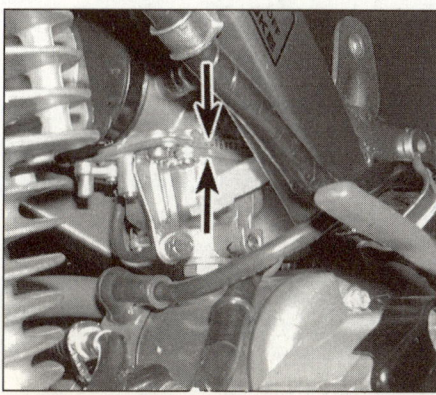

15.1b Inspect the seam (arrows) between the carburetor housing and the float bowl for leaks

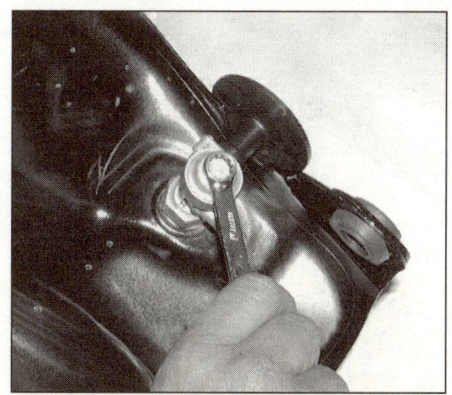

15.5a Remove the fuel tap strainer housing bolt . . .

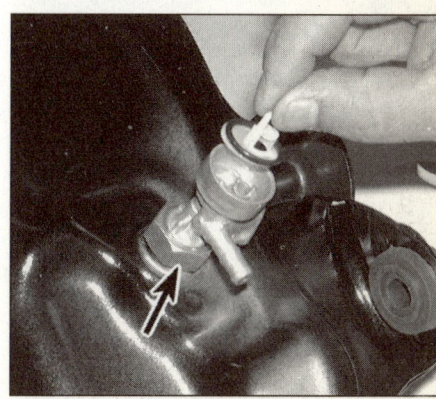

15.5b . . . the strainer and O-ring; unscrew the tap from the tank with a wrench on the hex (arrow)

6   Clean the strainer and the filter element with solvent and let them dry; don't try to blow these parts dry with compressed air. Inspect the fuel tap **(see illustration)** and replace any damaged parts. Reassemble the fuel tap and install the strainer. Be sure to use a new O-ring. Install the strainer cover and tighten the cover bolt securely, but do not over-tighten it. If you do, the O-ring will be distorted, which will result in fuel leaks.

7   Install the fuel tap. Tighten the fuel tap nut securely.

8   Install the fuel tank (see Chapter 3).

9   After installation, run the engine and check for fuel leaks.

10   Anytime the vehicle is going to be stored for a month or more, remove and drain the fuel tank. Also loosen the float chamber drain screw and drain the fuel from the carburetor.

11   Inspect the condition of the crankcase breather hose. Replace it if it's cracked, torn or deteriorated.

---

16   Spark plug - inspection, cleaning and gapping

Refer to illustrations 16.1a, 16.1b, 16.5a and 16.5b

1   Twist the spark plug cap **(see illustration)** to break it free from the plug, then pull it off. If available, use compressed air to blow any accumulated debris from around the spark plug. Remove the plug with a spark plug socket **(see illustration)**.

2   Inspect the electrodes for wear. Both the center and side elec-trodes should have square edges and the side electrode should be of uniform thickness. Look for excessive deposits and evidence of a cracked or chipped insulator around the center electrode. Compare your spark plugs to the color spark plug reading chart on the inside back cover of this manual. Check the threads, the washer and the ceramic insulator body for cracks and other damage.

3   If the electrodes are not excessively worn, and if the deposits can be easily removed with a wire brush, the plug can be regapped and reused (if no cracks or chips are visible in the insulator). If in doubt con-cerning the condition of the plug, replace it with a new one, as the expense is minimal.

4   Cleaning the spark plug by sandblasting is permitted, provided you clean the plug with a high flash-point solvent afterwards.

5   Before installing a new plug, make sure it is the correct type and heat range. Check the gap between the electrodes, as it is not preset. For best results, use a wire-type gauge rather than a flat gauge to check the gap **(see illustration)**. If the gap must be adjusted, bend the side electrode only and be very careful not to chip or crack the insula-tor nose **(see illustration)**. Make sure the washer is in place before installing the plug.

6   Since the cylinder head is made of aluminum, which is soft and easily damaged, thread the plug into the head by hand. Slip a short length of hose over the end of the plug to use as a tool to thread it into place. The hose will grip the plug well enough to turn it, but will start to

15.6 Fuel tap details

1   *Fuel tap strainer cover/bolt*
2   *Fuel tap strainer O-ring*
3   *Fuel tap strainer (clean with solvent)*
4   *Fuel tap*
5   *Fuel tap filter (clean with solvent)*

16.1a To remove the spark plug cap, twist and pull at the same time

16.1b Unscrew the plug with a spark plug socket

16.5a Spark plug manufacturers recommend using a wire type gauge when checking the gap - if the wire doesn't slide between the electrodes with a slight drag, adjustment is required

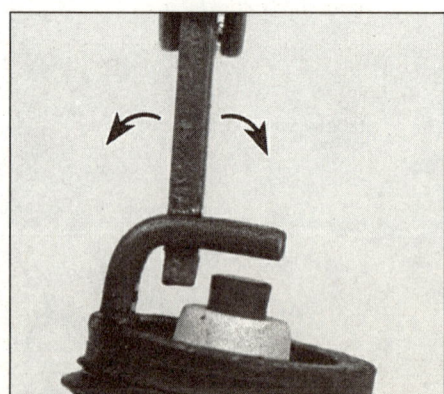

16.5b To change the gap, bend the side electrode only, as indicated by the arrows, and be very careful not to crack or chip the ceramic insulator surrounding the center electrode

slip if the plug begins to cross-thread in the hole - this will prevent damaged threads and the accompanying repair costs.

7   Once the plug is finger tight, the job can be finished with a socket. If a torque wrench is available, tighten the spark plug to the torque listed in this Chapter's Specifications. If you do not have a torque wrench, tighten the plug finger tight (until the washer bottoms on the cylinder head) then use a spark plug socket to tighten it an additional _ turn. Regardless of the method used, do not over-tighten it.

8   Reconnect the spark plug cap.

17 Engine oil and filter - change

Refer to illustrations 17.5a, 17.5b, 17.5c, 17.5d, 16.6a, 16.6b, 16.6c and 17.6d

1   Consistent routine oil and filter changes are the single most important maintenance procedure you can perform on a vehicle. The oil not only lubricates the internal parts of the engine, transmission and clutch, but it also acts as a coolant, a cleaner, a sealant and a protectant. Because of these demands, the oil takes a terrific amount of abuse and should be replaced often with new oil of the recommended grade and type.

2   Before changing the oil and filter, warm up the engine so the oil will drain easily. Be careful when draining the oil, as the exhaust pipe, the engine and the oil itself can cause severe burns.

3   Park the vehicle over a clean drain pan.

4   Remove the dipstick/oil filler cap to vent the crankcase and act as a reminder that there is no oil in the engine.

5   Remove the drain plug and washer from the engine and allow the

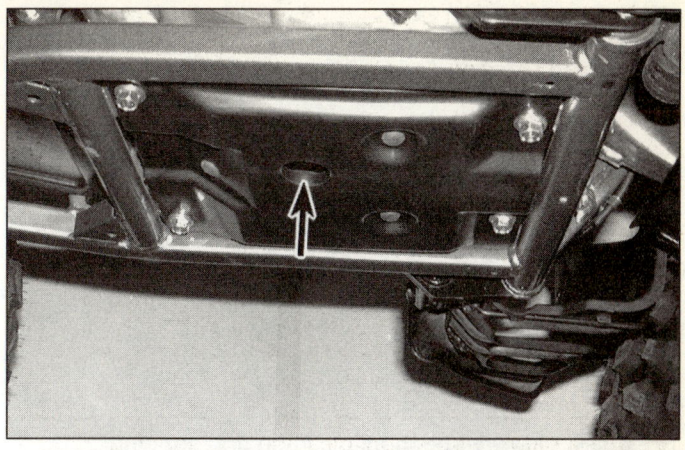

17.5a The oil drain plug on TRX300EX models is accessible through this hole (arrow) in the skid plate

oil to drain into the pan. On TRX300EX models, the drain plug can be reached through a hole **(see illustration)** in the skid plate; it's not necessary to remove the skid plate to remove the drain plug. On TRX400EX models, there are two drain plugs: one is located on the left side of the engine **(see illustration)**, just below the gearshift pedal, and the other is at the bottom of the oil tank **(see illustration)**. Discard the old drain plug washer **(see illustration)**.

6   Remove the oil filter cover bolts, then remove the cover and filter element **(see illustrations)**. Remove and inspect the O-rings; if they're

17.5b On TRX400EX models, there are two oil drain plugs: one is on the left side of the engine, below the gearshift pedal (arrow) . . .

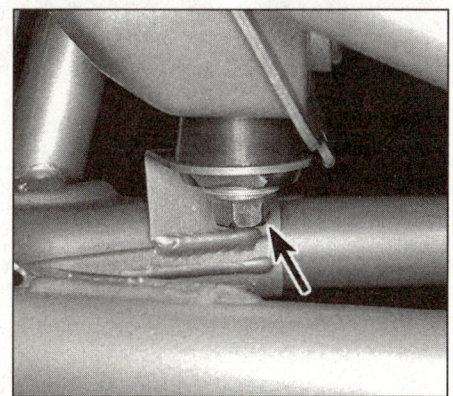

17.5c . . . and the other is on the bottom of the oil tank (arrow)

17.5d Be sure to discard the old drain plug washer(s) and install a new one when reinstalling the drain plug(s)

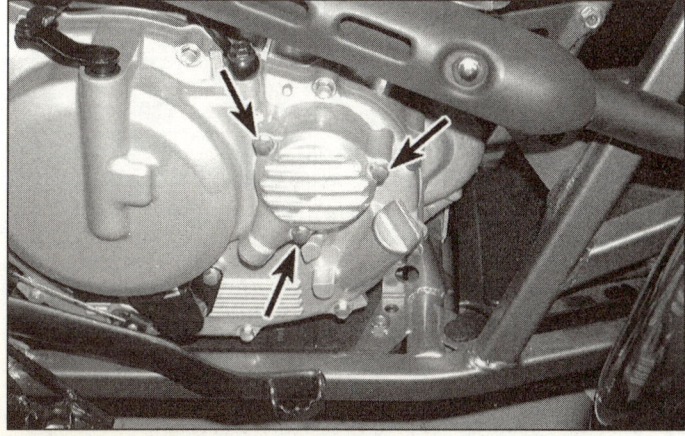

17.6a Remove the three filter cover bolts (arrows) on TRX300EX models, . . .

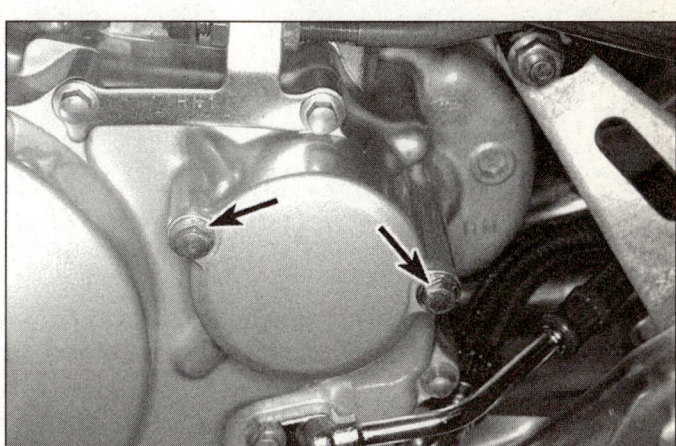

17.6b . . . or two bolts (arrows) on TRX400EX models . . .

1

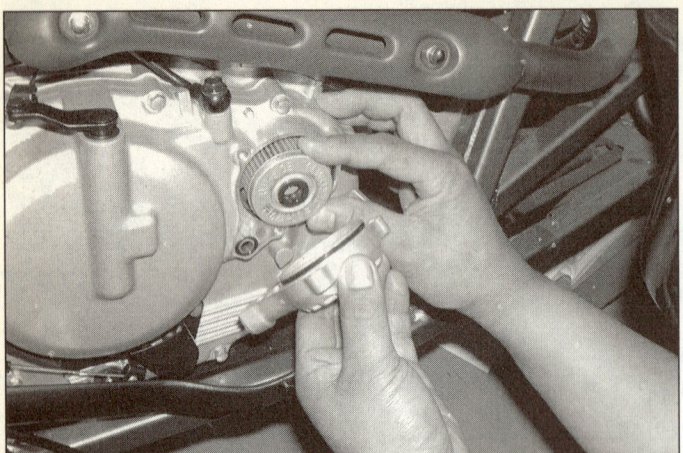

17.6c . . . lift off the cover and pull out the filter element; be sure the marked side of the element faces outward on installation

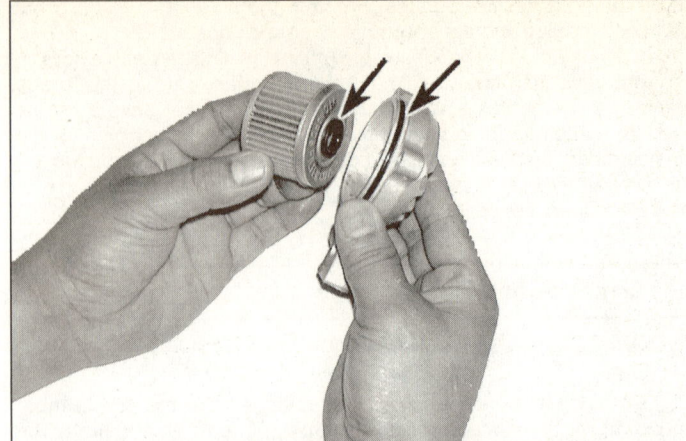

17.6d There are two O-rings: the smaller one (left arrow) goes on the filter and the larger one (right arrow) goes on the filter cover

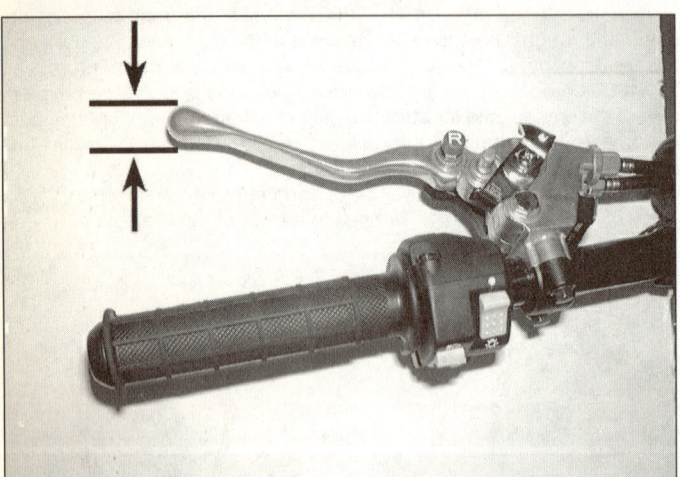

18.2 Measure clutch lever freeplay at the lever tip

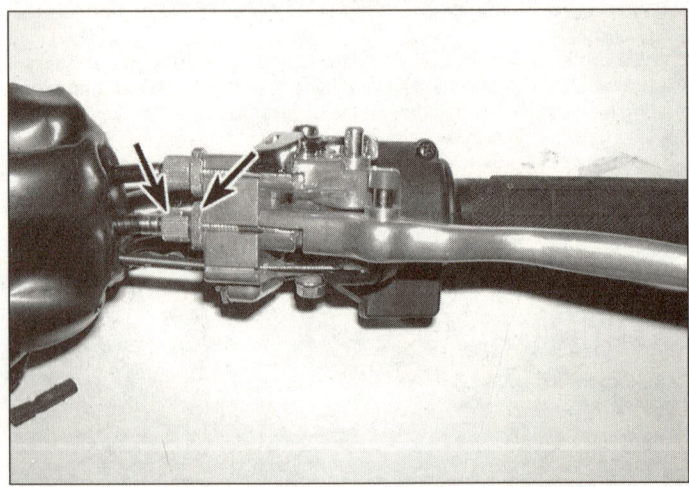

18.3 Peel back the dust cover, loosen the clutch cable locknut (right arrow), then turn the adjuster (left arrow) to adjust clutch lever freeplay

damaged or worn, replace them. If you're planning to do any other maintenance, this would be a good time to do so, while the oil is allowed to drain completely.

7   Wipe any remaining oil out of the filter housing area of the crankcase and make sure the oil passage is clear.

8   Inspect the drain plug threads; if they're stripped, replace the drain plug.

9   Install the filter element. **Caution:** *The filter must be installed facing the correct direction or oil starvation may cause severe engine damage.*

10   Install the cover with a new O-ring and tighten the bolts to the torque listed in this Chapter's Specifications.

11   Install the drain plug and tighten it to the torque listed in this Chapter's Specifications. Avoid overtightening, as damage to the engine case will result. Don't forget to install a new drain plug washer.

12   Before refilling the engine, check the old oil carefully. If the oil was drained into a clean pan, small pieces of metal or other material can be easily detected. If the oil is very metallic colored, then the engine is experiencing wear from break-in (new engine) or from insufficient lubrication. If there are flakes or chips of metal in the oil, then something is drastically wrong internally and the engine will have to be disassembled for inspection and repair.

13   If there are pieces of fiber-like material in the oil, the clutch is undergoing excessive wear and should be checked.

14   If the inspection of the oil turns up nothing unusual, refill the crankcase to the proper level with the recommended oil and install the dipstick/filler cap.

15   Start the engine and let it run for two or three minutes. Shut it off,

wait a few minutes, and then check the oil level. If necessary, add more oil to bring the level up to the upper level mark on the dipstick. Check around the drain plug and filter cover for leaks.

16   The old oil drained from the engine must be disposed of properly. Check with your local refuse disposal company, recycling facility or environmental agency to see if they will accept the oil for recycling. Don't pour used oil into drains or onto the ground. After the oil has cooled, it can be drained into a suitable container (capped plastic jugs, topped bottles, milk cartons, etc.) for transport to an appropriate disposal site.

18 Clutch cable - check and adjustment

1   Clutch cable freeplay is usually adjusted at the clutch lever. Freeplay can also be adjusted at the crankcase end of the clutch cable, but this is normally unnecessary.

Adjusting freeplay at the handlebar

Refer to illustrations 18.2 and 18.3

2   Operate the clutch lever and check freeplay at the lever tip **(see illustration)**.

3   If freeplay isn't within the range listed in this Chapter's Specifications, pull off the rubber dust cover, loosen the adjuster locknut, turn the adjuster nut to set freeplay and tighten the locknut **(see illustration)**.

18.4 On TRX300EX models, loosen the locknut (right arrow) and
turn the adjuster nut (left arrow) until the clutch lifter arm
has no freeplay

18.5 On TRX400EX models, loosen the locknut (left arrow) and
turn the adjuster nut (right arrow) until the clutch lifter
arm has no freeplay

19.3 To adjust the idle speed, turn the throttle stop screw (arrow)
in to increase idle speed or out to decrease it

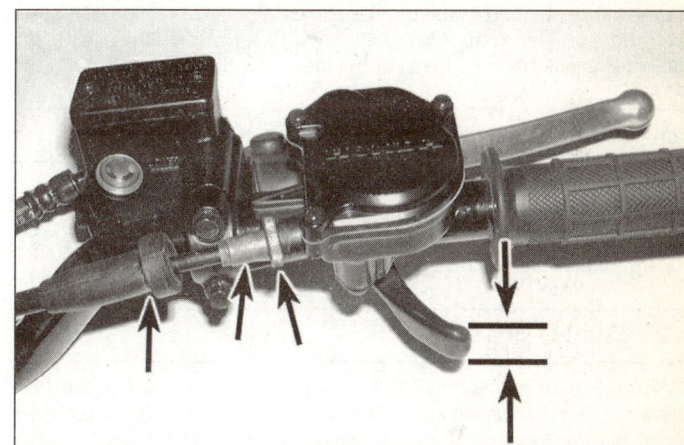

20.3 Check throttle lever freeplay at the lever tip; to adjust it, pull
back the dust boot (left arrow), loosen the locknut (right arrow)
and turn the adjuster (center arrow)

Adjusting freeplay at the crankcase

Refer to illustrations 18.4 and 18.5

Note: *Adjusting freeplay at the crankcase is normally unnecessary
unless you are replacing the clutch cable or have disconnected the
cable to remove or service the engine.*

4   On TRX300EX models, loosen the clutch cable locknut at the
cable guide on top of the crankcase **(see illustration)**. Turn the
adjuster nut in (clockwise, from the left side of the machine) to remove
all freeplay from the clutch lifter arm, then tighten the locknut.

5   On TRX400EX models, loosen the clutch cable locknut at the
cable guide on the right side of the crankcase **(see illustration)**. Turn
the adjuster nut in to remove all freeplay from the clutch lifter arm, then
tighten the locknut.

6   Adjust clutch lever freeplay at the handlebar (see above).

19 Idle speed - check and adjustment

Refer to illustration 19.3

1   Before adjusting the idle speed, make sure the spark plug gap is
correct (see Section 16). Also, turn the handlebars back-and-forth and
note whether the idle speed changes. If it does, the throttle cable may be
incorrectly routed. Be sure to correct this problem before proceeding.

2   Start the engine and warm it up to its normal operating temperature. Make sure the transmission is in Neutral, then hook up an induc-

tive-type tachometer.

3   Turn the idle stop screw **(see illustration)** to bring the idle speed
within the range listed in this Chapter's Specifications. Turning the
screw in increases the idle speed; backing it out decreases the idle
speed.

4   Snap the throttle open and shut a few times, then recheck the idle
speed. If necessary, repeat the adjustment procedure.

5   If a smooth, steady idle can't be achieved, the air/fuel mixture
may be incorrect (see Chapter 3).

20 Throttle cable - check and adjustment

Refer to illustrations 20.3

1   Before proceeding, check and, if necessary, adjust the idle speed
(see Section 19).

2   Make sure the throttle lever moves easily from fully closed to fully
open with the front wheel turned at various angles. The lever should
return automatically from fully open to fully closed when released. If
the throttle sticks, check the throttle cable for cracks or kinks in the
housing. Also, make sure the inner cable is clean and well-lubricated.

3   Measure freeplay at the throttle lever **(see illustration)**. If it's
within the range listed in this Chapter's Specifications, no adjustment
is necessary. If not, adjust it as follows.

23.6a The TRX300EX valve adjuster covers are secured by bolts (arrows)

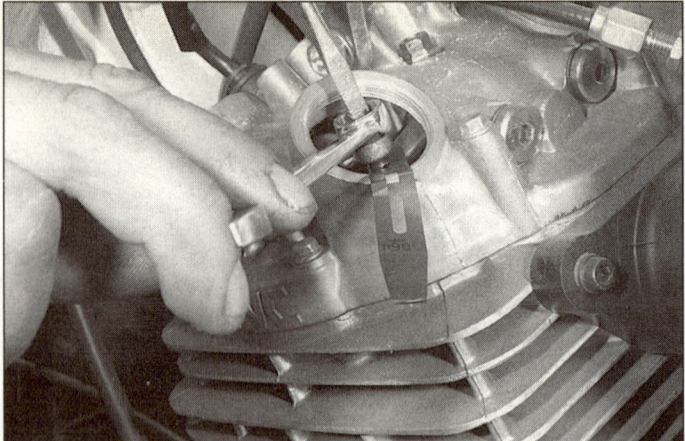

23.6b The TRX400EX valve adjuster covers are screwed into the valve cover

4   Pull back the rubber boot **(see illustration 20.3)** from the adjuster at the handlebar end of the throttle cable.
5   Loosen the adjuster locknut **(see illustration 20.3)**. Turn the adjuster to set freeplay, then tighten the locknut.

21   Choke - operation check

1   Operate the choke lever and note whether it operates smoothly.
2   If the choke lever doesn't operate smoothly, inspect the choke lever assembly (see "Carburetor - disassembly, cleaning and inspection" and "Carburetor - reassembly and float height check" in Chapter 3).

22   Battery - check

Warning: *Be extremely careful when handling or working around the battery. The electrolyte is very caustic and an explosive gas (hydrogen) is given off when the battery is charging.*
Note: *The original-equipment battery is a sealed maintenance-free unit. This procedure applies only to aftermarket replacement batteries with filler caps.*
1   This procedure only applies to aftermarket batteries with removable filler caps which can be removed to add water to the battery. Unless the original equipment battery has been replaced by an aftermarket unit, the electrolyte can't be topped up.
2   Remove the seat and rear fenders (see Chapter 7).
3   To remove the battery, disconnect the hold-down bolt and lift the battery up partially. The electrolyte level is visible through the translucent battery case - it should be between the Upper and Lower level marks.
4   If the electrolyte is low, remove the cell caps and fill each cell to the upper level mark with distilled water. Do not use tap water (except in an emergency) and do not overfill. The cell holes are quite small, so it may help to use a plastic squeeze bottle with a small spout to add the water. If the level is within the marks on the case, additional water is not necessary.
5   Next, check the specific gravity of the electrolyte in each cell with a small hydrometer made especially for motorcycle batteries. These are available from most dealer parts departments or motorcycle accessory stores.
6   Remove the caps, draw some electrolyte from the first cell into the hydrometer, and then note the specific gravity. Compare the reading to the value listed in the hydrometer manufacturer's instructions.
Note: Add 0.004 points to the reading for every 10-degrees F above 68-degrees F (20-degrees C) - subtract 0.004 points from the reading for every 10-degrees below 68-degrees F (20-degrees C).
7   Return the electrolyte to the appropriate cell and repeat the check for the remaining cells. When the check is complete, rinse the hydrom-

eter thoroughly with clean water.
8   If the specific gravity of the electrolyte in each cell is as specified, the battery is in good condition and is apparently being charged by the machine's charging system.
9   If the specific gravity is low, the battery is not fully charged. This may be due to corroded battery terminals, a dirty battery case, a malfunctioning charging system, or loose or corroded wiring connections. On the other hand, it may be that the battery is worn out, especially if the machine is old, or that infrequent use of the machine prevents normal charging from taking place.
10   Be sure to correct any problems and charge the battery if necessary. Refer to Chapter 8 for additional battery maintenance and charging procedures.
11   Install the battery cell caps, tightening them securely. Reconnect the cables to the battery, attaching the positive cable first and the negative cable last. Make sure to install the insulating boot over the positive terminal.
12   Install all components removed for access and make sure the battery vent tube is routed correctly. Be very careful not to pinch or otherwise restrict the tube, as the battery may build up enough internal pressure during normal charging system operation to explode.
13   If the vehicle will be stored for an extended time, fully charge the battery, then remove it. Disconnect the negative cable and remove the battery retainer strap. Disconnect the positive cable and vent tube and lift the battery out. **Warning:** Always disconnect the negative cable first and reconnect it last to avoid sparks which could cause a battery explosion.
14   Store the battery in a cool dark place. Check specific gravity at least once a month and recharge the battery if it's low.

23   Valve clearance - check and adjustment

Refer to illustrations 23.6a, 23.6b, 23.7, 23.8, 23.10 and 23.11
1   The engine must be cool to the touch for this maintenance procedure, so if possible let the machine sit overnight before beginning.
2   Remove the seat and rear fenders (see Chapter 7).
3   Disconnect the cable from the negative terminal of the battery (see "Battery – removal and installation" in Chapter 8).
4   Remove the front fender (see Chapter 7) and the fuel tank (see Chapter 3).
5   Remove the spark plug (see Section 16). This will make it easier to turn the crankshaft.
6   Remove the valve adjuster covers **(see illustrations)**.
7   Remove the timing hole plug and the crankshaft hole plug **(see illustration)**.
8   Position the piston at Top Dead Center (TDC) on the compression stroke as follows: Turn the crankshaft until the mark on the flywheel is

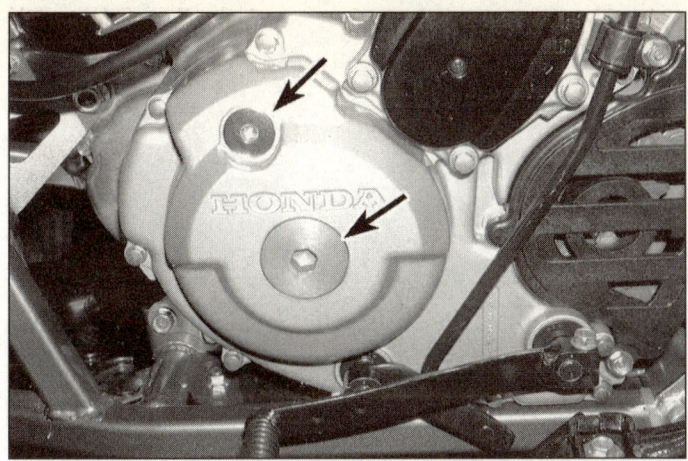

23.7 Remove the timing hole plug and the crankshaft hole plug (arrows)

23.8 Align the "T" mark on the flywheel with the notch in the left crankcase cover

23.10 Measure valve clearance with a feeler gauge of the specified thickness

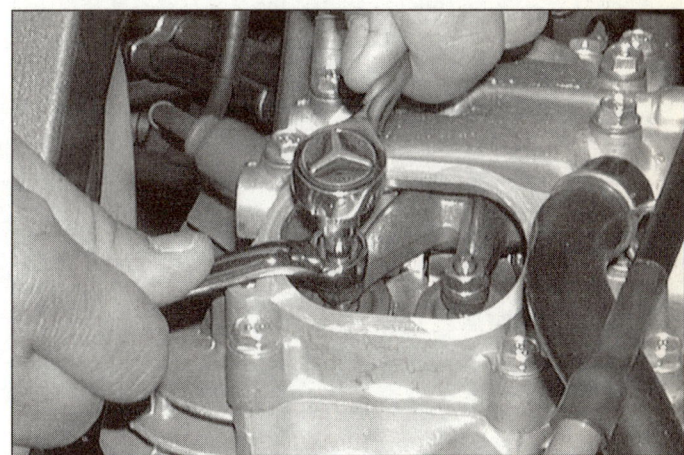

23.11 To adjust the valve clearance, loosen the locknut and turn the adjuster

aligned with the timing notch on the crankcase **(see illustration)**. Wiggle the rocker arms - there should be some play. If the rocker arms are tight, the engine is at TDC on the exhaust stroke. Rotate it one full turn, until the timing mark and notch are aligned again. Recheck to make sure the rocker arms are now loose.

9   With the engine in this position, all four valves can be checked. Start with the intake valves.

10   To check, insert a feeler gauge of the thickness listed in this Chapter's Specifications between the valve stem and rocker arm **(see illustration 23.6b and the accompanying illustration)**. Pull the feeler gauge out slowly - you should feel a slight drag. If there's no drag, the clearance is too loose. If there's a heavy drag, the clearance is too tight.

11   If the clearance is incorrect, loosen the adjuster locknut with a box-end wrench. Turn the adjusting screw **(see illustration)** until the clearance is correct, then tighten the locknut. For the TRX300EX, Honda manufactures a special wrench that allows you to hold the locknut and turn the adjuster with one hand while holding the feeler gauge with the other hand, but a box wrench and socket (or two wrenches) will also work. Use a box wrench and screwdriver for TRX400EX models.

12   After adjusting, recheck the clearance with the feeler gauge to make sure it wasn't changed when the locknut was tightened.

13   Now measure the other valve, following the same procedure you used for the first valve. Make sure to use a feeler gauge of the specified thickness.

14   After the clearances have been checked and adjusted for the intake valves, proceed to the exhaust valves.

15   When the clearances are correct for all four valves, install the valve adjusting hole covers. Be sure to use new O-rings on the covers if the old ones are hardened, deteriorated or damaged.

16   Apply silicone sealant to the threads of the timing hole and crankshaft hole plugs and install both plugs. Tighten the plugs to the torque listed in this Chapter's Specifications.

17   The remainder of installation is the reverse of removal.

24   Cylinder compression - check

Refer to illustration 24.5

1   Among other things, poor engine performance may be caused by leaking valves, incorrect valve clearances, a leaking head gasket, or worn piston, rings and/or cylinder wall. A cylinder compression check will help pinpoint these conditions and can also indicate the presence of excessive carbon deposits in the cylinder head.

2   The only tools required are a compression gauge and a spark plug wrench. Depending on the outcome of the initial test, a squirt-type oil can may also be needed.

3   Check valve clearance and adjust if necessary (see Section 23). Start the engine and allow it to reach normal operating temperature, then remove the spark plug (see Section 16). Work carefully - don't strip the spark plug hole threads and don't burn your hands.

4   Disable the ignition by disconnecting the primary wires from the coil (see Chapter 4). Be sure to mark the locations of the wires before detaching them.

24.5 A compression gauge with a threaded fitting for the spark plug hole is preferable to the type that requires hand pressure to maintain the seal

25.5a On TRX300EX models, remove these two bolts (arrows), the plate and gasket . . .

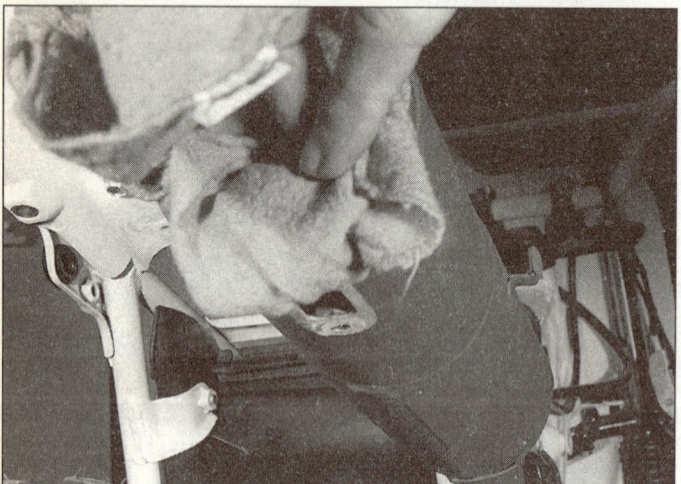

25.5b . . . then hold a rag against the muffler opening and rev the engine a few times to blow the carbon out of the spark arrester

25.5c DO NOT remove these two screws on TRX300EX models

5   Install the compression gauge in the spark plug hole **(see illustration)**. Hold or block the throttle wide open.
6   Crank the engine over a minimum of four or five revolutions (or until the gauge reading stops increasing) and observe the initial movement of the compression gauge needle as well as the final total gauge reading. Compare the results to the value listed in this Chapter's Specifications.
7   If the compression built up quickly and evenly to the specified amount, you can assume the engine upper end is in reasonably good mechanical condition. Worn or sticking piston rings and a worn cylinder will produce very little initial movement of the gauge needle, but compression will tend to build up gradually as the engine spins over. Valve and valve seat leakage, or head gasket leakage, is indicated by low initial compression which does not tend to build up.
8   To further confirm your findings, add a small amount of engine oil to the cylinder by inserting the nozzle of a squirt-type oil can through the spark plug hole. The oil will tend to seal the piston rings if they are leaking.
9   If the compression increases significantly after the addition of the oil, the piston rings and/or cylinder are definitely worn. If the compression does not increase, the pressure is leaking past the valves or the head gasket. Leakage past the valves may be due to insufficient valve clearances, burned, warped or cracked valves or valve seats or valves that are hanging up in the guides.

10   If compression readings are considerably higher than specified, the combustion chamber is probably coated with excessive carbon deposits. It is possible (but not very likely) for carbon deposits to raise the compression enough to compensate for the effects of leakage past rings or valves. Refer to Chapter 2, remove the cylinder head and carefully decarbonize the combustion chamber.

25   Exhaust system - inspection and spark arrester cleaning

Refer to illustrations 25.5a, 25.5b, 25.5c, 25.6a and 25.6b
1   Periodically, inspect the exhaust system for leaks and loose fasteners.
2   The exhaust pipe flange nuts at the cylinder head are especially prone to loosening, which could cause damage to the head. Check them frequently and keep them tight. If tightening the flange nuts fails to stop the leak, replace the exhaust pipe gasket(s) (see Chapter 3).
3   Inspect the heat shield fasteners, the clamp screw that secures the muffler to the exhaust pipe, and the muffler mounting bolts. Make sure all of these fasteners are tight.
4   A spark arrester in the muffler prevents the discharge of sparks which might start a fire. Clean the spark arrester at the specified interval to ensure that it functions effectively.
5   On TRX300EX models, remove the plate and gasket from the

25.6a On TRX400EX models, remove these three bolts (arrows) . . .

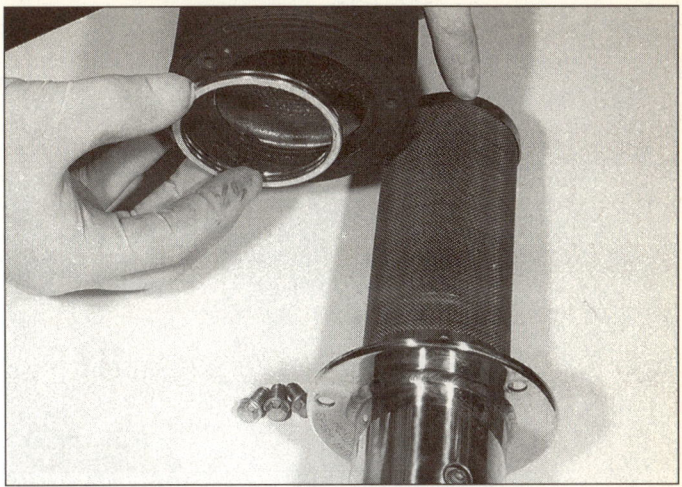

25.6b . . . and pull out the spark arrester for cleaning; install it with a new gasket

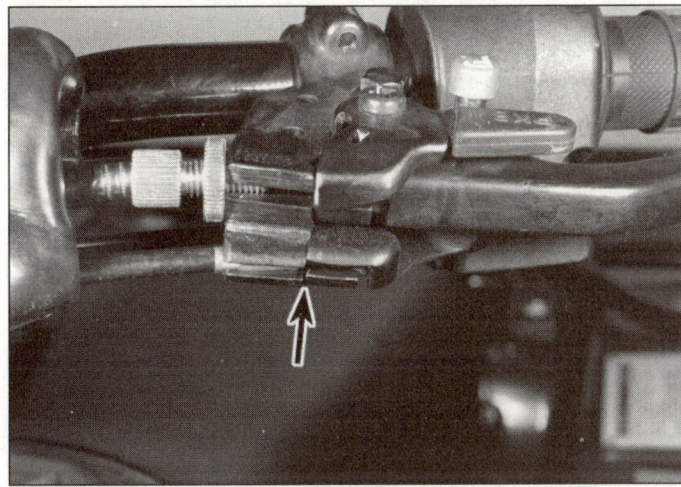

26.2 Measure reverse selector cable freeplay at the gap (arrow) between the reverse lock lever and the clutch and parking brake lever bracket . . .

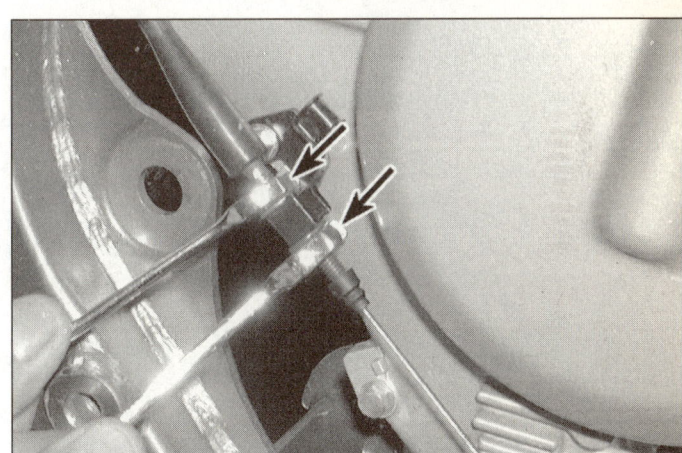

26.3 . . . to adjust freeplay, loosen the cable locknut (lower arrow) and turn the adjuster nut (upper arrow) to produce the correct gap

underside of the muffler **(see illustrations)**. Hold a rag firmly over the exhaust outlet at the rear end of the muffler **(see illustration)**, then have an assistant start the engine and rev it up a few times to blow carbon out of the plate hole. Shut off the engine. After the exhaust has cooled, install the plate and gasket and tighten the bolts securely. **Caution:** *Do NOT remove the two screws* **(see illustration)** *in the end of the muffler; these screws secure the exhaust baffle to the muffler/spark arrester and must be installed for the spark arrester to function properly.*
6   On TRX400EX models, remove the three bolts from the end of the muffler **(see illustration)**, pull out the spark arrester **(see illustration)**, discard the old gasket and clean the spark arrester with a wire brush. Install the spark arrester and a new gasket. Tighten the bolts securely.

26   Reverse selector cable (TRX300EX models) - check and adjustment

Refer to illustrations 26.2 and 26.3
1   Trace the reverse selector cable from its lever on the handlebar (below the clutch lever) to the reverse selector arm at the rear of the right crankcase cover. Look for kinks, bends, loose retainers or other problems and correct them as necessary.
2   To determine cable freeplay, measure the gap between the reverse selector lever and the clutch and parking brake lever bracket **(see illustration)**. If it's not within the range listed in this Chapter's Specifications, adjust it as follows.
3   Loosen the locknut at the cable bracket on the right crankcase cover **(see illustration)**. Turn the adjusting nut to produce the correct gap at the handlebar, then tighten the locknut securely.

Notes

Chapter 2 Part A
TRX300EX engine, clutch and transmission

Contents

2A

Specifications

General

Bore	74 mm (2.91 inches)
Stroke	65.5 mm (2.58 inches)
Displacement	281.7 cc (17.2 cubic inches)

Rocker arms

Rocker arm inside diameter
Standard	11.988 to 12.006 mm (0.4720 to 0.4727 inch)
Limit	12.038 mm (0.4739 inch)

Rocker shaft outside diameter
Standard	11.966 to 11.984 mm (0.4711 to 0.4718 inch)
Limit	11.92 mm (0.469 inch)

Shaft-to-arm clearance
Standard	0.004 to 0.040 mm (0.0002 to 0.0016 inch)
Limit	0.068 mm (0.0027 inch)

Camshaft

Lobe height

Intake
Standard	35.751 mm (1.4075 inches)
Limit	35.571 mm (1.4004 inches)

Exhaust
Standard	35.764 mm (1.4080 inches)
Limit	35.584 mm (1.4009 inches)

Cylinder head, valves and valve springs

Cylinder head warpage limit	0.10 mm (0.004 inch)
Valve stem diameter	
Intake	
Standard	5.480 to 5.490 mm (0.2157 to 0.2161 inch)
Limit	5.45 mm (0.215 inch)
Exhaust	
Standard	5.460 to 5.470 mm (0.2150 to 0.2154 inch)
Limit	5.43 mm (0.214 inch)
Valve guide inside diameter (intake and exhaust)	
Standard	5.500 to 5.512 mm (0.2165 to 0.2170 inch)
Limit..	5.525 mm (0.2175 inch)
Stem-to-guide clearance	
Intake	
Standard	0.010 to 0.032 mm (0.0004 0.0013 inch)
Limit	0.2 mm (0.005 inch)
Exhaust	
Standard	0.030 to 0.052 mm (0.0012 to 0.0020 inch)
Limit	0.14 mm (0.006 inch)
Valve seat width (intake and exhaust)	
Standard..	1.2 mm (0.05 inch)
Limit..	1.5 mm (0.06 inch)
Valve spring free length	
Inner spring (intake and exhaust)	
Standard	
Yellow paint..	40.78 mm (1.606 inches)
Pink paint..	40.68 mm (1.602 inches)
Limit	37.8 mm (1.49 inches)
Outer spring (intake and exhaust)	
Standard	
Yellow paint..	42.76 mm (1.683 inches)
Pink paint..	42.78 mm (1.684 inches)
Limit	39.8 mm (1.57 inches)

Cylinder

Bore diameter	
Standard..	74.00 to 74.01 mm (2.913 to 2.914 inches)
Limit..	74.1 mm (2.92 inches)
Taper and out-of-round limits..	0.10 mm (0.004 inch)
Warpage	0.05 mm (0.002 inch)

Pistons

Piston diameter (measured at a point 20 mm, or 0.8 inch, from the bottom)	
Standard..	73.965 to 73.985 mm (2.9120 to 2.9128 inches)
Limit..	73.90 mm (2.909 inches)
Piston-to-cylinder clearance	
Standard..	0.015 to 0.045 mm (0.0006 to 0.0018 inch)
Limit..	0.010 mm (0.004 inch)
Piston pin bore	
Standard..	17.002 to 17.008 mm (0.6694 to 0.6696 inch)
Limit..	17.04 m (0.671 inch)
Piston pin outside diameter	
Standard..	16.994 to 17.000 (0.6691 to 0.6693 inch)
Limit..	16.96 (0.668 inch)
Piston pin-to-piston bore clearance	
Standard..	0.002 to 0.014 mm (0.0001 to 0.0006 inch)
Limit..	0.02 mm (0.001 inch)
Connecting rod-to-piston pin clearance	
Standard..	0.016 to 0.040 mm (0.0006 to 0.0016 inch)
Limit..	0.14 mm (0.006 inch)
Ring side clearance	
Top	
Standard	0.030 to 0.060 mm (0.0012 to 0.0024 inch)
Limit	0.11 mm (0.004 inch)
Second	
Standard	0.015 to 0.045 mm (0.0006 to 0.0018 inch)
Limit	0.095 mm (0.0037 inch)

Ring end gap
  Top
    Standard .... 0.15 to 0.30 mm (0.006 to 0.012 inch)
    Limit .... 0.5 mm (0.02 inch)
  Second
    Standard .... 0.30 to 0.45 mm (0.012 to 0.018 inch)
    Limit .... 0.5 mm (0.02 inch)
  Oil
    Standard .... 0.20 to 0.70 mm (0.008 to 0.028 inch)
    Limit .... not specified

Clutch

Clutch outer inside diameter
  Standard.. 28.000 to 28.021 mm (1.1024 to 1.1032 inches)
  Limit .... 28.05 mm (1.104 inches)
Clutch outer guide outside diameter
  Standard.. 27.959 to 27.980 mm (1.1007 to 1.1016 inches)
  Limit .... 27.92 mm (1.099 inches)
Clutch outer guide inside diameter
  Standard.. 22.000 t0 22.021 mm (0.8661 to 0.8670 inch)
  Limit .... 22.05 mm (0.868 inch)
Mainshaft outside diameter at clutch outer guide
  Standard.. 21.959 to 21.980 mm (0.8645 to 0.8654 inch)
  Limit .... 21.93 mm (0.863 inch)
Spring free length
  Standard.. 35.9 mm (1.41 inches)
  Limit .... 34.9 mm (1.37 inches)
Clutch friction disc thickness
  Standard.. 2.92 to 3.08 mm (0.115 to 0.121 inch)
  Limit .... 2.6 mm (0.10 inch)
Clutch metal plate warpage limit .... 0.20 mm (0.008 inch)

Oil pump

Outer rotor-to-pump body clearance
  Standard.. 0.15 mm to 0.21 mm (0.006 to 0.008 inch)
  Limit .... 0.25 mm (0.010 inch)
Inner rotor-to-outer rotor clearance
  Standard.. 0.15 mm (0.006 inch)
  Limit .... 0.20 mm (0.008 inch)
Side clearance (rotors-to-straightedge)
  Standard.. 0.02 to 0.08 mm (0.001 to 0.003 inch)
  Limit .... 0.10 mm (0.004 inch)

Shift forks

Shift fork inside diameter
  Standard.. 13.000 to 13.021 mm (0.5118 to 0.5126 inch)
  Limit .... 13.04 mm (0.513 inch)
Shift fork shaft outside diameter
  Standard.. 12.966 to 12.984 mm (0.5105 to 0.5112 inch)
  Limit .... 12.96 mm (0.510 inch)
Shift fork ear thickness
  Standard.. 4.93 to 5.00 mm (0.194 to 0.197 inch)
  Limit .... 4.50 mm (0.177 inch)

Transmission

Transmission gear inside diameters
  Mainshaft fourth, mainshaft fifth, countershaft second
    Standard .... 25.020 to 25.041 mm (0.9850 to 0.9859 inch)
    Limit .... 25.07 mm (0.987 inch)
  Countershaft first, countershaft third, reverse
    Standard .... 28.020 to 28.041 mm (1.1031 to 1.1040 inch)
    Limit .... 28.07 mm (1.105 inches)
  Reverse idle
    Standard .... 18.000 to 18.018 mm (0.7087 to 0.7094 inches)
    Limit .... 18.05 mm (0.711 inch)

2A

Transmission (continued)

Transmission gear bushing outside diameters
    Mainshaft fourth, mainshaft fifth, countershaft second
        Standard .... 24.979 to 25.000 mm (0.9834 to 0.9843 inch)
        Limit .... 24.93 mm (0.981 inch)
    Countershaft third, countershaft first, countershaft reverse
        Standard .... 27.979 to 28.000 mm (1.1015 to 1.1024 inch)
        Limit .... 27.93 mm (1.100 inch)
    Reverse idle
        Standard .... 17.966 to 17.984 mm (0.7073 to 0.7080 inch)
        Limit .... 17.93 mm (0.706 inch)
Transmission gear bushing inside diameters
    Mainshaft fifth, countershaft second
        Standard .... 22.000 to 22.021 mm (0.8661 to 0.8670 inch)
        Limit .... 22.05 mm (0.868 inch)
    Countershaft third
        Standard .... 25.000 to 25.021 mm (0.9843 to 0.9851 inch)
        Limit .... 25.05 mm (0.986 inch)
    Reverse idle
        Standard .... 14.000 to 14.025 mm (0.5512 to 0.5522 inch)
        Limit .... 14.05 (0.553 inch)
Gear-to-bushing clearances
    At mainshaft fifth and countershaft second gears
        Standard .... 0.020 to 0.062 mm (0.0008 to 0.0024 inch)
        Limit .... 0.10 mm (0.004 inch)
    At countershaft third gear
        Standard .... 0.020 to 0.062 mm (0.0008 to 0.0024 inch)
        Limit .... 0.10 mm (0.004 inch)
    At reverse idler gear
        Standard .... 0.016 to 0.052 mm (0.0006 to 0.0020 inch)
        Limit .... 0.09 mm (0.004 inch)
Mainshaft outside diameter at mainshaft fifth gear bushing
    Standard.. 21.959 to 21.980 mm (0.8645 to 0.8654 inch)
    Limit.. 21.93 mm (0.863 inch)
Countershaft outside diameter at countershaft second gear bushing
    Standard .... 21.959 to 21.980 mm (0.8645 to 0.8654 inch)
    Limit .... 21.93 mm (0.863 inch)
Countershaft outside diameter at countershaft third gear bushing
    Standard .... 24.959 to 24.980 mm (0.9826 to 0.9835 inch)
    Limit .... 24.93 mm (0.981 inch)
Gear bushing-to-shaft clearance
    At mainshaft fifth, countershaft second and countershaft third gears
        Standard .... 0.020 to 0.062 mm (0.0008 to 0.0024 inch)
        Limit .... 0.10 mm (0.004 inch)
    At reverse idle
        Standard .... 0.016 to 0.052 mm (0.0006 to 0.0020 inch)
        Limit .... 0.09 mm (0.004 inch)
Reverse idler gear shaft outside diameter
    Standard.. 13.973 to 13.984 mm (0.5501 to 0.5506 inch)
    Limit.. 13.93 mm (0.548 inch)

Crankshaft

Connecting rod small end inside diameter
    Standard.. 17.016 to 17.034 mm (0.6699 to 0.6706 inch)
    Limit.. 17.10 mm (0.673 inch)
Connecting rod big end side clearance
    Standard.. 0.05 to 0.65 mm (0.002 to 0.026 inch)
    Limit.. 0.80 mm (0.031 inch)
Connecting rod big end radial clearance
    Standard.. 0.006 to 0.018 mm (0.0002 to 0.0007 inch)
    Limit.. 0.05 mm (0.002 inch)
Runout limit.. 0.05 mm (0.002 inch)

Torque specifications

Bearing retainer plate (on right crankcase)............................ 12 Nm (108 in-lbs) (3)
Breather separator Allen bolt (on right crankcase) .... 12 Nm (108 in-lbs) (3)
Cam chain tensioner
    Tensioner mounting bolts... 10 Nm (84 in-lbs)
    Tensioner sealing bolt .... 10 Nm (84 in-lbs)

Cam sprocket bolts	20 Nm (168 in-lbs)
Clutch	
Clutch center locknut	110 Nm (80 ft-lbs) (1)
Clutch lifter plate bolts	12 Nm (108 in-lbs)
Crankcase cover bolts (left or right)............................	10 Nm (84 in-lbs)
Crankcase bolts..	12 Nm (108 in-lbs)
Cylinder base bolts	10 Nm (84 in-lbs)
Cylinder head cover	
Flange bolts (13)	12 Nm (108 in-lbs)
Shorter flange bolt	10 Nm (84 in-lbs)
Cylinder head	
Allen bolts (2)	40 Nm (29 ft-lbs)
Cap nuts (4)	40 Nm (29 ft-lbs)
Engine mounting bolt nuts	
Front nuts (upper and lower)	60 Nm (43 ft-lbs)
Rear nut...	75 Nm (54 ft-lbs)
Oil pipe bolts	
6 X 12 mm bolt	9 Nm (78 in-lbs
7 X 18.5 mm bolt	12 Nm (108 in-lbs)
Gearshift linkage	
Gearshift return spring pin	22 Nm (16 ft-lbs) (3)
Gearshift drum shifter pin	23 Nm (17 ft-lbs) (3)
Gearshift pedal pinch bolt	16 Nm (144 in-lbs)
Reverse switch rotor/reverse shifter cam bolts	12 Nm (108 in-lbs) (3)
Reverse shifter rod/lever pinch bolt............................	16 Nm (144 in-lbs)

1 *Use non-permanent thread locking agent on the threads. Turn counterclockwise (anti-clockwise) to tighten. Stake the locknut.*
2 *Use non-permanent thread locking agent on the threads. Stake the locknut.*
3 *Use non-permanent thread locking agent on the threads.*

1 General information

The engine/transmission unit is of the air-cooled, single-cylinder four-stroke design. The four valves are operated by an overhead camshaft which is chain driven by the crankshaft. The engine/transmission assembly is constructed from aluminum alloy. The crankcase is divided vertically.

The crankcase incorporates a wet sump, pressure-fed lubrication system which uses a gear-driven rotor-type oil pump and an oil filter.

Power from the crankshaft is routed to the transmission via a wet multi-plate clutch. The transmission has five forward speeds and reverse.

2 Operations possible with the engine in the frame

The components and assemblies listed below can be removed without having to remove the engine from the frame. If, however, a number of areas require attention at the same time, removal of the engine is recommended.

Carburetor
Cylinder head cover/rocker arm assembly
Cam chain tensioner
Camshaft
Cylinder head and valve spring assembly
Cylinder and piston
Starter motor
Starter reduction gears
Alternator rotor and stator
Starter clutch
Gear selector mechanism external components
Clutch assembly
Primary drive gear
Oil pump

3 Operations requiring engine removal

2A

It is necessary to remove the engine/transmission assembly from the frame and separate the crankcase halves to gain access to the following components:

Crankshaft and connecting rod
Balancer shaft
Transmission shafts
Shift drum and forks

4 Major engine repair - general note

1   It is not always easy to determine when or if an engine should be completely overhauled, as a number of factors must be considered.
2   High mileage is not necessarily an indication that an overhaul is needed, while low mileage, on the other hand, does not preclude the need for an overhaul. Frequency of servicing is probably the single most important consideration. An engine that has regular and frequent oil and filter changes, as well as other required maintenance, will most likely give many miles of reliable service. Conversely, a neglected engine, or one which has not been broken in properly, may require an overhaul very early in its life.
3   Exhaust smoke and excessive oil consumption are both indications that piston rings and/or valve guides are in need of attention. Make sure oil leaks are not responsible before deciding that the rings and guides are bad. Refer to Chapter 1 and perform a cylinder compression check to determine for certain the nature and extent of the work required.
4   If the engine is making obvious knocking or rumbling noises, the connecting rod and/or main bearings are probably at fault.
5   Loss of power, rough running, excessive valve train noise and high fuel consumption rates may also point to the need for an overhaul, especially if they are all present at the same time. If a complete tune-up does not remedy the situation, major mechanical work is the only solution.

5.5 Disconnect the breather hose from the cylinder head cover (right arrow) and, on later models (as shown here), from the air intake duct (left arrow)

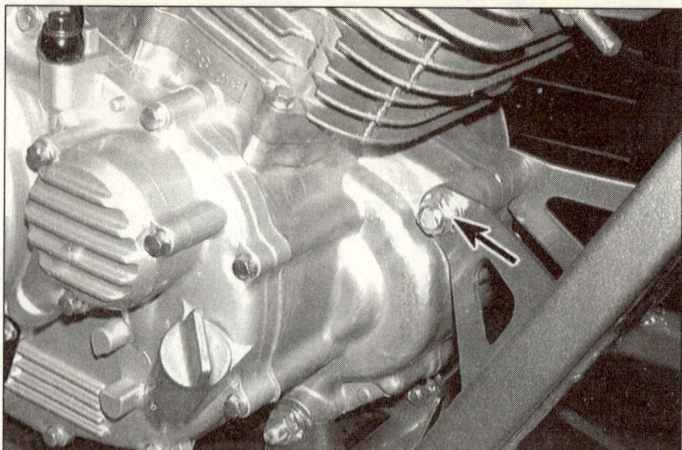

5.14a Remove the nut (arrow) from the upper front engine mount through-bolt . . .

5.14b . . . remove the upper front mount through-bolt (lower arrow), remove the engine hanger plate bolts (upper arrows; left bolts not shown) . . .

6   An engine overhaul generally involves restoring the internal parts to the specifications of a new engine. During an overhaul the piston rings are replaced and the cylinder walls are bored and/or honed. If a rebore is done, then a new piston is also required. The crankshaft and connecting rod are permanently assembled, so if one of these components needs to be replaced both must be. Generally the valves are serviced as well, since they are usually in less than perfect condition at this point. While the engine is being overhauled, other components such as the carburetor and the starter motor can be rebuilt also. The end result should be a like-new engine that will give as many trouble-free miles as the original.

7   Before beginning the engine overhaul, read through all of the related procedures to familiarize yourself with the scope and requirements of the job. Overhauling an engine is not all that difficult, but it is time consuming. Plan on the vehicle being tied up for a minimum of two (2) weeks. Check on the availability of parts and make sure that any necessary special tools, equipment and supplies are obtained in advance.

8   Most work can be done with typical shop hand tools, although a number of precision measuring tools are required for inspecting parts to determine if they must be replaced. Often a dealer service department or repair shop will handle the inspection of parts and offer advice concerning reconditioning and replacement. As a general rule, time is the primary cost of an overhaul so it doesn't pay to install worn or substandard parts.

9   As a final note, to ensure maximum life and minimum trouble from a rebuilt engine, everything must be assembled with care in a spotlessly clean environment.

5   Engine - removal and installation

Note: *Engine removal and installation should be done with the aid of an assistant to avoid damage or injury that could occur if the engine is dropped. A hydraulic floor jack should be used to support and lower the engine if possible (they can be rented at low cost).*

Removal

Refer to illustrations 5.5, 5.14a, 5.14b, 5.14c and 5.14d

1   Drain the engine oil (see Chapter 1).

2   Remove the seat/rear fender assembly, the front fender and the left footpeg assembly (see Chapter 7).

3   Disconnect the negative battery cable (see Chapter 8).

4   Remove the fuel tank (see Chapter 3).

5   Disconnect the crankcase breather tube from the cylinder head cover and, on later models, from the air intake duct **(see illustration)**. Note the routing of the breather tube assembly, then remove it.

6   Remove the carburetor (see Chapter 3). It's not necessary to disconnect the choke and throttle cables. Plug the carburetor intake opening with a rag.

7   Remove the exhaust system (see Chapter 3).

8   Label and disconnect the following wires (refer to Chapters 4 or 8 for component location if necessary):

   Spark plug wire (see Chapter 1)
   Ignition pulse generator (green three-pin connector) (see Chapter 4)
   Alternator (three-pin connector) (see Chapter 8)
   Reverse and neutral switches (see Chapter 8)
   Starter cable (see Chapter 8)

9   Disconnect the reverse cable (see Section 19).

10   Remove the shift pedal (see Section 22).

11   Disconnect the clutch cable from the lifter arm and detach the cable from the bracket on top of the crankcase (see Section 17). Set the clutch cable aside.

12   Remove the drive sprocket cover, remove the drive sprocket and pull the drive chain back so that it doesn't interfere with engine removal (see Chapter 5).

13   Support the engine securely from below with a floor jack. Place a block of wood between the jack head and the engine to protect the aluminum cases.

14   Remove the engine mounting bolts, nuts and brackets at the upper front, upper rear, lower front and lower rear **(see illustrations)**.

15   Have an assistant help you lift the engine and remove it from the left side.

16   Slowly lower the engine to a suitable work surface.

5.14c . . . remove the nuts (arrows) from the lower front mount, lower rear mount and upper rear mount through-bolts . . .

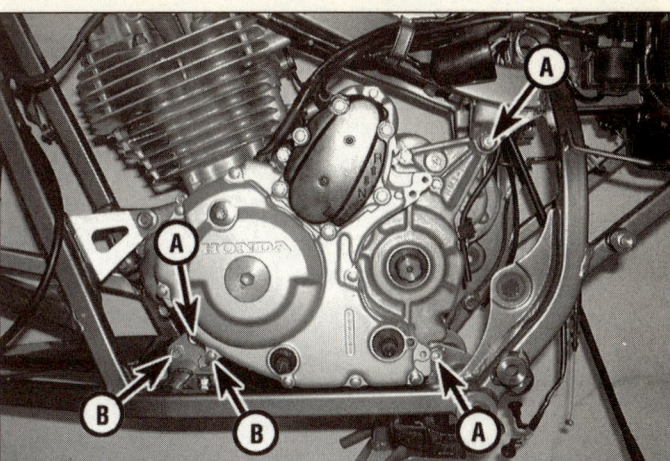

5.14d . . . remove the lower front, lower rear and upper rear through-bolts (A), the lower front hanger plate bolts (B) and the lower front hanger plate

Installation

17   Lift the engine up into its installed position in the frame.

18   Place a floor jack under the engine. Again, be sure to protect the engine from the jack head with a block of wood. Lift the engine to align the mounting bolt holes, then install the brackets, bolts and nuts. Tighten them to the torques listed in this Chapter's Specifications.

19   The remainder of installation is the reverse of the removal steps, with the following additions:

a) Use new gaskets at all exhaust pipe connections.
b) Adjust the clutch cable and the throttle cable (see Chapter 1).
c) Fill the engine with oil (see Chapter 1). Run the engine and check for leaks.

6   Engine disassembly and reassembly - general information

Refer to illustrations 6.2 and 6.3

1   Before disassembling the engine, clean the exterior with a degreaser and rinse it with water. A clean engine will make the job easier and prevent the possibility of getting dirt into the internal areas of the engine.

2   In addition to the precision measuring tools mentioned earlier, you will need a torque wrench, a valve spring compressor, oil gallery brushes **(see illustration)**, a piston ring removal and installation tool, a piston ring compressor and a clutch holder tool (described in Section 17). Some new, clean engine oil of the correct grade and type, some engine assembly lube (or moly-based grease) and a tube of RTV (silicone) sealant will also be required.

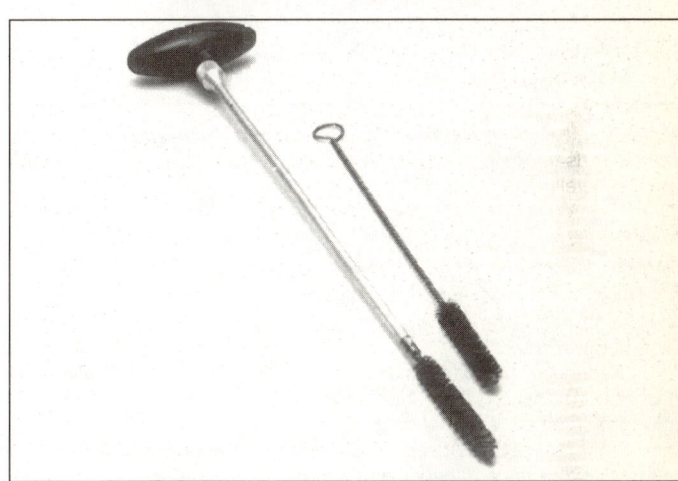

6.2 A selection of brushes is required for cleaning holes and passages in the engine components

3   An engine support stand made from short lengths of 2 x 4's bolted together will facilitate the disassembly and reassembly procedures **(see illustration)**. If you have an automotive-type engine stand, an adapter plate can be made from a piece of plate, some angle iron and some nuts and bolts.

4   When disassembling the engine, keep "mated" parts together (including gears, drum shifter pawls, etc.) that have been in contact with each other during engine operation. These "mated" parts must be reused or replaced as an assembly.

5   Engine/transmission disassembly should be done in the following general order with reference to the appropriate Sections.

Remove the cylinder head cover and rocker assembly
Remove the cam chain tensioner and camshaft
Remove the cylinder head
Remove the cylinder
Remove the piston
Remove the clutch
Remove the oil pump
Remove the external shift mechanism
Remove the alternator rotor
Separate the crankcase halves
Remove the shift drum/forks
Remove the transmission shafts/gears
Remove the crankshaft and connecting rod

6   Reassembly is accomplished by reversing the general disassembly sequence.

6.3 An engine stand can be made from short lengths of lumber and lag bolts or nails

2A

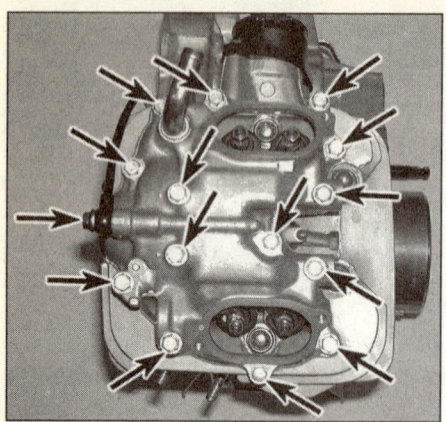

7.3 Loosen the cylinder head cover bolts (arrows) in two or three stages, in a criss-cross pattern

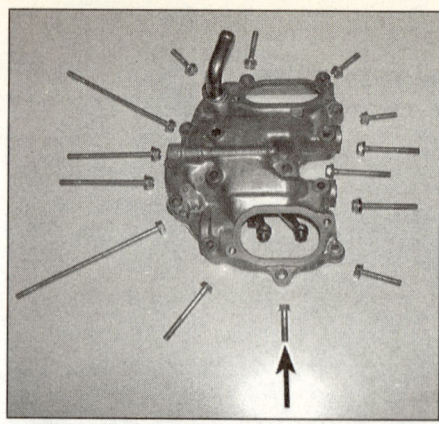

7.5 Note the various bolt lengths; the bolt in front of the exhaust valve adjuster hole (arrow) is tightened to a lower torque than the others

7.6 If the cylinder head cover is stuck on the cylinder head, gently tap it loose with a plastic or rubber mallet

7   Cylinder head cover and rocker arms - removal, inspection and installation

Note: *The cylinder head cover can be removed with the engine in the frame. If the engine has been removed, ignore the steps that don't apply.*

Removal

Refer to illustrations 7.3, 7.5, 7.6 and 7.7

1   Remove the fuel tank (see Chapter 3).
2   Disconnect the oil breather hose from the cylinder head cover **(see illustration 5.5)**.
3   Disconnect the external oil pipe from the cylinder head cover **(see illustration)**. Discard the old sealing washers (see Section 16 for more information on the external oil pipe)
4   Place the piston at top dead center on the compression stroke (see *Valve clearances - check and adjustment* in Chapter 1).
5   Loosen the cylinder head cover bolts in two or three stages, in a criss-cross pattern **(see illustration 7.3)**; it isn't necessary to remove the valve adjuster covers to remove the cylinder head cover, as shown in the illustration, but you will have to check the valve clearances after reinstalling the cover, and this would be a good time to inspect the cover O-rings. Remove the cover bolts and note the locations of the bolts of various lengths **(see illustration)**.

6   Remove the cover and cover bolts from the cylinder head. If the head is stuck, don't try to pry it off; tap around the sides of it with a plastic hammer to dislodge it **(see illustration)**.
7   Remove the cover dowels **(see illustration)**. They may be in the cylinder head or they may have come off with the cover. Stuff clean rags into the cam chain openings so small parts or tools can't fall into them.

Inspection

Refer to illustrations 7.9, 7.10a, 7.10b, 7.11, 7.12, 7.15 and 7.16

8   Inspect of the rocker arms for wear at the camshaft contact surfaces and the tips of the valve adjusting screws **(see illustration 7.7)**. Try to twist the rocker arms from side-to-side on the shafts. If they're loose on the shafts or if there's visible wear, remove them as described below.
9   Using a pair of diagonal cutters, lever the rocker arm dowel pins out of the cylinder head cover **(see illustration)**.
10   Using a suitable tool (such as a small L-shaped Allen key), push the rocker arm shafts out of the head far enough to get a hand on them, then pull the rocker shafts out of the cover and remove the rocker arms **(see illustrations)**. Do NOT mix up the rocker arms and rocker arm shafts. Mark the intake or the exhaust side with a laundry marker or paint to prevent confusion during reassembly.
11   Remove the old O-rings from the rocker arm shafts **(see illustration)**.

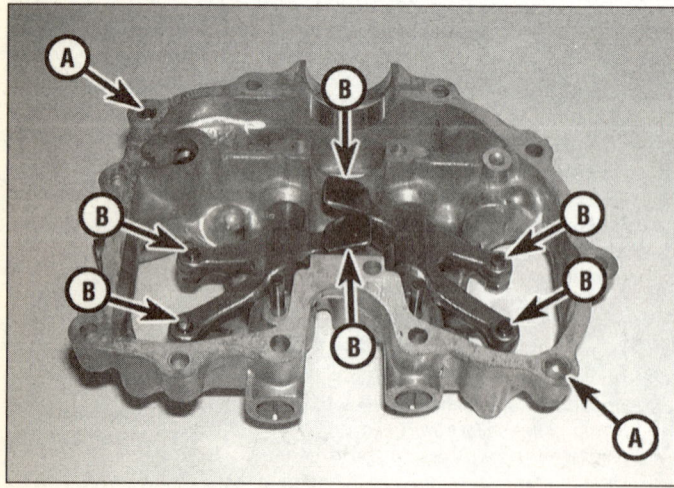

7.7 Cylinder head cover dowel locations (A) and rocker arm inspection points (B)

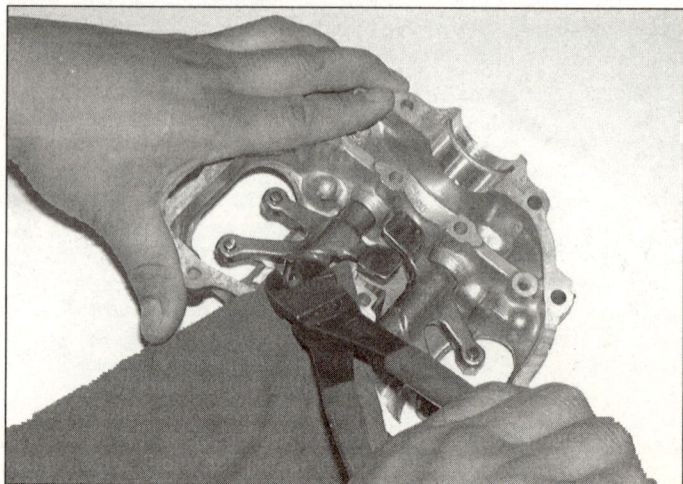

7.9 Using a rag to protect the cylinder head cover, lever the rocker arm shaft dowel pins out with a pair of diagonal cutters

7.10a Push each rocker arm shaft out of the cover far enough to pull it out the rest of the way by hand . . .

7.10b . . . then remove each rocker arm

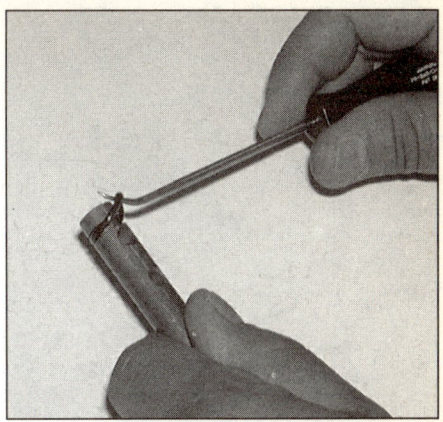

7.11 Remove the O-rings from their grooves in the rocker arm shafts; coat new O-rings with clean engine oil

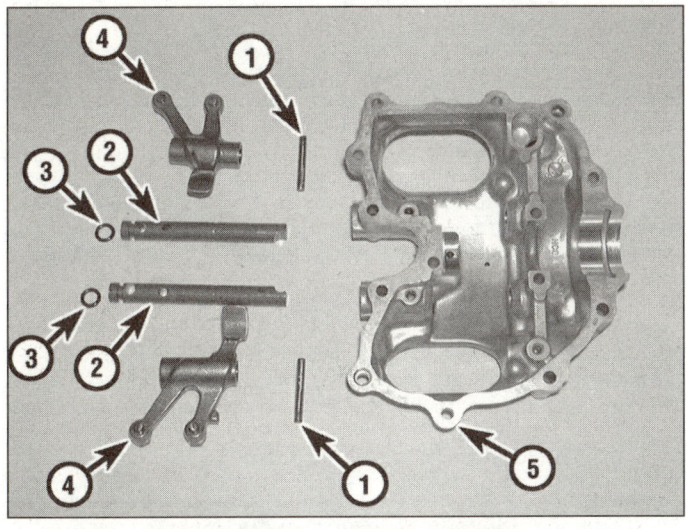

7.12 Cylinder head cover details

1	Rocker arm shaft dowel pins
2	Rocker arm shafts
3	Rocker arm shaft O-rings
4	Rocker arms
5	Cylinder head cover

12   Wash all parts in clean solvent and lay them out for inspection (see illustration).

13   Measure the outer diameter of each rocker shaft with a micrometer. Measure the inner diameter of the rocker arms with a small-hole gauge and a micrometer. Compare your measurements to the values listed in this Chapter's Specifications. If rocker arm-to-shaft clearance is excessive, replace the rocker arm and/or shaft, whichever is worn.

14   Coat new rocker shaft O-rings with clean engine oil and install them on the shafts.

15   Coat the rocker shafts and rocker arm bores with moly-based grease. Install the rocker shafts and rocker arms in the cylinder head cover (see illustration). Be sure to install the intake and exhaust rocker arms and shafts in the correct side of the cover. Align the cutouts in the ends of the shafts with the bolt holes in the far side of the cover and the dowel pin holes in the shafts with the dowel pin holes in the cover.

16   Coat the new dowel pins with clean engine oil and tap the dowels into the cylinder head cover and through the rocker arm shafts. If the dowel pin hole in either shaft is not precisely aligned with the dowel hole in the cover, use a screwdriver to turn the slot in the end of the rocker arm shaft to line it up precisely (see illustration). Make sure the dowel pins go all the way through the rocker shafts and seat in the cover.

2A

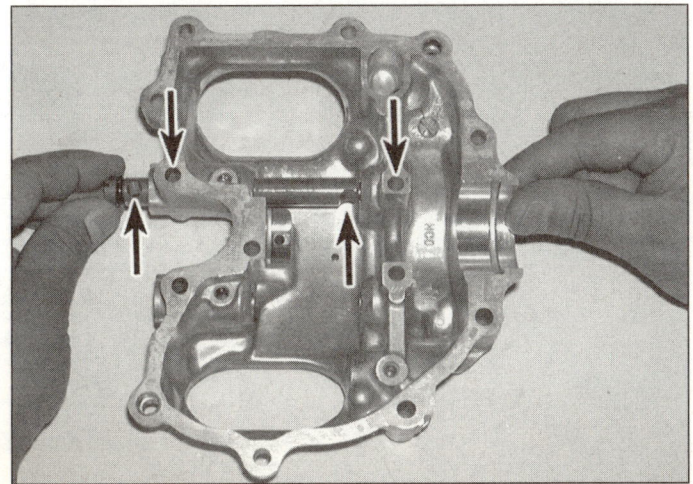

7.15 Align the dowel pin holes in the shaft and cover (left arrows) and the shaft cutout with the cover bolt hole (left arrows) (rocker arm removed for clarity)

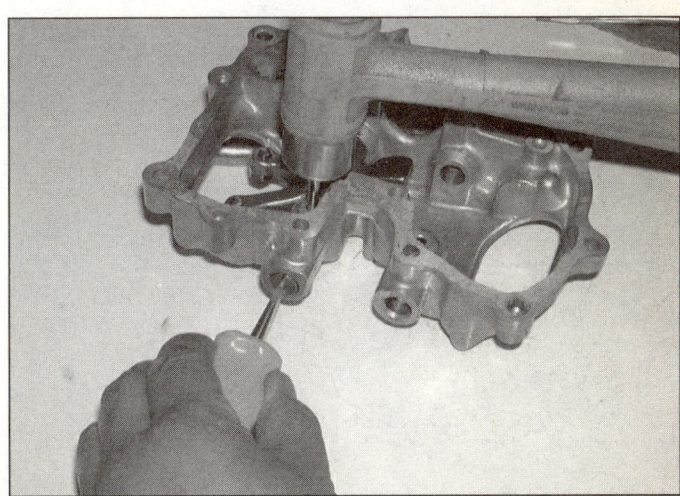

7.16 Use a screwdriver in the shaft slot to precisely align the holes before tapping in the dowel pin

7.17 Coat the circumference of the camshaft end cap (arrow) with RTV sealant

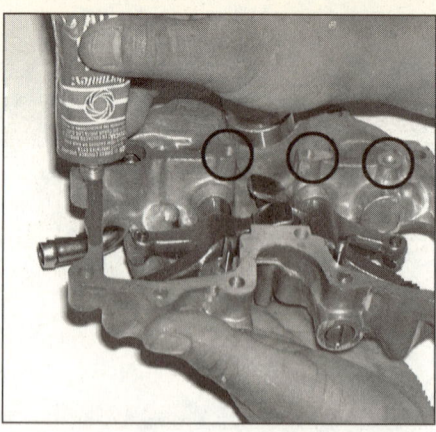

7.18 Apply sealant to the periphery of the cylinder head cover, but not to the three circled areas

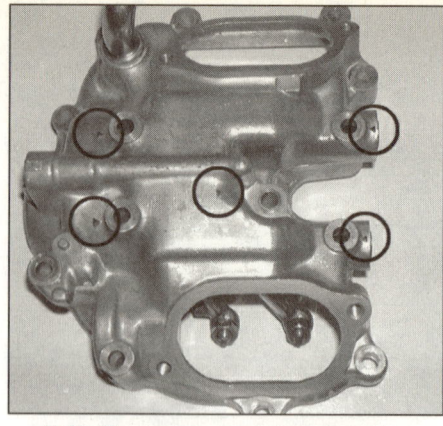

7.20 New copper washers must be installed at the bolt holes marked with a "triangle" (circled)

Installation

Refer to illustrations 7.17, 7.18 and 7.20

17 Inspect the camshaft end cap. If the cap is cracked, hardened or deteriorated, pull it off and install a new one **(see illustration)**. Before installing the new cap, apply RTV sealant to the circumference.

18 Clean the mating surfaces of the cylinder head and the valve cover with lacquer thinner, acetone or brake system cleaner. Apply a thin film of RTV sealant to the cover sealing surface (there's no gasket). Apply sealant only around the outer circumference of the cover; do

NOT apply sealant to the areas indicated in the accompanying photo **(see illustration)**.

19 Install the cylinder head cover on the cylinder head. Make sure the camshaft end cap is in position.

20 Install a new copper washer at the cover bolt holes marked with a "triangle" **(see illustration)**. Install the bolts and tighten them evenly in two or three stages (starting with the center bolt) to the torque listed in this Chapter's Specifications. Note that there are two types of bolts with different torque settings. There are 14 bolts altogether, 13 of which are tightened to the same torque; the 14th bolt (the one that's tightened to the lower torque value) is located in front of the exhaust valve adjuster cover.

21 Using new sealing washers, reattach the external oil pipe to the cylinder head cover and tighten the banjo bolt to the torque listed in this Chapter's Specifications (see Section 16 for more information on the external oil pipe).

22 Check and, if necessary, adjust the valve clearances (see Chapter 1).

23 The remainder of installation is the reverse of removal.

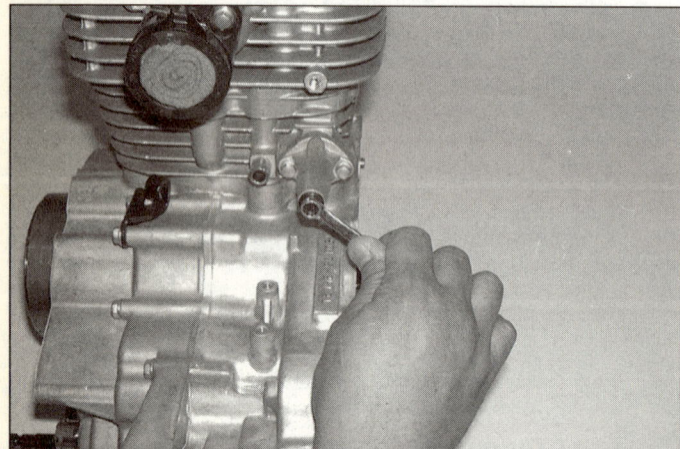

8.1a Remove the tensioner bolt . . .

8 Cam chain tensioner - removal and installation

Note: *This procedure can be performed with the engine in the frame.*

Removal

Refer to illustrations 8.1a, 8.1b, 8.2a, 8.2b and 8.4

1 Loosen the cam chain tensioner sealing bolt and washer **(see illustrations)**.

8.1b . . . and the sealing washer, then remove the tensioner mounting bolts (arrows)

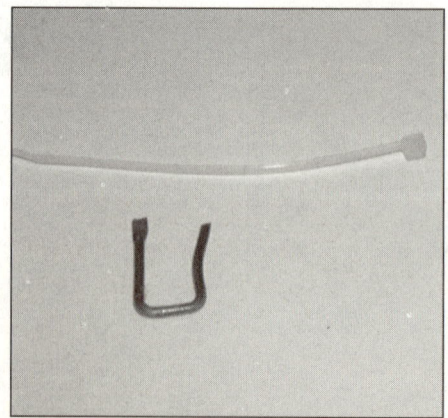

8.2a Bend a piece of coat hanger wire and flatten one end; you'll also need a cable tie long enough to wrap around the tensioner

8.2b Use the flattened end of the wire like a screwdriver to retract the lifter shaft, then secure it with a cable tie

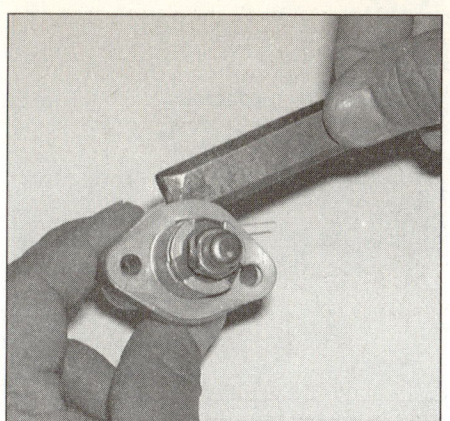

8.4 Remove all old gasket material from the tensioner and cylinder

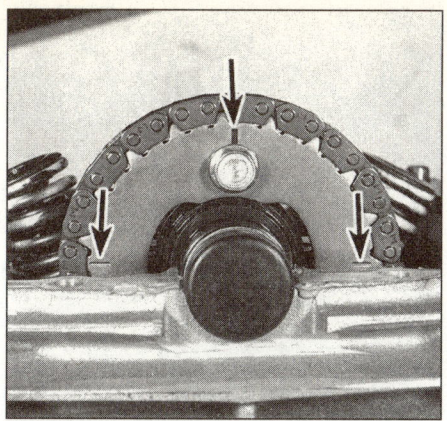

9.1 With the engine at TDC compression, the sprocket timing marks should be even with the cylinder head surface and the index mark above the upper bolt should be straight up

9.5 Turn the sprocket as needed to expose the bolts (arrows) so you can remove them

2   The cam chain tensioner lifter shaft must be retracted and locked into place before removing the tensioner. Fabricate a small U-shaped section of coat hanger wire and flatten one end of the wire **(see illustration)**. Insert the flattened end of the wire into the slot in the end of the lifter shaft, turn the lifter shaft clockwise until it's fully retracted and lock the lifter shaft into place with a cable tie **(see illustration)**.

3   Remove the tensioner mounting bolts **(see illustration 8.1b)** and detach the tensioner from the cylinder block. **Caution:** *Do NOT remove the coat hanger wire from the tensioner until after the tensioner has been reinstalled. If the tensioner piston is allowed to extend before the tensioner mounting bolts are retightened, the piston will be forced against the cam chain, damaging the tensioner or the chain.*

4   Clean all old gasket material from the tensioner body **(see illustration)** and from its mating surface on the cylinder block.

Installation

5   Lubricate the friction surfaces of the components with moly-based grease.

6   Install a new tensioner gasket on the cylinder. Position the tensioner body on the cylinder, install the tensioner bolts and tighten them to the torque listed in this Chapter's Specifications.

7   Cut the cable tie and remove the coat hanger wire. When released, the lifter shaft will automatically extend.

8   Install the tensioner sealing bolt and a new sealing washer and tighten the sealing bolt to the torque listed in this Chapter's Specifications.

9   Camshaft, sprocket and chain guides - removal, inspection and installation

Note: *This procedure can be performed with the engine in the frame.*

Removal

Refer to illustrations 9.1, 9.5, 9.6a, 9.6b and 9.8

1   Place the piston at top dead center on the compression stroke (see *Valve clearance - check and adjustment* in Chapter 1). Check the positions of the camshaft marks to make sure the engine is on the compression stroke, not the exhaust stroke **(see illustration)**.

2   Remove the cylinder head cover (see Section 7).

3   Remove the camshaft chain tensioner (see Section 8).

4   Remove the camshaft end cap **(see illustration 7.17)**.

5   Remove one of the sprocket bolts **(see illustration)**. Turn the crankshaft counterclockwise (viewed from the left side) one revolution, which will turn the camshaft one-half turn and expose the other sprocket bolt. Remove the other bolt.

6   Free the sprocket from the camshaft and remove the camshaft **(see illustrations)**.

7   Tie the chain to the engine with a piece of wire so it doesn't drop down into the cam chain tunnel. Engine damage could occur if the engine is rotated with the chain bunched up around the crank sprocket.

8   Remove the camshaft bearing stopper plate **(see illustration)**.

9   Cover the top of the cylinder head with a rag to prevent foreign objects from falling into the engine.

2A

9.6a Disengage the sprocket from the camshaft and from the cam chain . . .

9.6b . . . then, holding the sprocket and chain as shown, pull out the crankshaft

9.8 Remove the camshaft bearing stopper plate

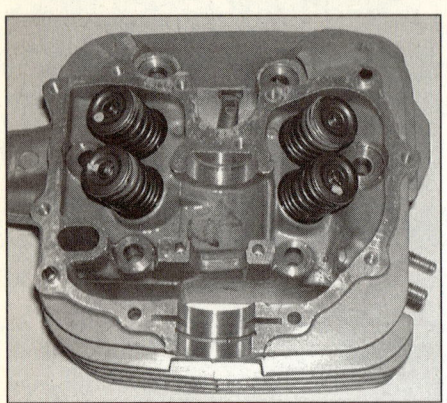

9.10 Inspect the cam bearing surfaces (arrows) for wear or damage

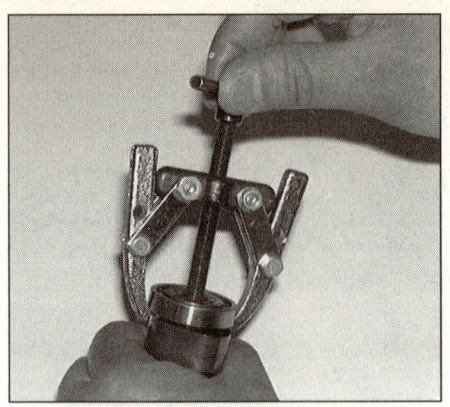

9.11 The camshaft bearings can be removed from the cam with a small puller

9.12a Check the cam lobes for wear - here's a good example of damage which will require replacement (or repair) of the camshaft

9.12b Measure the height of the cam lobes with a micrometer

9.15a The rear cam chain guide is a secured by a pivot bolt (left arrow); the front guide rests in a saddle (right arrow)

Inspection

Refer to illustrations 9.10, 9.11, 9.12a, 9.12b, 9.15a and 9.15b

Note: *Before replacing the camshaft because of damage, check with local machine shops specializing in ATV or motorcycle engine work. In the case of the camshaft, it may be possible for cam lobes to be welded, reground and hardened, at a cost lower than that of a new camshaft. If the bearing surfaces in the cylinder head or cover are damaged, it may be possible for them to be bored out to accept bearing inserts. Due to the cost of a new cylinder head it is recommended that all options be explored before condemning it as trash!*

10   Inspect the cam bearing surfaces in the cylinder head cover and in the cylinder head **(see illustration)**. Look for score marks, deep scratches and evidence of spalling (a pitted appearance).

11   Inspect the cam bearings for wear. If either bearing is blue (overheated due to lack of lubrication), stiff or noisy, replace both bearings. The bearings can be removed from the cam with a small puller **(see illustration)**. Tap new bearings onto the cam with a socket of sufficient diameter and depth (make sure that the socket pushes on the inner hub of the bearing, not on the bearing cage or outer race).

12   Check the camshaft lobes for heat discoloration (blue appearance), score marks, chipped areas, flat spots and spalling **(see illustration)**. Measure the height of each lobe with a micrometer **(see illustration)** and compare the results to the minimum lobe height listed in this Chapter's Specifications. If damage is noted or wear is excessive, the camshaft must be replaced. Also inspect the condition of the rocker arms (see Section 7).

13   Inspect the sprocket for wear, cracks and other damage, replacing it if necessary. If the sprocket is worn, the chain is also worn, and possibly the sprocket on the crankshaft. If wear this severe is appar-

ent, the entire engine should be disassembled for inspection.

14   Except in cases of oil starvation, the camshaft chain wears very little. If the chain has stretched excessively, the cam chain tensioner will be unable to maintain correct tension. If the chain has stretched excessively, replace it. To replace the chain, you'll need to remove the right crankcase cover and the clutch (see Section 18), the oil pump (see Section 20) and the primary drive gear (see Section 21).

15   Using a flashlight, inspect the chain guides for wear or damage. If

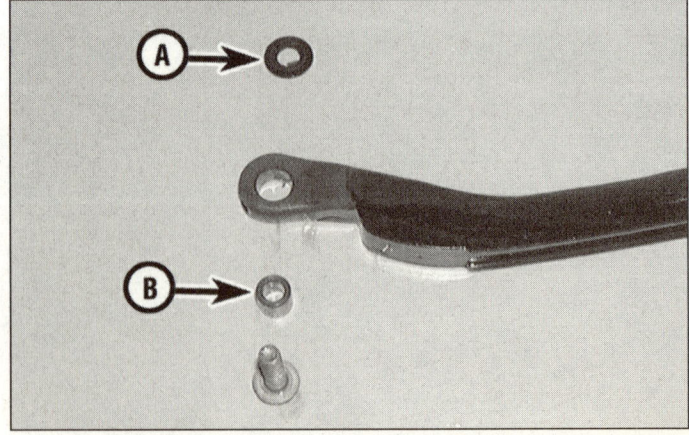

9.15b The rear cam chain guide washer (A) goes between the guide and the crankcase; the pivot bolt bushing (B) fits inside the guide

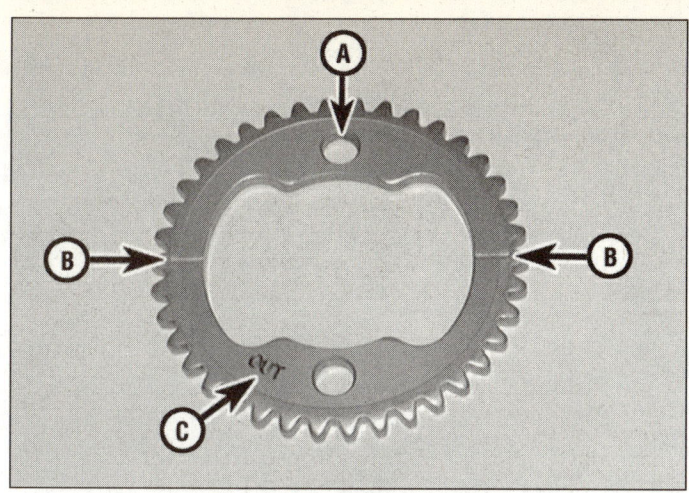

9.18 Camshaft sprocket marks

A   *Bolt hole index mark (install straight up)*
B   *Timing marks (align with the cylinder head mating surface)*
C   *OUT mark (faces away from the engine)*

they're worn or damaged, replace them. To replace the front guide, simply pull it straight up; when you install the new front guide, make sure it's correctly seated in the saddle in the crankcase **(see illustration)**. To replace the rear guide (the one that the tensioner pushes against), you'll have to remove the right crankcase cover and the clutch (see Section 18) to remove the pivot bolt from the lower end of the guide. Note that the rear guide pivot bolt uses a bushing and there's a washer between the guide and the crankcase **(see illustration)**. When installing the rear guide, don't forget to install the washer and the bushing.

Installation

Refer to illustrations 9.18 and 9.24

16   Install the cam chain guides, if removed (see Step 15).
17   Make sure that the T mark on the flywheel is still aligned with the index mark on the left crankcase cover (see *Valve clearance - check and adjustment* in Chapter 1).
18   Make sure the bearing surfaces in the cylinder head are clean. Lubricate the cam journals with molybdenum disulfide grease, then place the camshaft sprocket in position with the OUT mark facing out and the indexed bolt hole facing up **(see illustration)** and engage the sprocket teeth with the cam chain.
19   Place the camshaft in position on the cylinder head with the lobes pointing down.
20   Align the indexed bolt hole with the camshaft bolt hole and verify that the timing marks on the sprocket are aligned with the cylinder

head cover gasket surface. The index mark (and the sprocket bolt hole next to it) should be pointing straight up. Recheck the timing mark on the flywheel and make sure it's still aligned with the index mark on the left crankcase cover.
21   Apply a non-permanent thread-locking agent to the threads of the sprocket bolt. Install the bolt and tighten it to the torque listed in this Chapter's Specifications.
22   Coat the cam lobes with moly-based grease or engine assembly lube, then rotate the crankshaft to expose the other sprocket bolt hole. Coat the bolt threads with non-permanent thread-locking agent, install the bolt and tighten it to the torque listed in this Chapter's Specifications.
23   Rotate the engine back to the TDC position and recheck the timing marks. With the crankshaft mark in the TDC position, the timing marks and index mark should be correctly aligned (timing marks aligned with cylinder head gasket mating surface and index mark facing straight up). If this isn't the case, remove the sprocket and reposition it in the chain. **Caution:** *DO NOT run the engine with the marks out of alignment or severe engine damage could occur.*
24   Install the cam bearing stopper plate **(see illustration)**. Make sure it's fully seated into its groove in the cylinder head.
25   Install the cam chain tensioner (see Section 8).
26   Make sure that the bearing surfaces in the cylinder head cover are clean, then install the cylinder head cover (see Section 7).
27   Adjust the valve clearances (see *Valve clearance - check and adjustment* in Chapter 1).
28   The remainder of installation is the reverse of removal.

10   Cylinder head - removal and installation

Caution: *The engine must be completely cool before beginning this procedure, or the cylinder head may become warped.*

Removal

Refer to illustrations 10.3, 10.4, 10.5
1   Remove the carburetor and the exhaust system (see Chapter 3).
2   Remove the cylinder head cover, the cam chain tensioner and the camshaft (see Sections 7, 8 and 9, respectively). Remove the external oil pipe (see Section 20).
3   Loosen the cylinder head Allen bolts and cap nuts in two or three stages in the reverse of the tightening sequence **(see illustration)**. Remove the nuts, bolts and washers (use needle-nose pliers to remove the washers if necessary).
4   Lift the cylinder head off the cylinder **(see illustration)**. If the head is stuck, tap around the side of the head with a rubber mallet to jar it loose, or use two wooden dowels inserted into the intake or exhaust ports to lever the head off. Don't attempt to pry the head off by inserting a screwdriver between the head and the cylinder block - you'll damage the sealing surfaces.

2A

9.24 Install the cam bearing stopper plate; make sure it's fully seated into its groove in the head

10.3 Remove the cylinder head Allen bolts and cap nuts (arrows) gradually and evenly, in a criss-cross pattern

10.4 Carefully lift the cylinder head off the cylinder

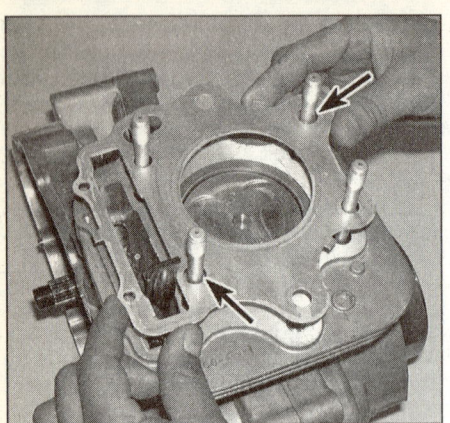

10.5 Remove the gasket and the dowels (arrows) (if the dowels aren't in the cylinder block, they're probably in the cylinder head)

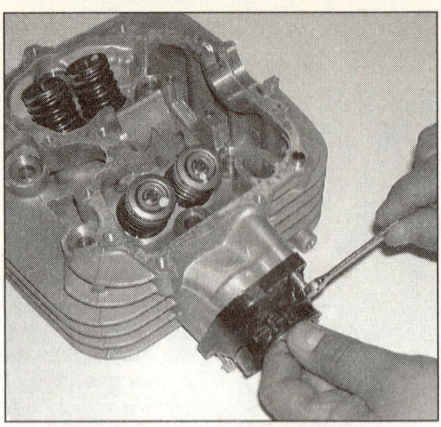

12.4a Remove the carburetor insulator bolts and detach the insulator from the cylinder head

12.4b Remove the old O-ring from the insulator; use a new one on installation

5   Remove the head gasket and the two dowel pins from the cylinder block **(see illustration)**. (The dowel pins might be in the cylinder head or in the cylinder block.) Support the cam chain so it won't drop into the cam chain tunnel, and stuff a clean rag into the cam chain tunnel to prevent the entry of debris.

6   Check the cylinder head gasket and the mating surfaces on the cylinder head and cylinder for leakage, which could indicate warpage. Check the flatness of the cylinder head (see Section 12).

7   Clean all traces of old gasket material from the cylinder head and cylinder. Be careful not to let any of the gasket material fall into the crankcase, the cylinder bore or the bolt holes.

Installation

8   Install the two dowel pins, then lay the new gasket in place on the cylinder block. Never re-use the old gasket and don't use any type of gasket sealant.

9   Carefully lower the cylinder head over the studs and dowels. It is helpful to have an assistant support the camshaft chain with a piece of wire so it doesn't fall and become kinked or detached from the crankshaft. When the head is resting on the cylinder, wire the cam chain to another component to keep tension on it.

10   Install the steel washers and cap nuts on the studs. Install new copper washers on the Allen bolts, then thread them lightly into their holes.

11   Tighten the two Allen bolts and four cap nuts in two or three stages to the torque listed in this Chapter's Specifications.

12   Install the external oil pipe (see Section 20).

13   Installation is otherwise the reverse of removal.

14   Change the engine oil (see Chapter 1).

11 Valves/valve seats/valve guides - servicing

1   Because of the complex nature of this job and the special tools and equipment required, servicing of the valves, the valve seats and the valve guides (commonly known as a valve job) is best left to a professional.

2   The home mechanic can, however, remove and disassemble the head, do the initial cleaning and inspection, then reassemble and deliver the head to a dealer service department or properly equipped repair shop for the actual valve servicing. Refer to Section 12 for those procedures.

3   The dealer service department will remove the valves and springs, recondition or replace the valves and valve seats, replace the valve guides, check and replace the valve springs, spring retainers and keepers (as necessary), replace the valve seals with new ones and reassemble the valve components.

4   After the valve job has been performed, the head will be in like-

new condition. When the head is returned, be sure to clean it again very thoroughly before installation on the engine to remove any metal particles or abrasive grit that may still be present from the valve servicing operations. Use compressed air, if available, to blow out all the holes and passages.

12 Cylinder head and valves - disassembly, inspection and reassembly

1   As mentioned in the previous Section, valve servicing and valve guide replacement should be left to a dealer service department or other repair shop. However, disassembly, cleaning and inspection of the valves and related components can be done (if the necessary special tools are available) by the home mechanic. This way no expense is incurred if the inspection reveals that service work is not required at this time.

2   To properly disassemble the valve components without the risk of damaging them, a valve spring compressor is absolutely necessary. If the special tool is not available, have a dealer service department or other repair shop handle the entire process of disassembly, inspection, service or repair (if required) and reassembly of the valves.

Disassembly

Refer to illustrations 12.4a, 12.4b and 12.8a through 12.8f, 12.10 and 12.12

3   Remove the cylinder head (see Section 10).

4   Remove the carburetor insulator and discard the old O-ring **(see illustrations)**.

5   Before the valves are removed, scrape away any traces of gasket material from the head gasket sealing surface. Work slowly and do not nick or gouge the soft aluminum of the head. Gasket removing solvents, which work very well, are available at most vehicle shops and auto parts stores.

6   Carefully scrape all carbon deposits out of the combustion chamber area. A hand held wire brush or a piece of fine emery cloth can be used once most of the deposits have been scraped away. Do not use a wire brush mounted in a drill motor, or one with extremely stiff bristles; the cylinder head material is soft and may be eroded away or scratched by the wire brush.

7   Before proceeding, arrange to label and store the valves along with their related components so they can be kept separate and reinstalled in the same valve guides they are removed from (again, plastic bags work well for this).

8   Compress the valve springs on the first valve with a valve spring compressor **(see illustration)**, then remove the keepers **(see illustration)**. Do not compress the springs any more than is absolutely necessary. Carefully release the valve spring compressor and remove the

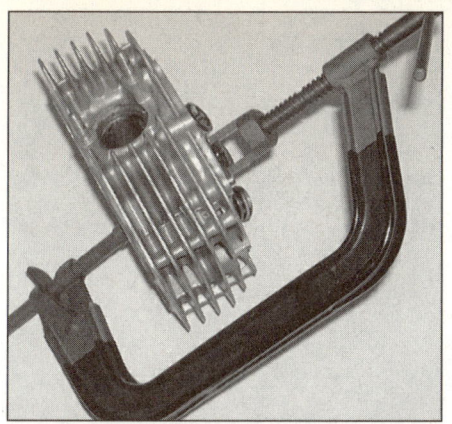

12.8a Install a valve spring compressor, compress the spring . . .

12.8b . . . remove the keepers (arrow), then release the spring compressor

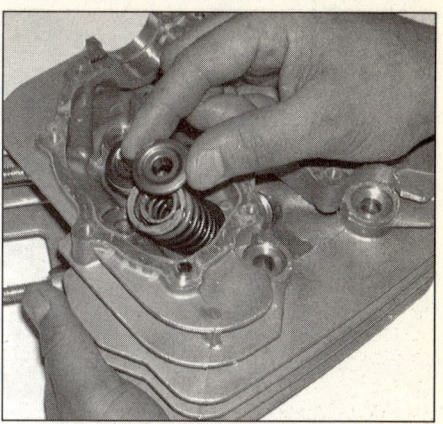

12.8c Remove the valve spring retainer . . .

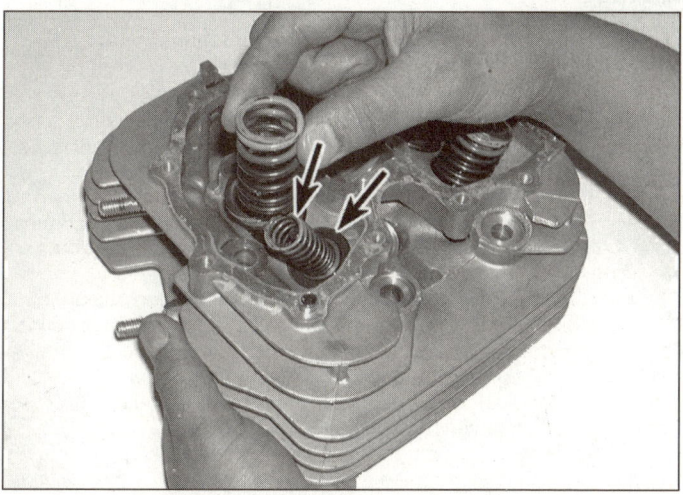

12.8d . . . the outer spring, inner spring and spring seat (arrows)

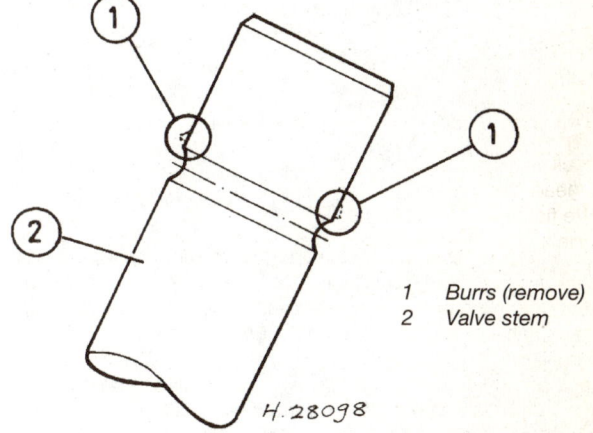

1 Burrs (remove)
2 Valve stem

H.28098

12.8e Inspect the area around the keeper groove for burrs and carefully remove them with a file

retainer, springs, spring seat and valve from the head **(see illustrations)**. Check the area around the keeper groove for any burrs that might damage the guide when the valve is pulled through. If there are any, remove them with a very fine file or whetstone **(see illustration)**.

9 Repeat the preceding step for each valve. Remember to keep the parts for each valve together so they can be reinstalled in the same location.

10 Once the valves have been removed and labeled, pull off the valve stem seals **(see illustration)** with pliers and discard them (the old seals must not be re-used).

11 Clean the cylinder head with solvent and dry it thoroughly. Compressed air will speed the drying process and ensure that all holes and recessed areas are clean.

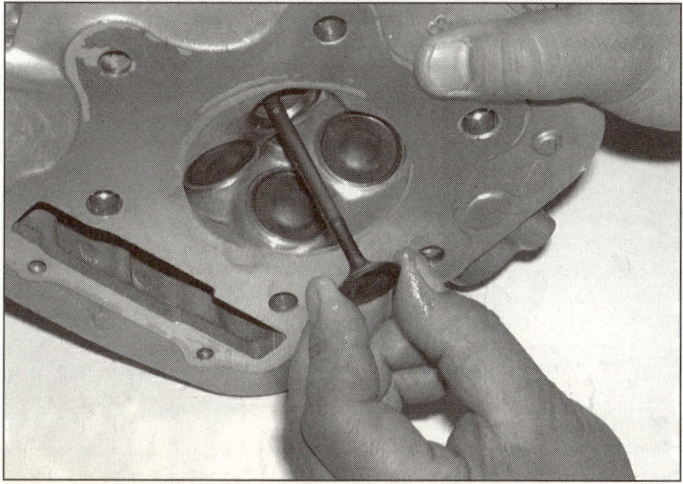

12.8f Pull the valve out of the guide

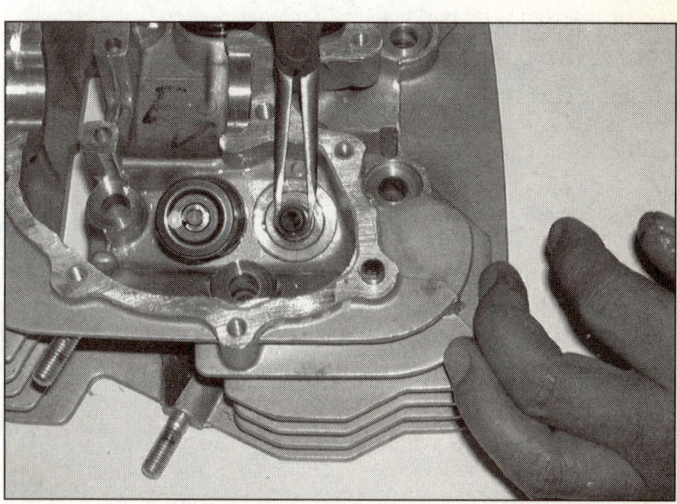

12.10 Remove the old valve guide seals with a pair of needle-nose pliers

2A

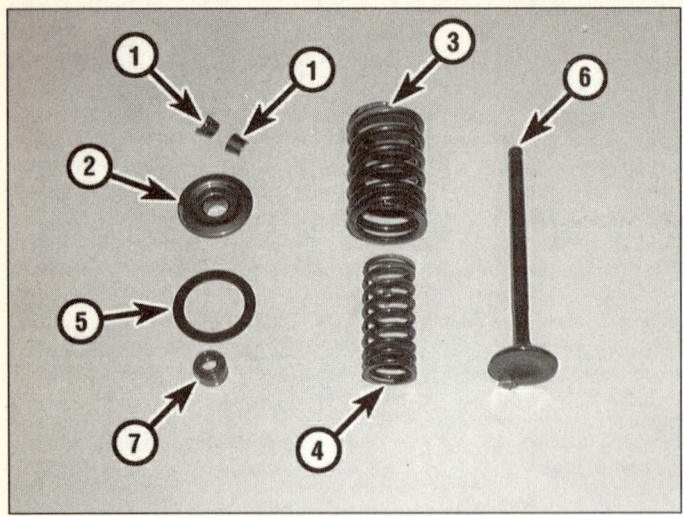

12.12 Valve components

1 Keepers	4 Inner valve spring (the
2 Valve spring retainer	tightly wound coils face
3 Outer valve spring (the	down, toward the spring
tightly wound coils face	seat)
down, toward the spring	5 Spring seat
seat)	6 Valve
	7 Valve guide oil seal

12   Clean all of the valve springs, keepers, retainers and spring seats with solvent and dry them thoroughly. Do the parts from one valve at a time so that no mixing of parts between valves occurs. It's a good idea to label each valve assembly **(see illustration)**, then put it in a plastic bag with the label.

13   Scrape off any deposits that may have accumulated on the valves then use a motorized wire brush to remove deposits from the valve heads and stems. Again, make sure the valves do not get mixed up.

Inspection

Refer to illustrations 12.15a, 12.15b, 12.16, 12.17, 12.18, 12.19a, 12.19b, 12.20a and 12.20b

14   Inspect the head very carefully for cracks and other damage. If cracks are found, a new head will be required. Check the cam bearing surfaces for wear and evidence of seizure. Check the camshaft for wear as well (see Section 9).

15   Using a precision straightedge and a feeler gauge, check the head gasket mating surface for warpage. Lay the straightedge length-wise, across the head and diagonally (corner-to-corner), intersecting

12.15a Check the gasket surface for flatness with a straightedge and feeler gauge . . .

the head bolt holes, and try to slip a feeler gauge under it, on either side of each combustion chamber **(see illustrations)**. The feeler gauge thickness should be the same as the cylinder head warpage limit listed in this Chapter's Specifications. If the feeler gauge can be inserted between the head and the straightedge, the head is warped and must either be machined or, if warpage is excessive, replaced with a new one.

16   Examine the valve seats in each of the combustion chambers. If they are pitted, cracked or burned, the head will require valve service that is beyond the scope of the home mechanic. Measure the valve seat width **(see illustration)** and compare it to this Chapter's Specifications. If it is not within the specified range, or if it varies around its circumference, valve service work is required.

17   Clean the valve guides to remove any carbon buildup, then measure the inside diameters of the guides (at both ends and the center of the guide) with a small-hole gauge and a micrometer **(see illustration)**. Record the measurements for future reference. The guides are measured at the ends and at the center to determine if they are worn in a bell-mouth pattern (more wear at the ends). If they are, guide replacement is an absolute must.

18   Carefully inspect each valve face for cracks, pits and burned spots. Check the valve stem and the keeper groove area for cracks **(see illustration)**. Rotate the valve and check for any obvious indication that it is bent. Check the end of the stem for pitting and excessive wear and make sure the bevel is the specified width. The presence of any of the above conditions indicates the need for valve servicing.

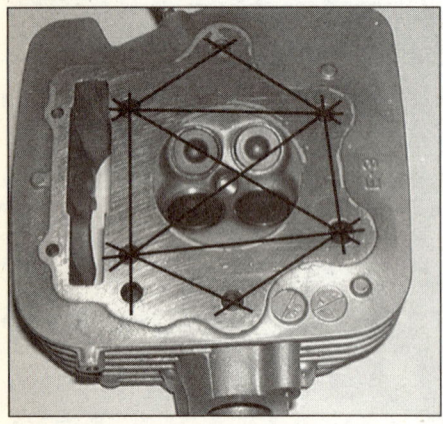

12.15b . . . measuring in the directions shown

12.16 Measuring valve seat width

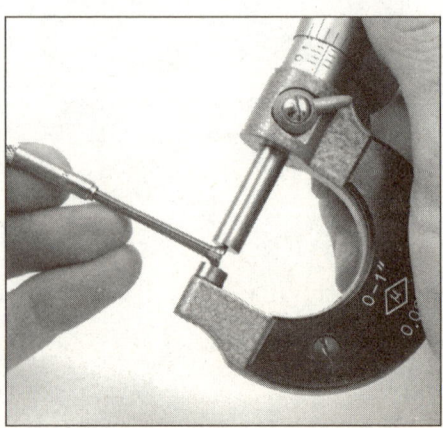

12.17 Measure the valve guide inside diameter with a small-hole gauge, then measure the gauge with a micrometer

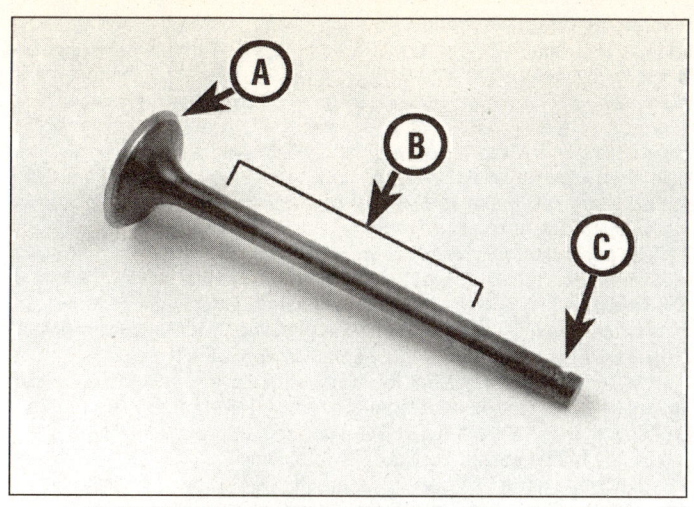

12.18 Check the valve face (A), stem (B) and keeper groove (C) for wear and damage

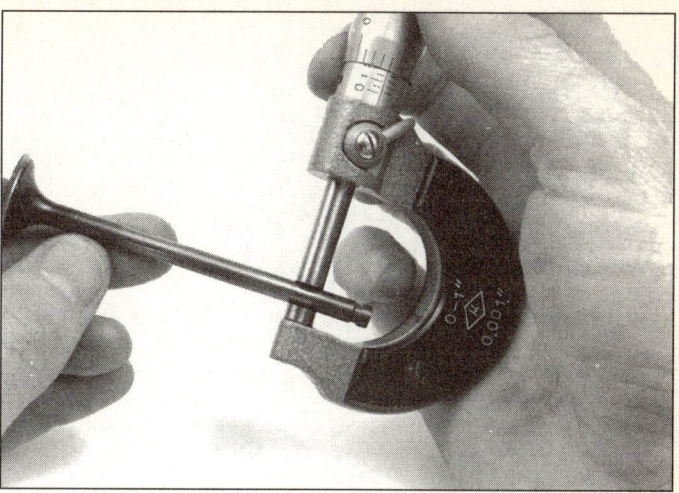

12.19a Measuring valve stem diameter

19   Measure the valve stem diameter **(see illustration)**. If the diameter is less than listed in this Chapter's Specifications, the valves will have to be replaced with new ones. Also check the valve stem for bending. Set the valve in a V-block with a dial indicator touching the middle of the stem **(see illustration)**. Rotate the valve and look for a reading on the gauge (which indicates a bent stem). If the stem is bent, replace the valve.

20   Inspect the end of each valve spring for wear and pitting. Measure the free length **(see illustration)** and compare it to this Chapter's Specifications. Any springs that are shorter than specified have sagged and should not be reused. Stand the spring on a flat surface and check it for squareness **(see illustration)**.

21   Inspect the spring retainers and keepers for obvious wear and cracks. Any questionable parts should not be re-used, as extensive damage will occur in the event of failure during engine operation.

22   If the inspection indicates that no service work is required, the valve components can be reinstalled in the head.

Reassembly

Refer to illustrations 12.24, 12.25a, 12.25b, 12.27 and 12.28

23   If the valve seats have been ground, the valves and seats should be lapped before installing the valves in the head to ensure a positive seal between the valves and seats. This procedure requires coarse and fine valve lapping compound (available at auto parts stores) and a valve lapping tool. If a lapping tool is not available, a piece of rubber or plastic hose can be slipped over the valve stem (after the valve has been installed in the guide) and used to turn the valve.

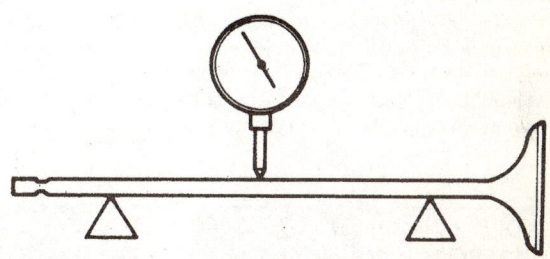

12.19b Check the valve stem for bends with a V-block (or V-blocks, as shown here) and a dial indicator

2A

24   Apply a small amount of coarse lapping compound to the valve face **(see illustration)**, then slip the valve into the guide. **Note:** *Make sure the valve is installed in the correct guide and be careful not to get any lapping compound on the valve stem.*

25   Attach the lapping tool (or hose) to the valve and rotate the tool between the palms of your hands. Use a back-and-forth motion rather than a circular motion. Lift the valve off the seat and turn it at regular intervals to distribute the lapping compound properly. Continue the lapping procedure until the valve face and seat contact area is of uniform width and unbroken around the entire circumference of the valve face and seat **(see illustrations)**. Once this is accomplished, wipe the valve and seat clean and lap the valves again with fine lapping compound.

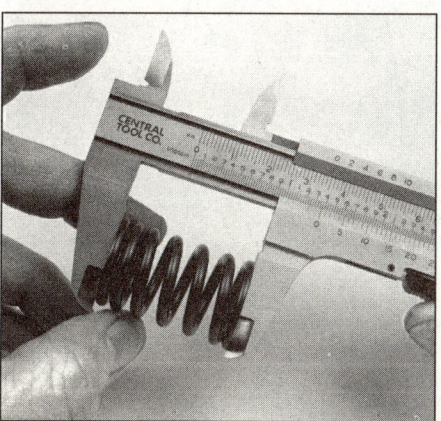

12.20a Measuring the free length of a valve spring

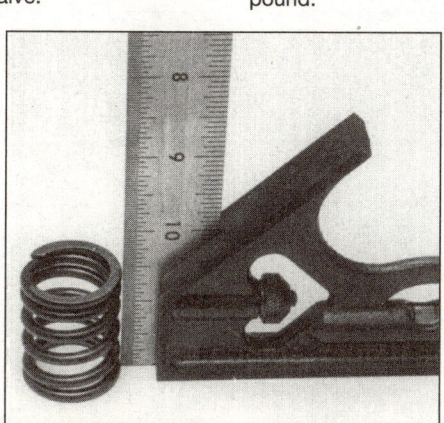

12.20b Checking a valve spring for squareness

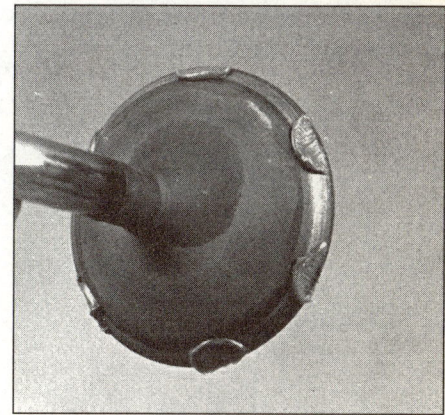

12.24 Apply the lapping compound very sparingly, in small dabs, to the valve face only

12.25a After lapping, the valve face should exhibit a uniform, unbroken contact pattern (arrow) . . .

26   Carefully remove the valve from the guide and wipe off all traces of lapping compound. Use solvent to clean the valve and wipe the seat area thoroughly with a solvent soaked cloth. Repeat the procedure for the remaining valves.

27   Lay the spring seat in place in the cylinder head, then install new valve stem seals on both of the guides **(see illustration)**. Use an appropriate size deep socket to push the seals into place until they are properly seated. Don't twist or cock them, or they will not seal properly against the valve stems. Also, don't remove them again or they will be damaged.

28   Coat the valve stems with assembly lube or moly-based grease, then install one of them into its guide. Next, install the spring seat, springs and retainers, compress the springs and install the keepers. **Note:** *Install the springs with the tightly wound coils at the bottom (next to the spring seat). When compressing the springs with the valve spring compressor, depress them only as far as is absolutely necessary to slip the keepers into place. Apply a small amount of grease to the keepers* **(see illustration)** *to help hold them in place as the pressure is released from the springs. Make certain that the keepers are securely locked in their retaining grooves.*

29   Support the cylinder head on blocks so the valves can't contact the workbench top, then very gently tap each of the valve stems with a soft-faced hammer. This will help seat the keepers in their grooves.

30   Once all of the valves have been installed in the head, check for proper valve sealing by pouring a small amount of solvent into each of the valve ports. If the solvent leaks past the valve(s) into the combustion chamber area, disassemble the valve(s) and repeat the lapping procedure, then reinstall the valve(s) and repeat the check. Repeat the procedure until a satisfactory seal is obtained.

31   Once the head has been reassembled, install the carburetor insulator with a new O-ring. Tighten the insulator bolts securely.

13   Cylinder - removal, inspection and installation

Removal

Refer to illustrations 13.2, 13.3, 13.4a, 13.4b and 13.5

1   Remove the external oil pipe (see Section 20), the cylinder head cover (see Section 7) and the cylinder head (see Section 10).

2   Lift out the cam chain front guide **(see illustration)**.

12.25b . . . and the seat (arrow) should be the specified width with a smooth, unbroken appearance

12.27 Push the oil seal onto the valve guide (arrow)

12.28 A small dab of grease will help hold the keepers in place on the valve while the spring compressor is released

13.2 Lift the front cam chain guide from its saddle

13.3 Remove the two bolts (arrows) that attach the base of the cylinder to the crankcase

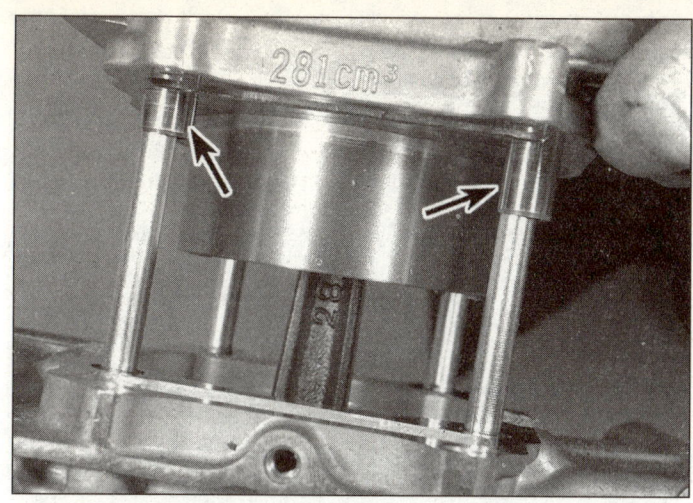

13.4a Lift the cylinder off and note the location of the dowels (arrows) (they may be in the crankcase)

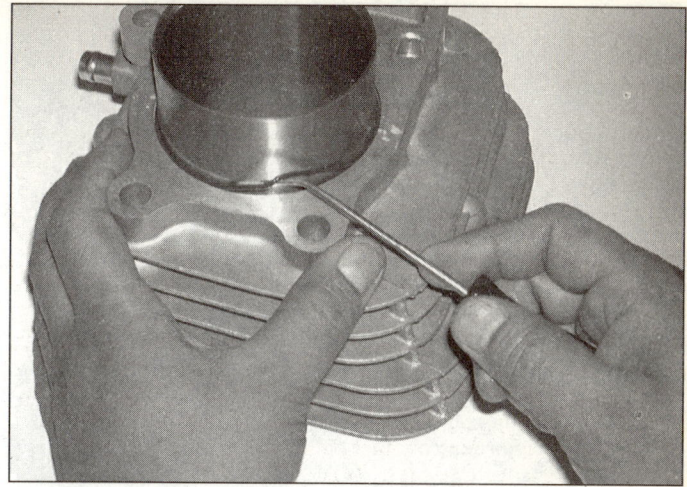

13.4b Remove the old O-ring from the cylinder liner and discard it

3   Remove two bolts that secure the base of the cylinder to the crankcase **(see illustration)**.

4   Lift the cylinder straight up to remove it **(see illustration)**. If it's stuck, tap around its perimeter with a soft-faced hammer, but don't tap on the cooling fins or they may break. Do NOT try to pry between the cylinder and the crankcase; this will ruin the sealing surfaces. Look for the dowel pins **(see illustration 13.5a)**. If they didn't come off with the cylinder, they're still in the crankcase. After removing the cylinder, remove the old O-ring from the cylinder liner **(see illustration)** and discard it.

5   Stuff rags around the piston **(see illustration)** and remove the cylinder base gasket and all traces of old gasket material from the surfaces of the cylinder and the crankcase.

Inspection

Refer to illustrations 13.6 and 13.8

Caution: *Don't attempt to separate the liner from the cylinder.*

6   Check the top surface of the cylinder for warpage, using the same method as for the cylinder head (see Section 12). Measure along the sides and diagonally across the stud holes **(see illustration)**.

7   Check the cylinder walls carefully for scratches and score marks.

8   Using the appropriate precision measuring tools, check the cylinder's diameter at the top, center and bottom of the cylinder bore, parallel to the crankshaft axis **(see illustration)**. Next, measure the cylinder's diameter at the same three locations across the crankshaft axis.

Compare the results to this Chapter's Specifications. If the cylinder walls are tapered, out-of-round, worn beyond the specified limits, or badly scuffed or scored, have the cylinder rebored and honed by a dealer service department or an ATV repair shop. If a rebore is done, oversize pistons and rings will be required as well. **Note:** *Honda supplies pistons in two oversizes.*

9   As an alternative, if the precision measuring tools are not available, a dealer service department or other repair shop can make the measurements and offer advice concerning servicing of the cylinder.

10   If it's in reasonably good condition and not worn to the outside of the limits, and if the piston-to-cylinder clearance can be maintained properly, then the cylinder does not have to be rebored; honing is all that is necessary.

11   To perform the honing operation you will need the proper size flexible hone with fine stones as shown in *Maintenance techniques, tools and working facilities* at the front of this book, or a "bottle brush" type hone, plenty of light oil or honing oil, some shop towels and an electric drill motor. Hold the cylinder block in a vise (cushioned with soft jaws or wood blocks) when performing the honing operation. Mount the hone in the drill motor, compress the stones and slip the hone into the cylinder. Lubricate the cylinder thoroughly, turn on the drill and move the hone up and down in the cylinder at a pace which will produce a fine crosshatch pattern on the cylinder wall with the crosshatch lines intersecting at approximately a 60-degree angle. Be sure to use plenty of lubricant and do not take off any more material

2A

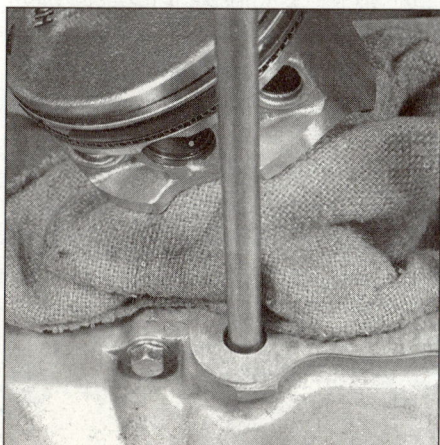

13.5 Stuff clean shop rags into the crankcase opening to keep out debris

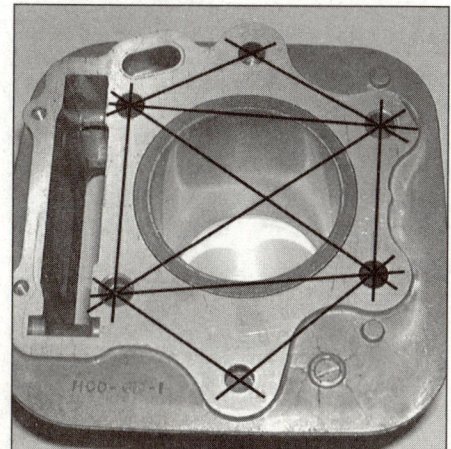

13.6 Check the cylinder top surface for warpage in the directions shown

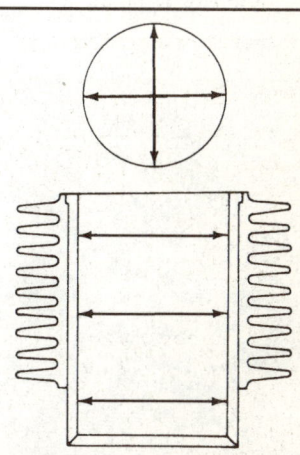

13.8 Measure the cylinder diameter in two directions, at the top, center and bottom of ring travel

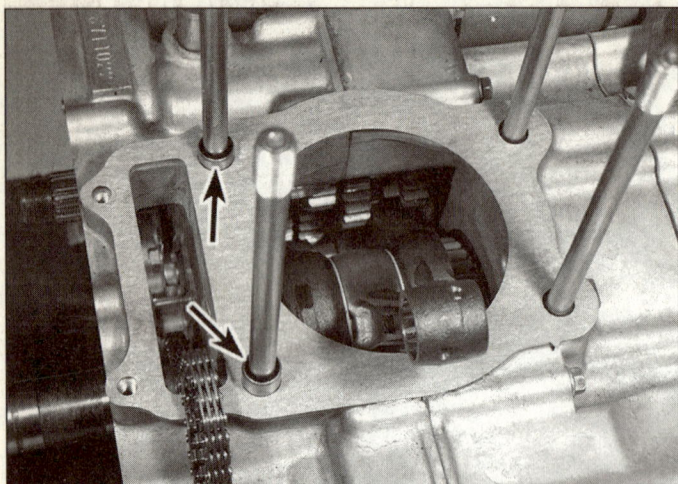

13.14 Install the cylinder base gasket and both dowels (arrows) on the crankcase (piston removed for clarity)

13.16 If you're experienced and very careful, the cylinder can be installed over the rings without a ring compressor, but a compressor is recommended

than is absolutely necessary to produce the desired effect. Do not withdraw the hone from the cylinder while it is running. Instead, shut off the drill and continue moving the hone up and down in the cylinder until it comes to a complete stop, then compress the stones and withdraw the hone. Wipe the oil out of the cylinder and repeat the procedure on the remaining cylinder. Remember, do not remove too much material from the cylinder wall. If you do not have the tools, or do not desire to perform the honing operation, a dealer service department or other repair shop will generally do it for a reasonable fee.

12   Next, the cylinder must be thoroughly washed with warm soapy water to remove all traces of the abrasive grit produced during the honing operation. Be sure to run a brush through the bolt holes and flush them with running water. After rinsing, dry the cylinder thoroughly and apply a coat of light, rust-preventative oil to all machined surfaces.

Installation

Refer to illustrations 13.14 and 13.16

13   Lubricate the cylinder bore with plenty of clean engine oil. Apply a thin film of moly-based grease to the piston skirt.

14   Install the dowel pins, then lower a new cylinder base gasket over them **(see illustration)**.

15   Attach a piston ring compressor to the piston and compress the piston rings. A large hose clamp can be used instead - just make sure it doesn't scratch the piston, and don't tighten it too much.

16   Install the cylinder over the studs and carefully lower it down until the piston crown fits into the cylinder liner **(see illustration)**. While doing this, pull the camshaft chain up, using a hooked tool or a piece

of stiff wire. Push down on the cylinder, making sure the piston doesn't get cocked sideways, until the bottom of the cylinder liner slides down past the piston rings. A wood or plastic hammer handle can be used to gently tap the cylinder down, but don't use too much force or the piston will be damaged.

17   Remove the piston ring compressor or hose clamp, being careful not to scratch the piston.

18   The remainder of installation is the reverse of the removal steps.

14   Piston - removal, inspection and installation

1   The piston is attached to the connecting rod with a piston pin that is a slip fit in the piston and rod.

2   Before removing the piston from the rod, stuff a clean shop towel into the crankcase hole, around the connecting rod. This will prevent the circlips from falling into the crankcase if they are inadvertently dropped.

Removal

Refer to illustrations 14.3a, 14.3b, 14.4a and 14.4b

3   The piston should have an IN mark on its crown that goes toward the intake (rear) side of the engine **(see illustration)**. If this mark is not visible due to carbon buildup, scribe an arrow into the piston crown before removal. Support the piston and pry the circlip out with a pointed tool **(see illustration)**.

14.3a The IN mark on top of the piston faces the intake (rear) side of the engine

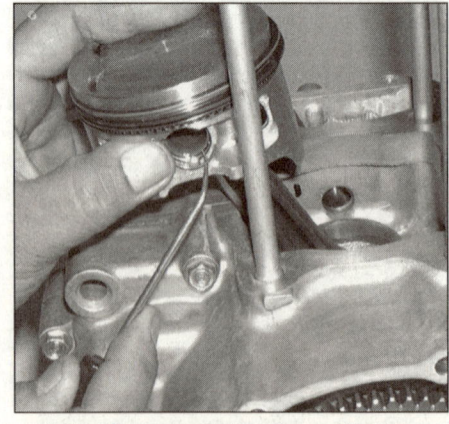

14.3b Wear eye protection and pry the circlip out of its groove with a pointed tool

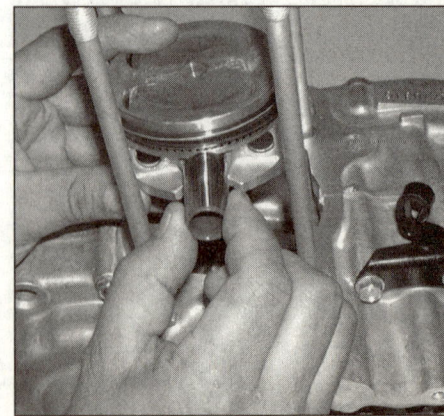

14.4a Push the piston pin part-way out, then pull it the rest of the way

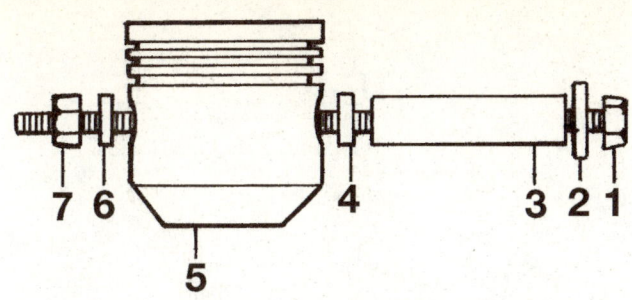

**14.4b The piston pin should come out with hand pressure -
if it doesn't, this removal tool can be fabricated from
readily available parts**

1	Bolt	7	Nut (B)
2	Washer	A)	Large enough for piston
3	Pipe (A)		pin to fit inside
4	Padding (A)	B)	Small enough to fit
5	Piston		through piston pin bore
6	Washer (B)		

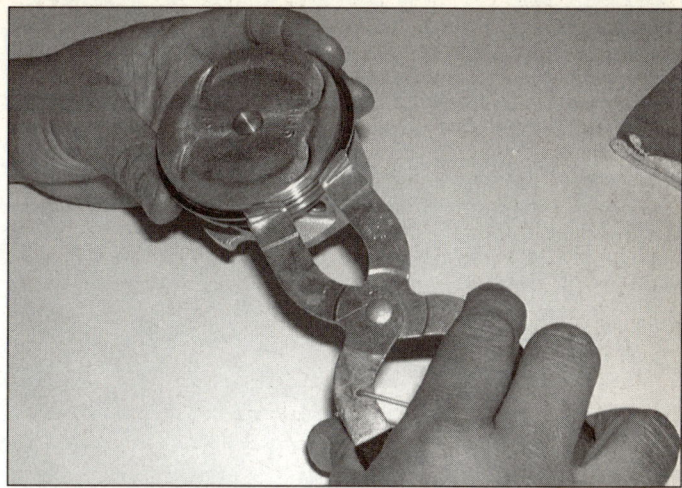

**14.6a Remove the top and second piston rings with a ring
removal and installation tool**

4   Push the piston pin out from the opposite end to free the piston from the rod **(see illustration)**. You may have to deburr the area around the groove to enable the pin to slide out (use a triangular file for this procedure). If the pin won't come out, you can fabricate a piston pin removal tool from a long bolt, a nut, and a piece of tubing and washers **(see illustration)**.

Inspection

Refer to illustrations 14.6a, 14.6b, 14.6c, 14.13, 14.14, 14.15 and 14.16

5   Before the inspection process can be carried out, the pistons must be cleaned and the old piston rings removed.

6   Using a piston ring removal and installation tool, carefully remove the top and second rings from the piston **(see illustration)**. Do not nick or gouge the pistons in the process. Remove the upper oil ring side rail, the oil ring spacer (the wavy part) and the lower oil ring side rail **(see illustrations)**.

7   Scrape all traces of carbon from the tops of the pistons. A hand-held wire brush or a piece of fine emery cloth can be used once the majority of the deposits have been scraped away. Do not, under any circumstances, use a wire brush mounted in a drill motor to remove deposits from the pistons; the piston material is soft and will be eroded away by the wire brush.

8   Use a piston ring groove cleaning tool to remove any carbon deposits from the ring grooves. If a tool is not available, a piece broken off the old ring will do the job. Be very careful to remove only the carbon deposits. Do not remove any metal and do not nick or gouge the

sides of the ring grooves.

9   Once the deposits have been removed, clean the piston with solvent and dry them thoroughly. Make sure the oil return holes below the oil ring grooves are clear.

10   If the piston is not damaged or worn excessively and if the cylinder is not rebored, a new piston will not be necessary. Normal piston wear appears as even, vertical wear on the thrust surfaces of the piston and slight looseness of the top ring in its groove. New piston rings, on the other hand, should always be used when an engine is rebuilt.

11   Carefully inspect each piston for cracks around the skirt, at the pin bosses and at the ring lands.

12   Look for scoring and scuffing on the thrust faces of the skirt, holes in the piston crown and burned areas at the edge of the crown. If the skirt is scored or scuffed, the engine may have been suffering from overheating and/or abnormal combustion, which caused excessively high operating temperatures. The oil pump should be checked thoroughly. A hole in the piston crown, an extreme to be sure, is an indication that abnormal combustion (pre-ignition) was occurring. Burned areas at the edge of the piston crown are usually evidence of spark knock (detonation). If any of the above problems exist, the causes must be corrected or the damage will occur again.

13   Measure the piston ring-to-groove clearance (side clearance) by laying a new piston ring in the ring groove and slipping a feeler gauge in beside it **(see illustration)**. Check the clearance at three or four locations around the groove. Be sure to use the correct ring for each groove; they are different. If the clearance is greater than specified, new pistons will have to be used when the engine is reassembled.

2A

**14.6b Remove the upper oil ring
side rail . . .**

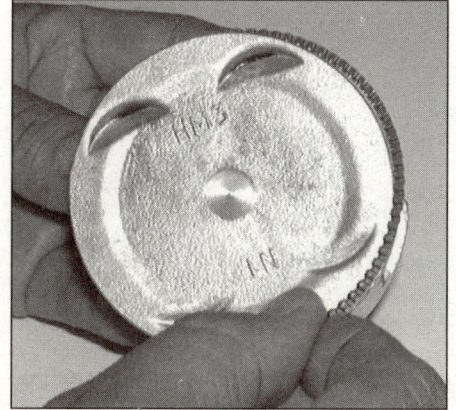

**14.6c . . . remove the expander, then
remove the lower side rail**

**14.13 Measure the piston ring-to-groove
clearance with a feeler gauge**

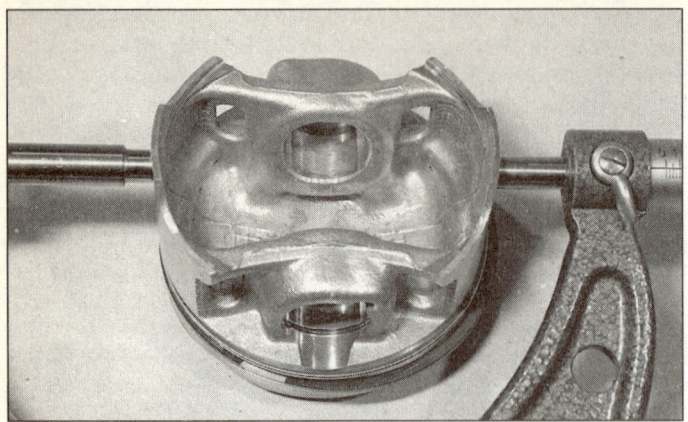

14.14 Measure the piston diameter with a micrometer

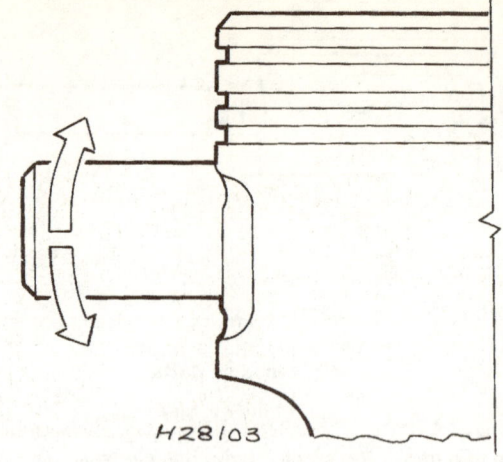

14.15 Slip the pin into the piston and try to wiggle it back-and-forth; if it's loose, replace the piston and pin

14   Check the piston-to-bore clearance by measuring the bore (see Section 13) and the piston diameter **(see illustration)**. Measure the piston across the skirt on the thrust faces at a 90-degree angle to the piston pin, at the specified distance up from the bottom of the skirt. Subtract the piston diameter from the bore diameter to obtain the clearance. If it is greater than specified, the cylinder will have to be rebored and a new oversized piston and rings installed. If the appropriate precision measuring tools are not available, the piston-to-cylinder clearance can be obtained, though not quite as accurately, using feeler gauge stock. Feeler gauge stock comes in 12-inch lengths and various thicknesses and is generally available at auto parts stores. To check the clearance, slip a piece of feeler gauge stock of the same thickness as the specified piston clearance into the cylinder along with thee piston. The cylinder should be upside down and the piston must be positioned exactly as it normally would be. Place the feeler gauge between the piston and cylinder on one of the thrust faces (90-degrees to the piston pin bore). The piston should slip through the cylinder (with the feeler gauge in place) with moderate pressure. If it falls through, or slides through easily, the clearance is excessive and a new piston will be required. If the piston binds at the lower end of the cylinder and is loose toward the top, the cylinder is tapered, and if tight spots are encountered as the piston/feeler gauge is rotated in the cylinder, the cylinder is out-of-round. Be sure to have the cylinder and piston checked by a dealer service department or a repair shop to confirm your findings before purchasing new parts.

15   Apply clean engine oil to the pin, insert it into the piston and check for freeplay by rocking the pin back-and-forth **(see illustration)**. If the pin is loose, a new piston and possibly a new pin must be installed.

16   Repeat Step 15, this time inserting the piston pin into the con-

necting rod **(see illustration)**. If the pin is loose, measure the pin diameter and the pin bore in the rod (or have this done by a dealer or repair shop). A worn pin can be replaced separately; if the rod bore is worn, the rod and crankshaft must be replaced as an assembly.

17   Refer to Section 15 and install the rings on the pistons.

Installation

Refer to illustration 14.18

18   Install the piston with its IN mark toward the intake side (rear) of the engine. Lubricate the pin and the rod bore with moly-based grease. Install a new circlip in the groove in one side of the piston (don't reuse the old circlips). Push the pin into position from the opposite side and install another new circlip. Compress the circlips only enough for them to fit in the piston. Make sure the circlips are fully seated in their grooves in the piston pin bore, and make sure that the end gaps of the circlips are not aligned with the notch in the pin bore **(see illustration)**.

15 Piston rings - installation

Refer to illustrations 15.2, 15.4, 15.7a, 15.7b, and 15.7c

1   Before installing the new piston rings, the ring end gaps must be checked.

2   Insert the top (No. 1) ring into the bottom of the first cylinder and square it up with the cylinder walls by pushing it in with the top of the piston. The ring should be about one-half inch above the bottom edge

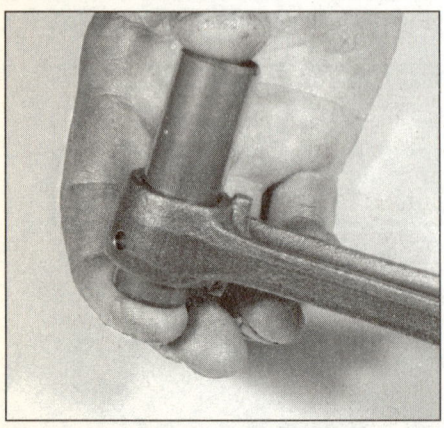

14.16 Slip the piston pin into the rod and try to rock it back-and-forth to check for looseness

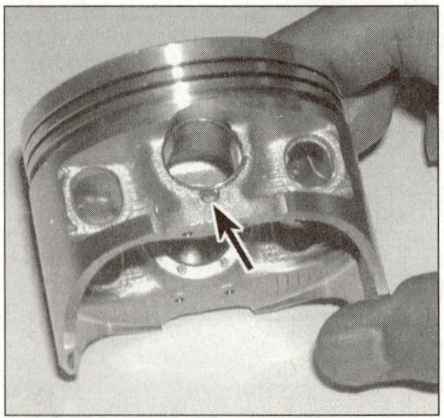

14.18 Seat both piston pin circlips securely in the piston grooves and make sure that their end gaps are away from the removal notch (arrow)

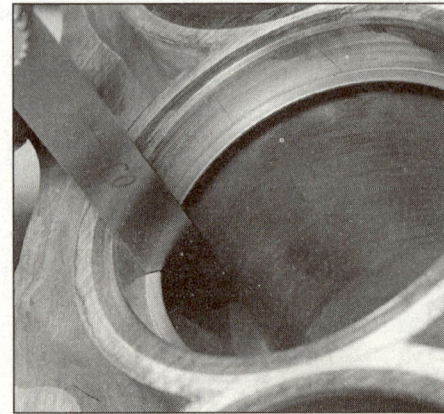

15.2 Check the piston ring end gap with a feeler gauge at the bottom of the cylinder

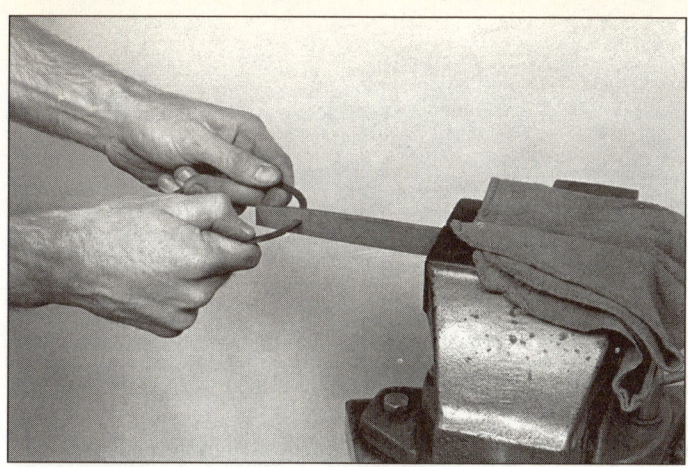

15.4 If the end gap is too small, clamp a file in a vise and file the ring ends (from the outside in only) to enlarge the gap slightly

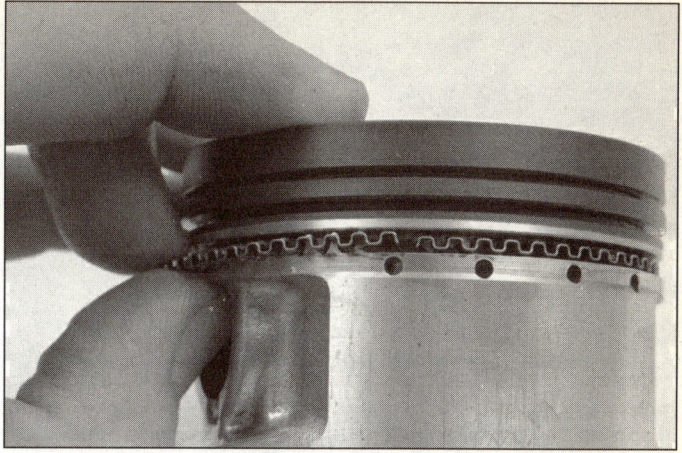

15.7a When installing the oil ring expander, make sure the ends don't overlap

of the cylinder. To measure the end gap, slip a feeler gauge between the ends of the ring **(see illustration)** and compare the measurement to the Specifications.

3   If the gap is larger or smaller than specified, double check to make sure that you have the correct rings before proceeding.

4   If the gap is too small, it must be enlarged or the ring ends may come in contact with each other during engine operation, which can cause serious damage. The end gap can be increased by filing the ring ends very carefully with a fine file **(see illustration)**. When performing this operation, file only from the outside in.

5   Repeat the procedure for the second compression ring (ring gap is not specified for the oil ring rails or spacer).

6   Once the ring end gaps have been checked/corrected, the rings can be installed on the piston.

7   The oil control ring (lowest on the piston) is installed first. It is composed of three separate components. Slip the spacer into the groove, then install the upper side rail **(see illustrations)**. Do not use a piston ring installation tool on the oil ring side rails as they may be damaged. Instead, place one end of the side rail into the groove between the spacer expander and the ring land. Hold it firmly in place and slide a finger around the piston while pushing the rail into the groove (taking care not to cut your fingers on the sharp edges). Next, install the lower side rail in the same manner.

8   After the three oil ring components have been installed, check to make sure that both the upper and lower side rails can be turned smoothly in the ring groove.

9   Install the no. 2 (middle) ring next. It can be readily distinguished from the top ring by its cross-section shape **(see illustration 15.7c)**. Do not mix the top and middle rings.

10   To avoid breaking the ring, use a piston ring installation tool and make sure that the identification mark is facing up **(see illustration 15.7c)**. Fit the ring into the middle groove on the piston. Do not expand the ring any more than is necessary to slide it into place.

11   Finally, install the no. 1 (top) ring in the same manner. Make sure the identifying mark is facing up. Be very careful not to confuse the top and second rings. Besides the different profiles, the top ring is narrower than the second ring.

12   Once the rings have been properly installed, stagger the end gaps, including those of the oil ring side rails **(see illustration 15.7c)**.

16   External oil pipe - removal and installation

Removal

Refer to illustration 16.2a, 16.2b and 16.2c

1   Drain the engine oil (see Chapter 1).

2   Remove the banjo bolts at the upper and lower end of the oil pipe and the retaining bolt in the center of the pipe **(see illustrations)**.

2A

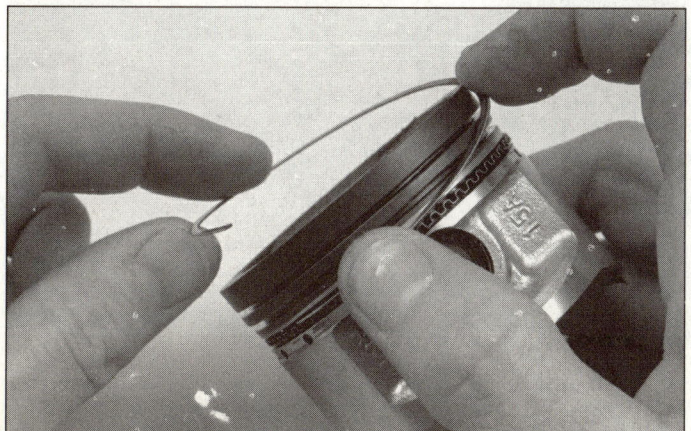

15.7b Don't use a ring installation tool to install the oil ring side rails; do it by hand

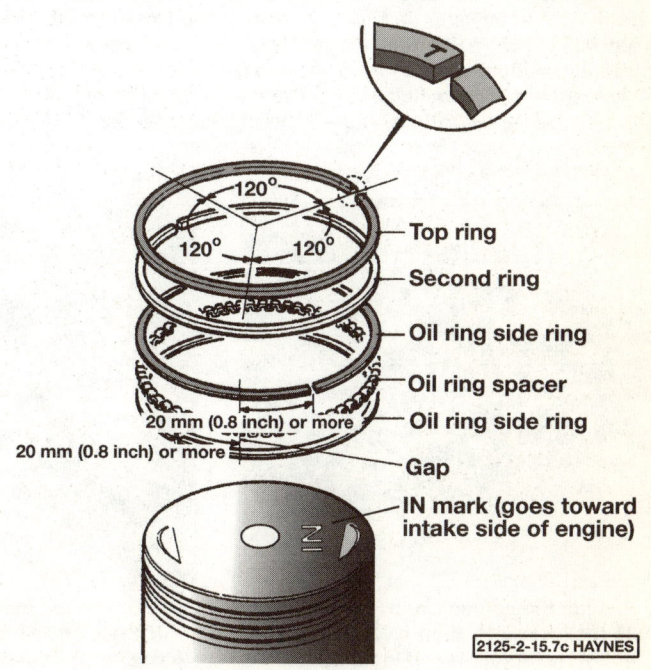

- Top ring
- Second ring
- Oil ring side ring
- Oil ring spacer
- Oil ring side ring
- Gap
- IN mark (goes toward intake side of engine)

120° 120° 120°

20 mm (0.8 inch) or more

20 mm (0.8 inch) or more

2125-2-15.7c HAYNES

15.7c Ring details

16.2a Remove the banjo bolt (arrow) from the lower end of the external oil pipe . . .

16.2b . . . on installation, use a new sealing washer on each side of the fitting

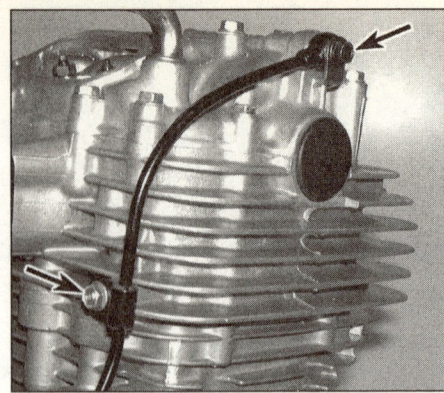

16.2c Remove the bracket bolt and upper banjo bolt (arrows) from the cylinder head cover, then lift off the oil line

Installation

Refer to illustration 16.3

3   Installation is the reverse of the removal steps, with the following additions:

a) *Replace the sealing washers for the upper and lower banjo bolts* **(see illustration)**.
b) *Position the lower end of the pipe against the cast lug on the crankcase so the pipe won't twist clockwise when the union bolt is tightened.*
c) *Install the metal rotation stopper on the upper end of the pipe, again so the pipe won't twist clockwise when the union bolt is tightened.*
d) *Tighten the union bolts and the retaining bolt to the torques listed in this Chapter's Specifications.*

17 Clutch cable - removal and installation

Refer to illustrations 17.2, 17.3, 17.4 and 17.7

1   At the handlebar, peel back the rubber dust cover, back off the locknut and unscrew the clutch cable adjuster **(see illustration 18.3 in Chapter 1)**.
2   To disengage the upper end of the cable from the clutch lever, rotate the cable adjuster and locknut so that the slots in both are aligned with the cable, then slip the cable out of the adjuster and locknut **(see illustration)** and disengage the end plug from the lever.
3   Trace the clutch cable down to the cable guide on top of the crankcase **(see illustration)**, loosen the locknut and back off the adjuster (see Section 18 in Chapter 1, if necessary), and unbolt the cable guide from the engine.

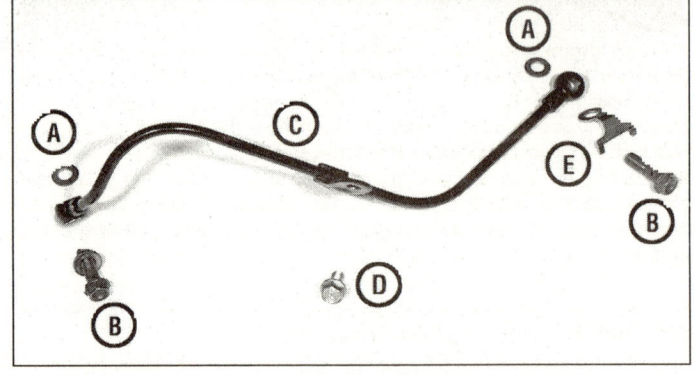

16.3 Oil pipe details

A	Sealing washers	D	Bracket bolt
B	Banjo bolts	E	Rotation stopper
C	Oil pipe		

4   Disengage the lower end of the cable from the clutch lifter arm **(see illustration)**. Note the routing of the clutch cable, then remove the cable.
5   Slide the cable back and forth in the housing and make sure it moves freely. If it doesn't, try lubricating it as described in Chapter 1. If that doesn't help, replace the cable.
6   To replace the clutch lever, simply remove the pivot bolt nut and pull out the pivot bolt.
7   To remove the clutch/parking brake lever bracket, disconnect the parking brake cable (see Chapter 6), unplug the electrical connector

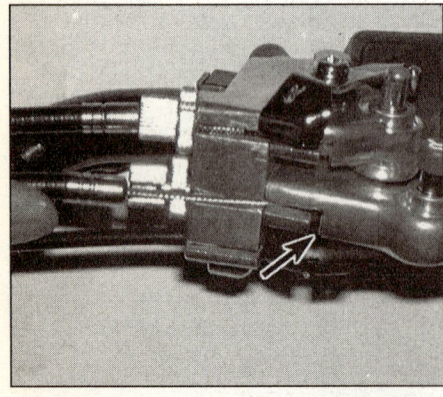

17.2 Align the adjuster and locknut slots with the lever slot, then pivot the cable out of the adjuster and locknut and lower the cable end plug (arrow) from the clutch lever

17.3 Loosen the clutch cable locknut (center arrow), back off the adjuster (upper arrow) and remove the two cable guide bolts (lower arrows)

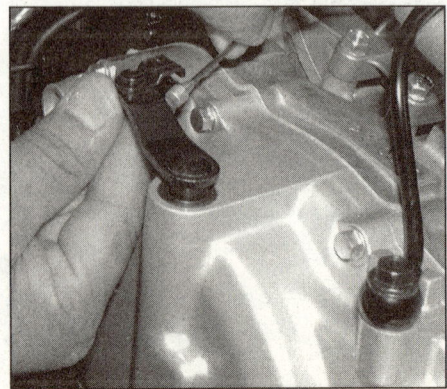

17.4 Push the clutch lifter arm toward the engine and disengage the cable end plug from the lifter arm clevis

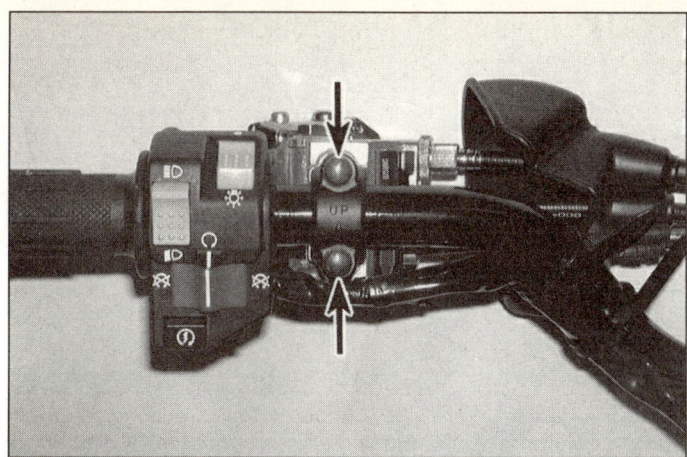

17.7 The clutch/parking brake lever bracket is secured
by two bolts (arrows)

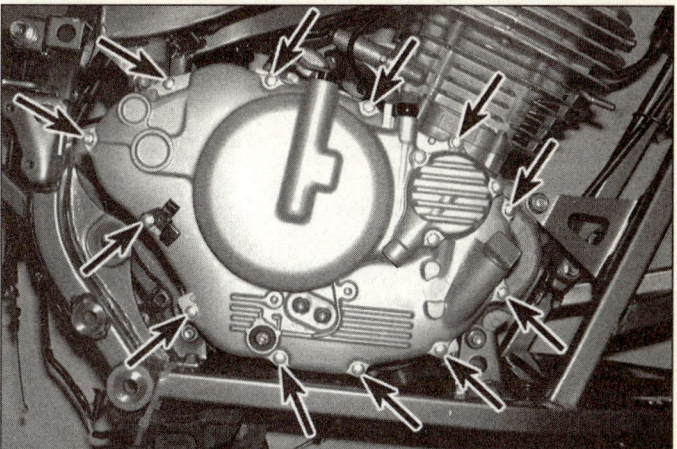

18.6a Right crankcase cover bolts (arrows); one bolt secures the
reverse selector cable guide (circled)

from the clutch switch (see Chapter 8), remove the two clamp bolts **(see illustration)** and remove the lever assembly.

8   Installation is the reverse of removal. If you removed the clutch/parking brake lever bracket, make sure that the UP arrow on the bracket clamp faces up **(see illustration 17.7)**, and be sure to adjust the parking brake cable (see "Parking brake - adjustment" in Chapter 1). Make sure that the parking brake cable is correctly routed and adjust the clutch cable when you're done (see "Clutch cable - check and adjustment" in Chapter 1).

18 Clutch - removal, inspection and installation

Right crankcase cover and clutch release mechanism

Removal

Refer to illustrations 18.6a, 18.6b, 18.7, 18.8, 18.9, 18.10a, 18.10b and 18.10c

1   Drain the engine oil and remove the oil filter (see Chapter 1).
2   Disconnect the external oil pipe from the right crankcase cover (see Section 20).
3   Loosen the clutch cable and disconnect it from the clutch lifter arm (see Section 17).
4   Remove the neutral and reverse switch cover and unplug the electrical connectors from the two switches (see "Neutral and reverse switches - check and replacement" in Chapter 8).
5   Disconnect the reverse shifter cable from the reverse selector arm

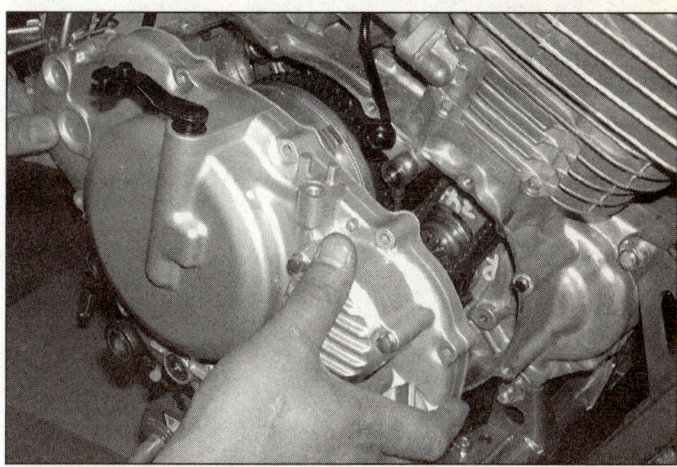

18.6b Carefully pull off the right crankcase cover; if it's stuck,
gently tap it loose with a plastic mallet

and remove the reverse selector arm (see Section 19).
6   Remove the bolts and the right crankcase cover **(see illustrations)**. Note that one of the rear bolts secures the reverse selector cable. Don't forget to install this part when reattaching the right cover.
7   Remove the dowel pins **(see illustration)** and put them in a plastic bag.
8   Remove the old gasket material from the right crankcase cover **(see illustration)** and from the cover mating surface on the crankcase.

18.7 Remove the two dowel pins (arrows) (if the dowels are not in
the crankcase, they're probably in the cover)

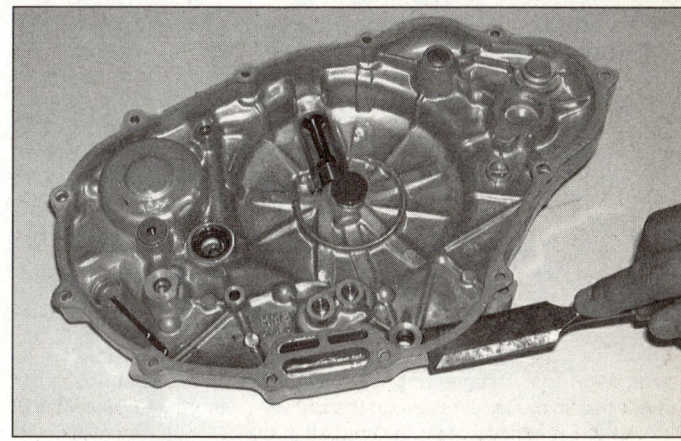

18.8 Remove the old gasket material from the right
crankcase cover and crankcase

2A

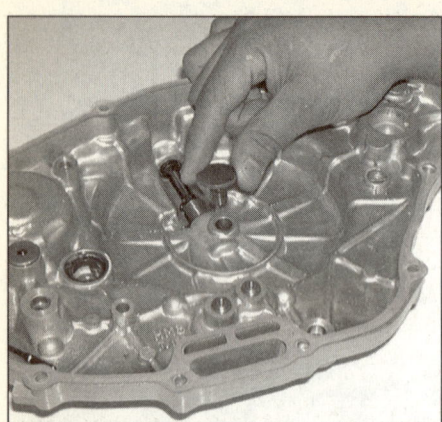

18.9 Remove the clutch lifter piece from the right crankcase cover

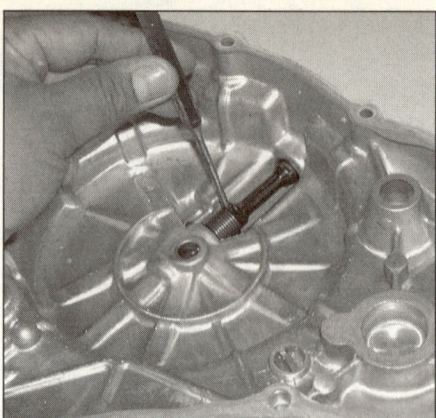

18.10a To remove the clutch lifter arm and shaft, drive in the spring pin until it's flush with the shaft . . .

18.10b . . . then pull out the shaft and remove the spring (arrow)

18.10c Drive out the old spring pin and discard it; always use a new spring pin on assembly

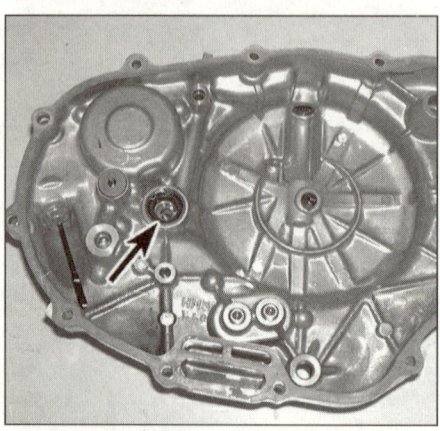

18.16a The crankshaft seal (arrow) is secured in the right crankcase cover by a snap-ring

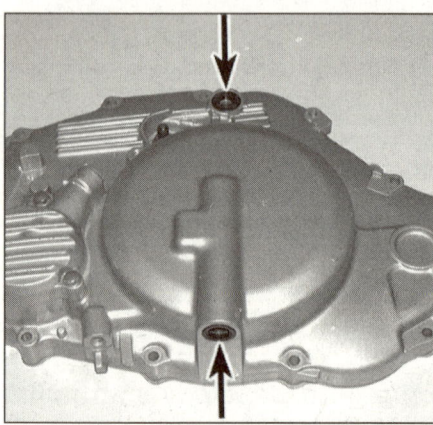

18.16b The clutch lifter arm seal (lower arrow) and reverse selector shaft seal (upper arrow)

9   Remove the clutch lifter piece **(see illustration)**.
10   Using a 3 mm punch, drive the spring pin into the clutch lifter arm shaft until it's flush with the shaft, then remove the clutch lifter arm and shaft assembly **(see illustrations)**. Drive the spring pin the rest of the way out of the lifter arm shaft **(see illustration)** and discard it.
11   Remove the lifter arm shaft return spring **(see illustration 18.10b)**.

Inspection

Refer to illustrations 18.16a and 18.16b

12   Inspect the inner end of the lifter piece (the disc-shaped friction face that pushes against the lifter plate needle bearing) and the outer end (which is pushed by the cam surface of the lifter arm shaft). If either end of the lifter piece is excessively worn, replace it.
13   If the outer end of the lifter piece is worn, carefully inspect the cam surface of the clutch lifter arm for visible wear or damage at the contact point of the cam.
14   If the friction face of the lifter piece is excessively worn, look closely at the clutch lifter plate needle bearing, which is probably also worn **(see illustrations 18.25a and 18.25b)**.
15   Make sure that the return spring is neither bent nor distorted. Replace any parts that show problems.
16   Inspect the right crankcase cover seals **(see illustrations)**. If any of them are leaking or excessively worn, replace them. The crankcase seal is retained by a snap-ring that must be removed before the seal can be replaced. To remove an old seal, pry it out of the cover with a screwdriver. To install a new seal, drive it into place with a small socket with an outside diameter (O.D.) slightly smaller than the O.D. of the seal.

Installation

Refer to illustration 18.18

17   Install the return spring and the lifter arm shaft. Drive a new spring pin into the hole in the lifter arm shaft until the pin height (the amount the pin protrudes from the shaft is 2.5 to 3.5 mm (0.1 to 0.14 inch).
18   Before installing the right crankcase cover, remove the oil strainer screen **(see illustration)** and inspect it carefully. Aluminum-colored sediment in the oil on the screen is no real cause for alarm; it's just the

18.18 Pull out the oil strainer screen and inspect it for fragments of aluminum, steel or clutch material

18.22a Remove these four bolts (arrows), remove the lifter plate (arrow) . . .

18.22b . . . and remove the clutch springs

18.22c Bend back the staked portion of the locknut

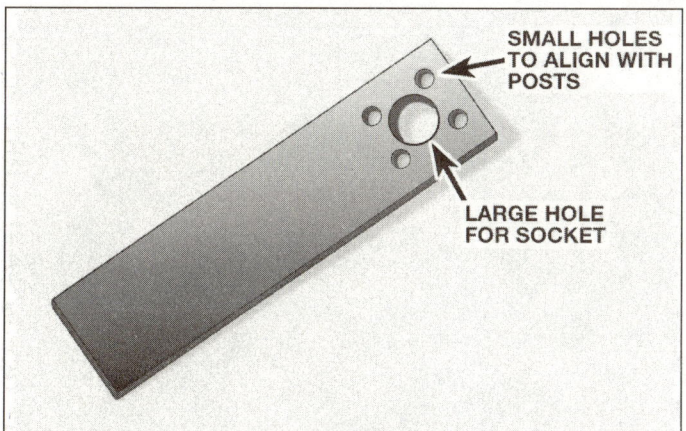

18.22d If you don't have the special Honda clutch holder tools, make your own out of flat stock

SMALL HOLES TO ALIGN WITH POSTS

LARGE HOLE FOR SOCKET

18.22e Remove the lockwasher; the OUT SIDE mark faces away from the engine on installation

2A

residue deposited by aluminum parts as they wear. But any bits or chunks of any material are likely indicators of abnormal internal engine wear. If the pieces are clutch material, it's time to overhaul the clutch. If there are pieces of metal on the screen, it's probably time to split the cases and inspect the crankshaft and transmission assemblies. When you're through with your inspection, wash the screen in clean solvent and install it.

19   Installation is otherwise the reverse of removal.

20   Refill the engine oil and adjust the clutch cable (see Chapter 1).

Clutch

Removal

Refer to illustrations 18.22a through 18.22k

21   Remove the right crankcase cover (see above).

22   Refer to the accompanying illustrations to remove the clutch components **(see illustrations)**. If an air wrench is not available, the clutch must be immobilized while removing the locknut. The Honda service tool (part no. 07JMB-MN50300) resembles a steering wheel

18.22f Remove the clutch center

18.22g Remove the discs and plates; if you plan to reuse them, keep them in order and don't flip them over

18.22h Remove the pressure plate from the clutch housing

18.22i Remove the thrust washer from the mainshaft

18.22j Remove the clutch housing

18.22k Remove the clutch housing guide bushing

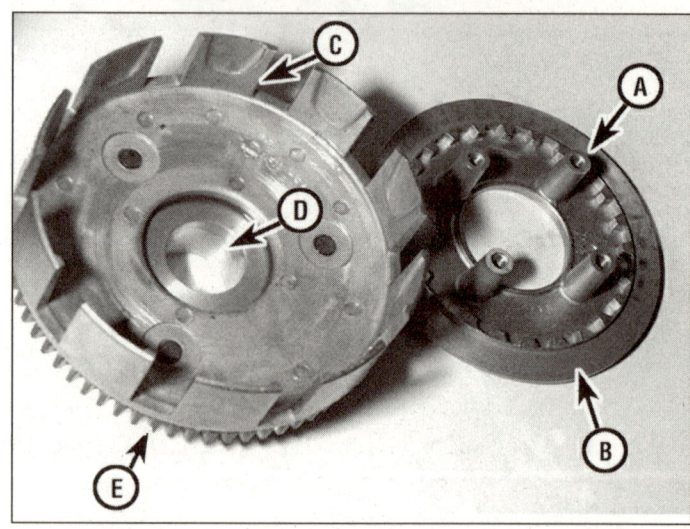

18.23 Clutch inspection points

A Pressure plate posts	D Clutch housing bearing
B Pressure plate friction	surface
surface	E Primary driven gear
C Clutch housing slots	

puller and is attached to the spring posts on the pressure plate. It is then held with a breaker bar while the nut is loosened with a socket. A similar tool can be fabricated from metal stock **(see illustration 18.22d)**. **Note:** *If you plan to reuse the friction plates and discs, keep them in order and don't flip them over.*

Inspection

Refer to illustrations 18.23, 18.25a, 18.25b, 18.26, 18.27 and 18.28

23   Inspect the bolt posts and the friction surface on the pressure plate for damaged threads, scoring or wear **(see illustration)**. Replace the pressure plate if any defects are found.

24   Inspect the edges of the slots in the clutch housing for indentations made by the friction plate tabs **(see illustration 18.23)**. If the indentations are deep they can prevent clutch release, so the housing should be replaced with a new one. If the indentations can be removed easily with a file, the life of the housing can be prolonged to an extent. Also, check the driven gear teeth for cracks, chips and excessive wear and the springs on the back side for breakage. If the gear is worn or damaged or the springs are broken, the clutch housing must be replaced with a new one. Check the bearing surface in the center of the clutch housing for score marks, scratches and excessive wear.

25   Inspect the clutch lifter plate for wear and damage. Rotate the inner race of the needle bearing and make sure it's not rough, loose or noisy. If the bearing is excessively worn, replace it **(see illustrations)**.

26   Measure the free length of the clutch springs **(see illustration)** and compare the results to this Chapter's Specifications. If the springs have sagged, or if cracks are noted, replace them with new ones as a set.

27   If the lining material of the friction plates smells burnt or if it is glazed, new parts are required. If the metal clutch plates are scored or

discolored, they must be replaced with new ones. Measure the thickness of the friction plates **(see illustration)** and replace with new parts any friction plates that are worn. **Note:** *If any friction plates are worn beyond the minimum thickness, it's a good idea to replace them all.*

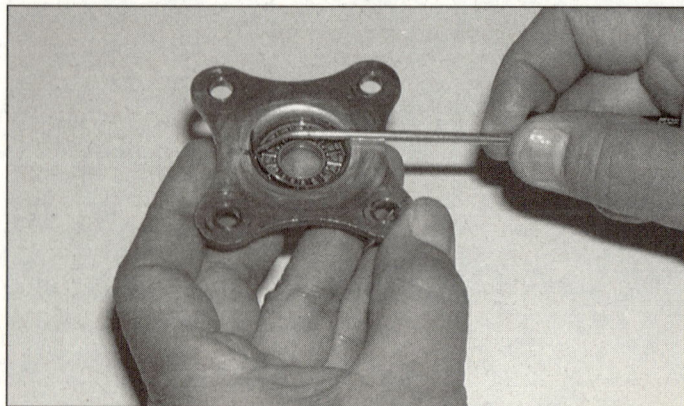

18.25a To replace the needle bearing in the lifter plate, pry out this stop-ring

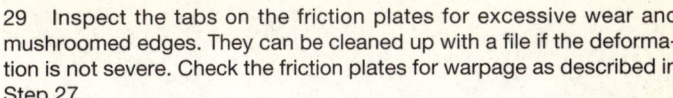

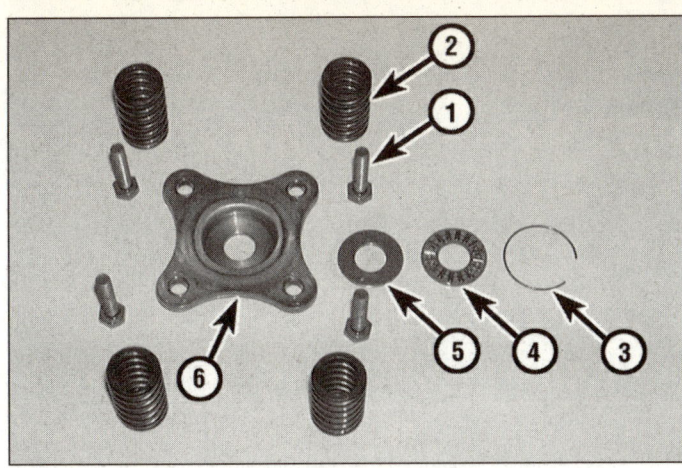

18.25b Lifter plate details

1	Clutch bolts (4)	4	Needle bearing
2	Clutch springs (4)	5	Thrust washer
3	Stop-ring	6	Lifter plate

28   Lay the metal plates, one at a time, on a perfectly flat surface (such as a piece of plate glass) and check for warpage by trying to slip a feeler gauge between the flat surface and the plate **(see illustration)**. The feeler gauge should be the same thickness as the maximum warpage listed in this Chapter's Specifications. Do this at several places around the plate's circumference. If the feeler gauge can be slipped under the plate, it is warped and should be replaced with a new one.

29   Inspect the tabs on the friction plates for excessive wear and mushroomed edges. They can be cleaned up with a file if the deformation is not severe. Check the friction plates for warpage as described in Step 27.

30   Inspect the clutch housing guide bushing for score marks, heat discoloration and evidence of excessive wear. Measure its inner and outer diameter and compare them to the values listed in this Chapter's Specifications (a dealer can do this if you don't have precision measuring equipment). Replace the guide bushing if it's worn beyond the specified limits. Also measure the end of the transmission mainshaft where the guide bushing rides; if it's worn to less than the limit listed in this Chapter's Specifications, replace the mainshaft (see Section 25).

31   Inspect the splines of the clutch center for wear or damage and replace the clutch center if problems are found.

Installation

Refer to illustrations 18.34a, 18.34b, 18.35a, 18.35b and 18.38

32   Lubricate the inner and outer surfaces of the clutch housing guide with moly-based grease and install it on the mainshaft **(see illustration 18.22k)**.

33   Install the clutch housing and thrust washer on the mainshaft **(see illustrations 18.22j and 18.22i)**.

34   Coat the friction plates with engine oil. Install a friction plate on the disc, followed by a metal plate, then alternate the remaining friction and metal plates until they're all installed (there are six friction plates and five metal plates). Friction plates go on first and last, so the friction material contacts the metal surfaces of the clutch center and the pressure plate **(see illustrations)**.

35   Install the clutch center over the posts, then install the assembly in the clutch housing **(see illustrations)**.

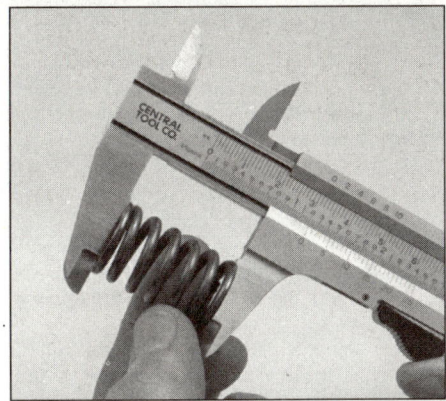

18.26 Measure the clutch spring free length

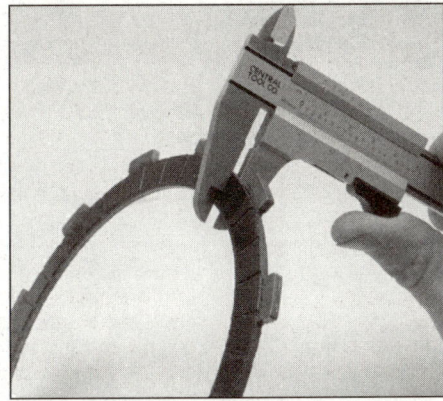

18.27 Measure the thickness of the friction plates

18.28 Check the metal plates for warpage

2A

18.34a Install a friction plate first, then a metal plate (shown), then alternate the remaining friction and metal plates . . .

18.34b . . . a friction plate goes on last

18.35a Install the clutch center . . .

18.35b . . . then install the assembly
on the mainshaft

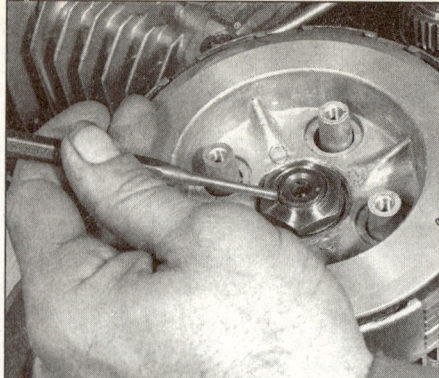

18.38 Stake the new locknut

19.1 Remove the guide bolt (upper arrow),
the pinch bolt (lower arrow) and slide the
reverse shifter rod arm off the shaft

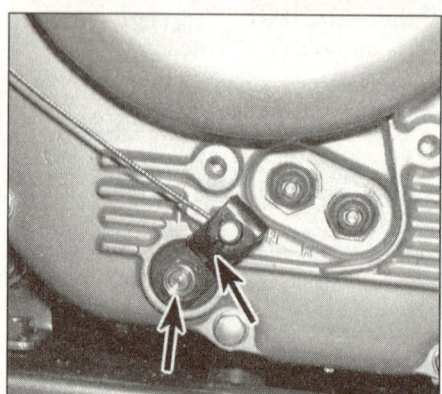

19.3 The reverse selector arm nut (lower
arrow) and cable removal slot
(upper arrow)

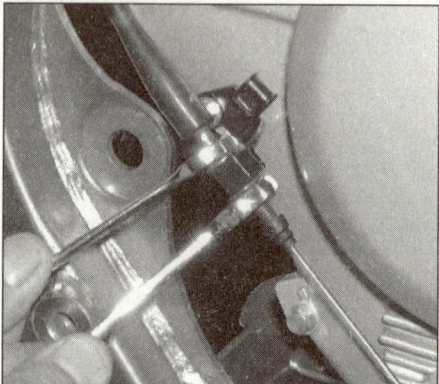

19.4 Back off the locknut (lower) and the
adjuster nut (upper) and detach the
cable from the bracket

36   Install the washer on the mainshaft with its OUT SIDE mark facing away from the engine **(see illustration 18.22e)**.
37   Coat the threads of a new locknut with non-hardening thread locking agent and install it on the mainshaft. Hold the clutch with one of the methods described in Step 8 and tighten the locknut to the torque listed in this Chapter's Specifications.
38   Stake the locknut with a hammer and punch **(see illustration)**.
39   Install the clutch springs and the lifter plate **(see illustrations 18.22b and 18.22a)**. Tighten the bolts to the torque listed in this Chapter's Specifications in two or three stages, in a criss-cross pattern.
40   The remainder of installation is the reverse of the removal steps.

19   Reverse selector cable and reverse mechanism - removal and installation

Reverse shifter rod

Refer to illustration 19.1

1   Remove the gearshift pedal (see Section 22). Remove the reverse shifter rod guide bolt and pinch bolt **(see illustration)** and remove the reverse shifter rod.
2   Installation is the reverse of removal. Make sure that there's 1 to 10 mm (0.04 to 0.39 inch) clearance between the reverse shifter rod and the left crankcase cover.

Reverse selector cable

Refer to illustrations 19.3, 19.4 and 19.5

3   Remove the reverse selector arm nut, pull the arm off the reverse selector shaft, disengage the reverse selector cable from the arm and remove the arm **(see illustration)**.
4   At the reverse selector cable bracket, loosen the locknut and

adjusting nut **(see illustration)**, then slip the cable out of the bracket.
5   At the handlebar, pull the reverse selector cable housing out of the clutch and parking brake lever bracket, then align the cable with its slot (the lowest of the three cable slots in the bracket), slip the cable out of the bracket and disengage the cable end plug from the reverse selector lever **(see illustration)**.
Before removing the old reverse selector cable, note how it's routed from the handlebar to the engine. Route the new cable exactly the same way.
6   Reconnect the ends of the reverse selector cable the reverse selector lever at the handlebar, to the bracket and to the reverse selector arm. When you're done, adjust the reverse selector cable (see Chapter 1).

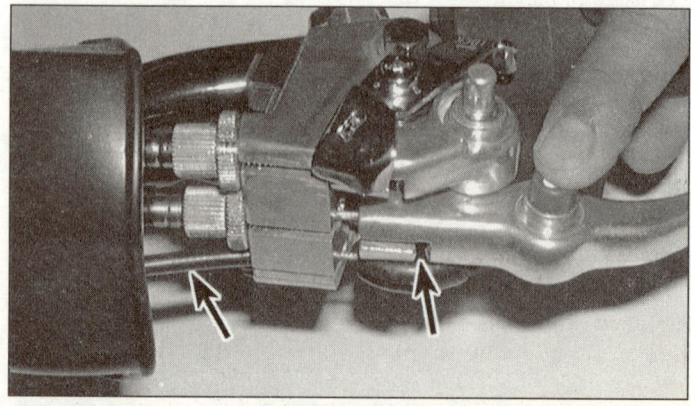

19.5 Pull the cable housing (left arrow) out of the bracket, pull the cable out of the horizontal slot and align it with the vertical slot (right arrow), then lower the cable end plug from the lever

19.8 Pull out the reverse selector shaft, noting how one end of the spring seats against the crankcase

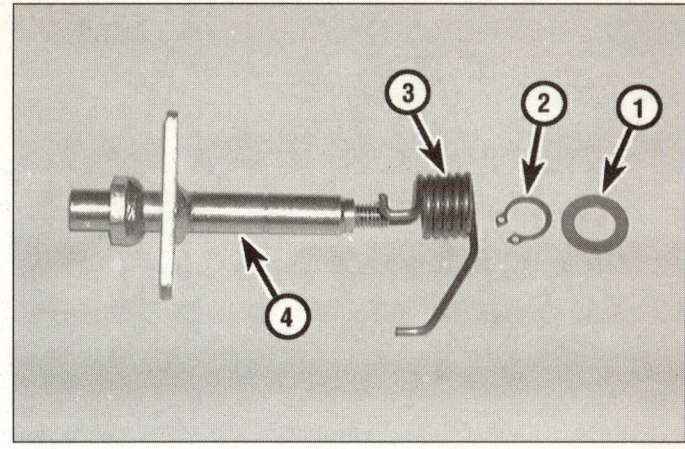

19.9 Reverse selector shaft details:

1	Thrust washer	3	Spring
2	Snap-ring	4	Reverse selector shaft

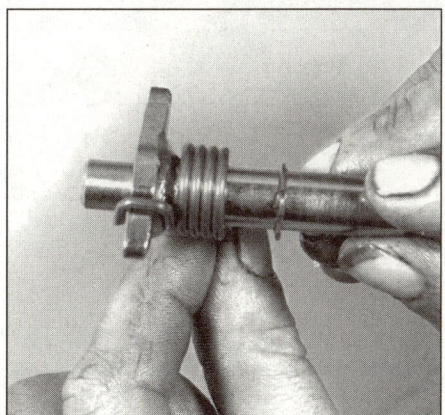

19.10 Hook the end of the spring into the lever notch

19.14a Note how the reverse shifter stopper arm (left arrow) engages the reverse shifter cam (upper arrow), then remove the bolt (lower arrow) . . .

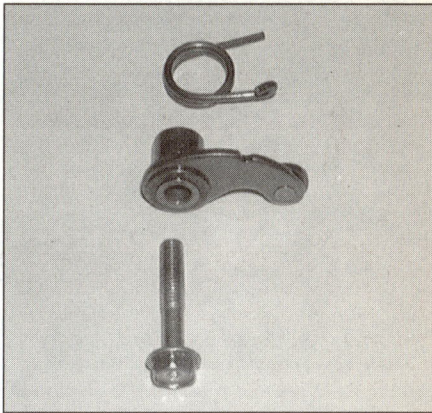

19.14b . . . and remove the stopper arm and return spring

2A

Reverse mechanism

Reverse selector shaft

Refer to illustrations 19.8, 19.9 and 19.10

7   Remove the right crankcase cover (see Section 18).

8   Note how the reverse selector shaft is installed in relation to the reverse switch rotor, then remove the reverse selector shaft assembly **(see illustration)**.

9   Slide the thrust washer off the reverse selector shaft and inspect the shaft. Leave the snap-ring on the shaft. Unless the shaft, snap-ring or spring has broken, it's not necessary to take apart the reverse selector shaft assembly. If you must disassemble it, pay close attention to how it's assembled before taking it apart **(see illustration)**.

10   Hook the spring into the notch in the shaft lever **(see illustration)**, lubricate the inner end of the shaft with oil, then install the reverse selector shaft and position the other end of the spring against the crankcase. Install the thrust washer on the shaft and slide it against the snap-ring.

11   Installation is otherwise the reverse of removal.

Reverse shifter assembly

Refer to illustrations 19.14a, 19.14b, 19.15a, 19.15b, 19.16a and 19.16b

12   Remove the right crankcase cover and the clutch (see Section 18).

13   Remove the reverse selector shaft assembly (see above).

14   Note how the reverse shifter stopper arm is installed in relation to the reverse shifter cam, then remove the reverse shifter stopper arm bolt and remove the stopper arm and return spring **(see illustrations)**.

15   Note the orientation of the neutral switch rotor arm, then remove the neutral switch rotor Allen bolt and washer and remove the neutral switch rotor **(see illustrations)**.

19.15a Note the orientation of the neutral switch rotor arm (right arrow), then remove the Allen bolt (left arrow) and washer . . .

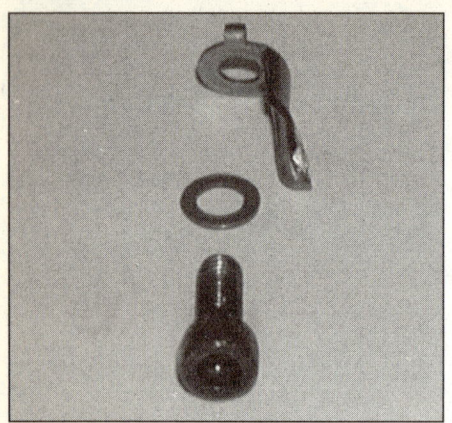

19.15b . . . and remove the neutral switch rotor

19.16a Note the installed relationship of the reverse switch rotor and the reverse shifter cam, then remove the reverse switch rotor bolts (arrows) . . .

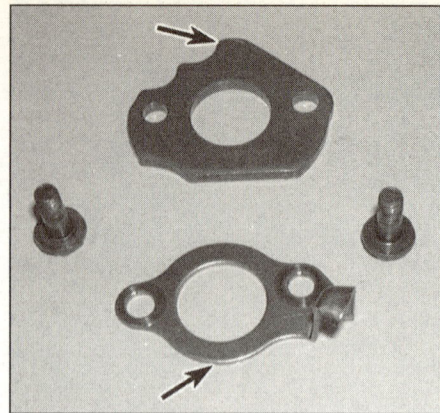

19.16b . . . and remove the reverse switch rotor (lower arrow) and reverse shifter cam (upper arrow)

20.2a Remove and discard the oil pipe O-ring; always use a new O-ring when installing the oil pipe

20.2b To detach the oil pipe from the crankcase, remove these two bolts (arrows)

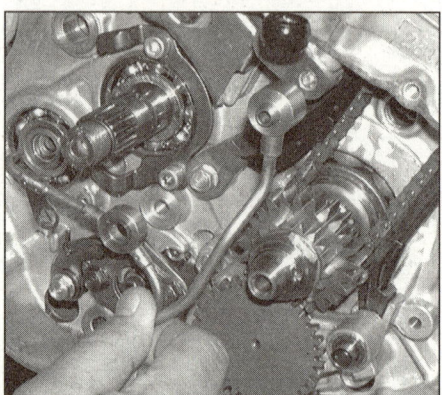

20.2c Carefully pull the oil pipe off the crankcase

16   Note the installed relationship between the reverse switch rotor and the reverse shifter cam, then remove the reverse switch rotor bolts and remove the reverse switch rotor and the reverse shifter cam **(see illustrations)**.

17   Installation is the reverse of removal. Be sure to tighten the reverse switch rotor/reverse shifter cam bolts to the torque listed in this Chapter's Specifications.

20 Oil pipe and pump - removal, inspection and installation

Note: *The oil pump can be removed with the engine in the frame.*

Removal

Refer to illustrations 20.2a, 20.2b, 20.2c, 20.3a, 20.3b, 20.3c and 20.3d

1   Remove the right crankcase cover and the clutch (see Section 18).

20.3a Remove this dowel pin and O-ring from the oil pump

20.3b Remove the oil pump mounting bolts (arrows) . . .

20.3c . . . and pull the pump off; note the locations of the dowels, which may stay in the crankcase or come off with the oil pump

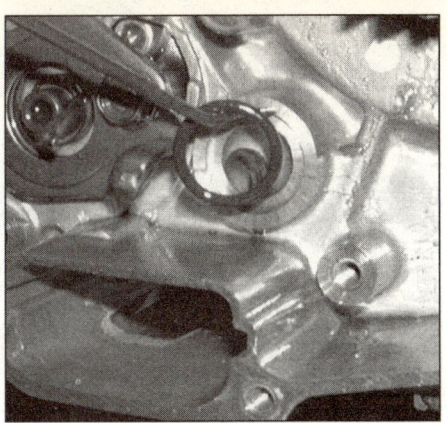

20.3d Remove and discard the O-ring; always use a new O-ring when installing the oil pump

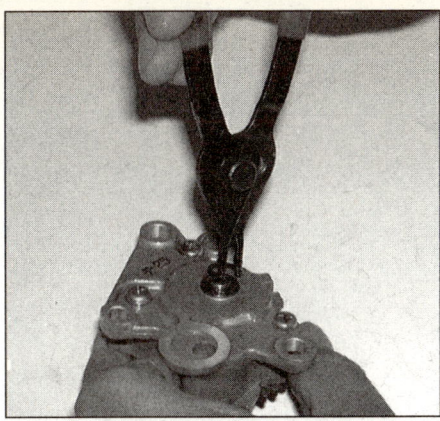

20.4a Remove the snap-ring and washer . . .

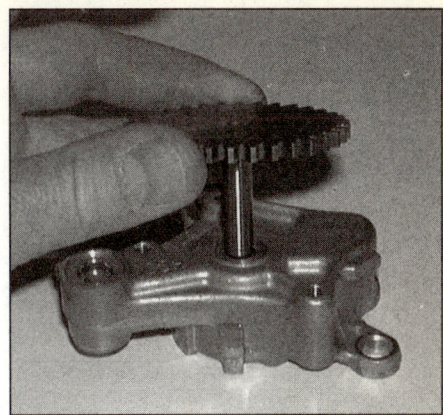

20.4b . . . and pull out the driven gear and shaft

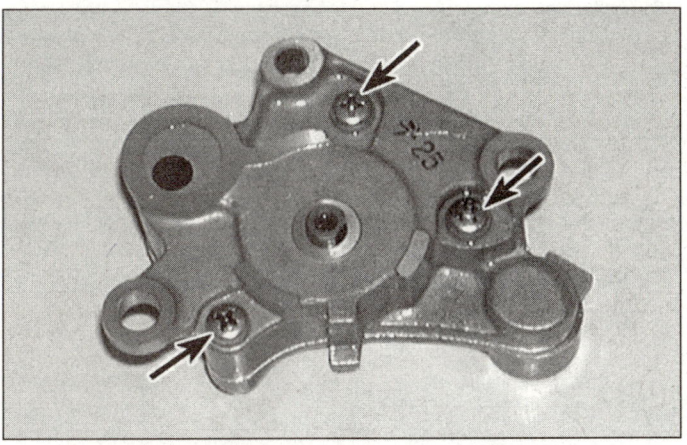

20.5a Remove the cover screws (arrows) . . .

20.5b . . . lift off the pump cover, note the location of the dowel (arrow), then remove it

2   Remove the O-ring from the oil pipe **(see illustration)**. Remove the banjo bolt and hex bolt that secure the oil pipe **(see illustration)**, then carefully work the pipe free of the engine **(see illustration)** and remove it. Discard the old O-ring.

3   Remove the O-ring and dowel pin from the oil pump **(see illustration)**, then remove the three pump mounting bolts and pull the pump off the engine **(see illustrations)**. Remove the old O-ring **(see illustration)** and discard it.

Inspection

Refer to illustrations 20.4a, 20.4b, 20.5a, 20.5b, 20.6a, 20.6b, 20.7a, 20.7b and 20.7c

4   Remove the snap-ring and washer and remove the pump driven gear and shaft **(see illustrations)**.

5   Remove the pump cover screws and separate the cover from the pump body **(see illustrations)**.

6   Remove the rotors **(see illustrations)**, wash all the components in solvent, and then dry them off. Check the pump body, the rotors and

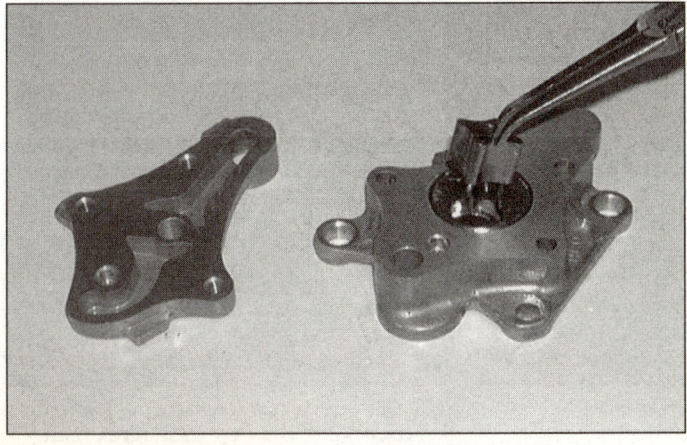

20.6a Remove the inner rotor . . .

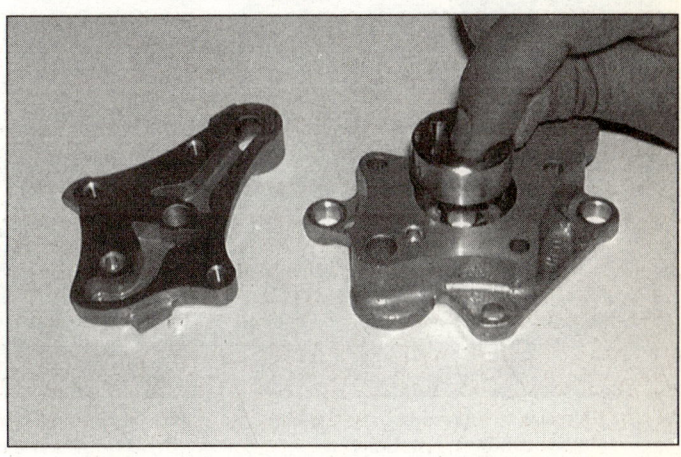

20.6b . . . and the outer rotor, then wash the pump body and rotors in fresh solvent

2A

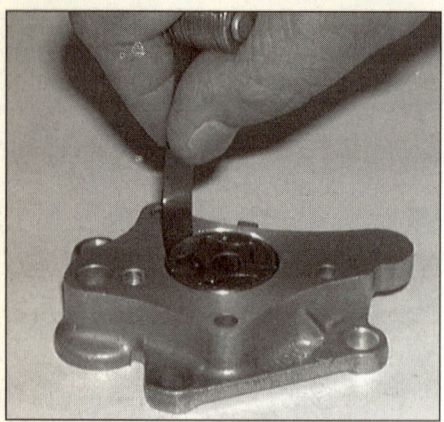

20.7a Measure the clearance between the outer rotor and the body . . .

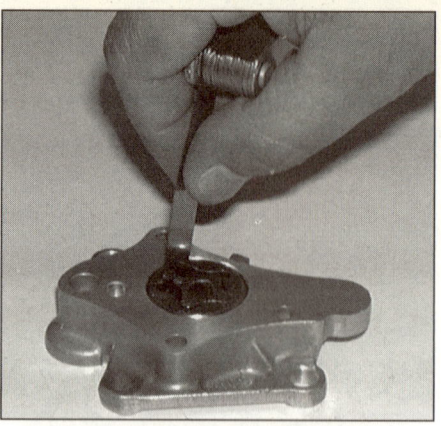

20.7b . . . between the inner and outer rotors . . .

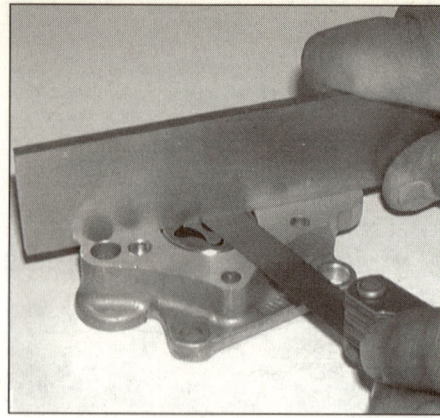

20.7c . . . and between the rotors and a straightedge laid across the pump body

the cover for scoring and wear. If any damage or uneven or excessive wear is evident, replace the pump. If you are rebuilding the engine, it's a good idea to install a new oil pump.

7   Place the rotors in the pump cover. Measure the clearance between the outer rotor and body, and between the inner and outer rotors, with a feeler gauge **(see illustrations)**. Place a straightedge across the pump body and rotors and measure the gap with a feeler gauge **(see illustration)**. If any of the clearances are beyond the limits listed in this Chapter's Specifications, replace the pump.

8   Reassemble the pump by reversing the disassembly steps, with the following additions:

a) *Before installing the cover, pack the cavities between the rotors with petroleum jelly - this will ensure the pump develops suction quickly and begins oil circulation as soon as the engine is started.*

b) *Make sure the cover dowel is in position* **(see illustration 20.5b)**.

c) *Tighten the cover screws securely, but don't strip them out.*

d) *When securing the driven gear shaft with the snap-ring, install the snap-ring with its chamfered edge toward the oil pump (against the washer).*

Installation

9   Installation is the reverse of removal, with the following additions:

a) *Make sure the two oil pump dowels are in position (the two upper pump bolts use dowels) and install a new O-ring* **(see illustration 20.3d)**.

b) *Tighten the oil pump mounting bolts securely, but don't over-tighten them.*

c) *Install the oil pipe bolts in the correct locations* **(see illustration 20.2b)**. *Tighten the bolts to the torque listed in this Chapter's Specifications. Install a new O-ring on the oil pipe* **(see illustration 20.2a)**.

21   Oil pump gear and primary drive gear - removal and installation

Refer to illustrations 21.2, 21.3, 21.5, 21.6 and 21.7

1   Remove the right crankcase cover (see Section 18).

2   Unstake the primary drive gear locknut **(see illustration)**.

3   To prevent the primary drive gear from turning when the locknut is loosened, wedge a penny or copper washer between the teeth of the primary drive gear and the driven gear on back of the clutch housing **(see illustration)**. Note that the penny goes where the two gears meet *underneath*, NOT on top. Hold the penny in place and turn the primary drive gear locknut *clockwise* to loosen it (it has left-hand threads).

4   Remove the clutch (see Section 18) and the oil pump (see Section 20).

5   Remove the primary drive gear locknut and lockwasher **(see illustration)**. Note that the OUT SIDE on the lockwasher faces out when the lockwasher is installed.

6   Remove the oil pump drive gear **(see illustration)**.

7   Slide the primary drive gear off the crankshaft **(see illustration)**.

8   Installation is the reverse of removal. Make sure that the OUT SIDE on the lock washer faces out, and be sure to stake the *new* locknut after tightening it to the torque listed in this Chapter's Specifications.

21.2 Unstake the primary drive gear locknut before loosening it

21.3 Wedge a penny between the primary drive gear and clutch driven gear teeth from below (arrow), then turn the primary drive gear locknut clockwise to loosen it

21.5 Remove the lockwasher from the crankshaft; the OUT SIDE mark on the lockwasher faces out when installed

21.6 Remove the oil pump drive gear

21.7 Remove the primary drive gear

22.2 Locate (or make) alignment marks on the shift shaft and pedal, then remove the pinch bolt (arrow) and pull the pedal off

22.6a Note how the gearshift spindle upper hole engages the shifter collar (upper arrow) and its lower hole centers on the return spring post (lower arrow) . . .

22.6b . . . then pull out the gearshift spindle and remove the washer

22.7a Note the relationship of the guide plate (upper arrow) to the drum shifter (lower arrow) . . .

2A

22   External shift mechanism - removal, inspection and installation

Shift pedal

Removal

Refer to illustration 22.2

1   Look for punch marks on the shift pedal and the end of the shift shaft. If there aren't any, make your own marks with a sharp punch.

2   Remove the shift pedal pinch bolt **(see illustration)** and slide the pedal off the shaft.

Inspection

3   Inspect the shift pedal for wear or damage such as bending. Check the splines on the shift pedal and shaft for stripping or step wear. Replace the pedal or shaft if these problems are found.

Installation

4   Install the shift pedal. Line up the punch marks and tighten the pinch bolt to the torque listed in this Chapter's Specifications.

External shift linkage

Removal

Refer to illustrations 22.6a, 22.6b, 22.7a, 22.7b, 22.7c, 22.8, 22.9a, 22.9b, 22.11a, 22.11b, 22.12a and 22.12b

5   Remove the left crankcase cover (see "Alternator stator coils and

rotor - check and replacement" in Chapter 8).

6   Note how the gearshift spindle assembly is installed **(see illustration)**, with its upper hole engaging the shifter collar and its lower hole centered on the return spring post, then remove the gearshift spindle and washer **(see illustration)**.

7   Note the relationship of the guide plate to the drum shifter **(see illustration)**, then remove the shifter collar and the guide plate bolts **(see illustration)** and lift off the guide plate **(see illustration)**.

22.7b . . . remove the shifter collar (center arrow) and the guide plate bolts (upper and lower arrows) . . .

22.7c . . . and lift off the guide plate

22.8 Remove the drum shifter from the shift drum center bolt and the drum center

22.9a Remove the two collars . . .

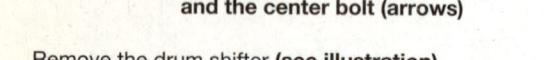

22.9b . . . the dowel pins, shift drum bearing stopper plates and the center bolt (arrows)

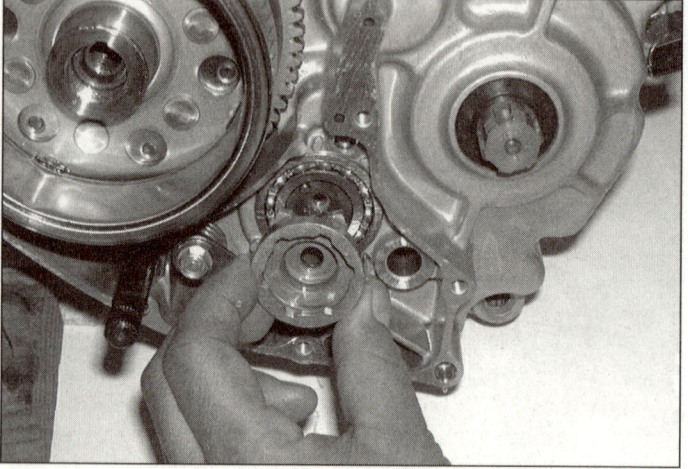

22.11a Remove the shift drum center . . .

8   Remove the drum shifter **(see illustration)**.
9   Remove the collars, dowel pins and bearing stopper plates **(see illustrations)**.
10   Remove the shift drum center bolt **(see illustration 22.9b)**.
11   Remove the shift drum center and dowel pin **(see illustrations)**.
12   Note how one end of the return spring hooks around the stopper arm, then unbolt and remove the arm and remove the spring **(see illustrations)**.

Inspection

Refer to illustrations 22.13a, 22.13b, 22.14a and 22.14b
13   Carefully note how the drum shifter parts go together, then remove the ratchet pawls, plungers and springs from the drum shifter **(see illustration)**. Inspect the parts and replace any that are worn or damaged. Reassembly is the reverse of disassembly **(see illustration)**.
14   Inspect the gearshift spindle for bends and damage to the splines. If the shaft is bent or the splines are damaged, replace the

22.11b . . . and the shift drum center dowel pin

22.12a Note how one end of the return spring hooks around the stopper arm (lower arrow), then remove the bolt (upper arrow) . . .

22.12b . . . and pull off the stopper arm and return spring

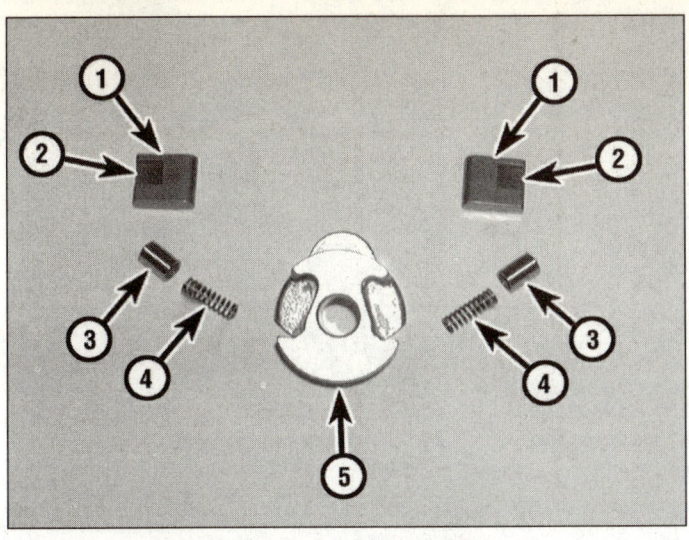

22.13a Drum shifter details:

1 Ratchet pawls	3 Plungers
2 Pawl slots (align these with the plungers)	4 Springs
	5 Drum shifter

gearshift spindle. Pry the return spring apart and remove it from the gearshift spindle **(see illustrations)**. Inspect the condition of the spring. If it's cracked, distorted or otherwise worn, replace it. If the spring and gearshift spindle are in good shape, reassemble them.

15 Make sure the return spring post isn't loose **(see illustration 22.6b)**. If it is, unscrew it, apply a non-hardening locking compound to the threads, then reinstall it and tighten it securely.

Installation

Refer to illustrations 22.17 and 22.19

16 Position the spring on the stopper arm, then install the stopper arm on the engine and tighten its bolt securely **(see illustrations 22.12c, 22.12b and 22.12a)**.

17 Pull down the stopper arm and install the drum center on the shift drum, making sure its dowel is located in the drum center notch **(see illustrations 22.11b and 22.11a)**. Apply non-permanent thread locking agent to the threads of drum center bolt, then install it and tighten it securely **(see illustration 22.10)**. Make sure that the stopper arm spring is correctly installed and the roller end of the stopper arm engages a notch in the drum center **(see illustration)**.

18 Install the dowels and bearing stopper plates in the crankcase, then install the collars over the dowels **(see illustrations 22.9c, 22.9b and 22.9a)**.

19 Assemble the drum shifter and guide plate, install the drum shifter/guide plate assembly over the dowels **(see illustration)**, then install the guide plate bolts **(see illustration 22.7b)**.

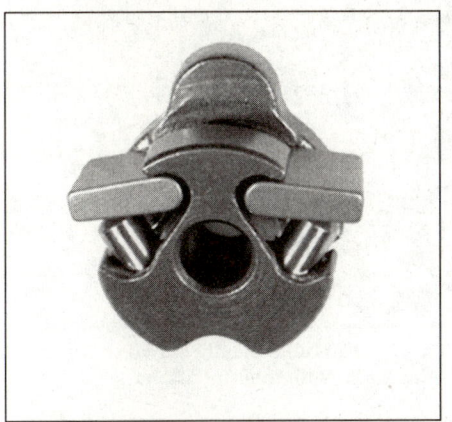

22.13b The rounded ends of the ratchet pawls face each other and the offset pawl slots align with the plungers

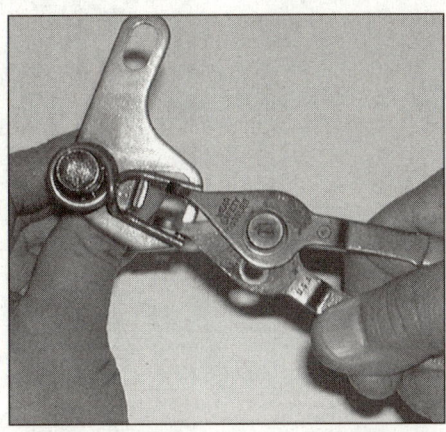

22.14a Pry apart the return spring with a pair of snap-ring pliers or a screwdriver . . .

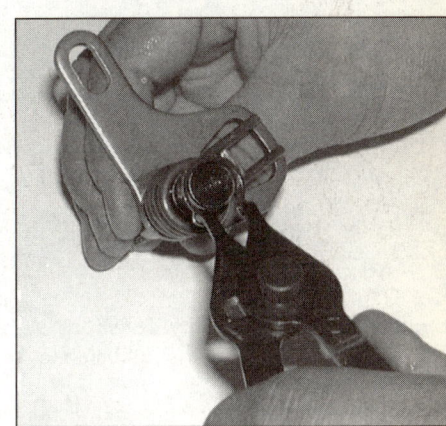

22.14b . . . then remove the spring from the gearshift spindle

22.17 The ends of the spring (lower arrows) rest against the crankcase and hook over the stopper arm; the roller end (upper arrow) engages a notch in the drum center

22.19 Place the guide plate over the dowels and against the collars

2A

23.12a Remove the crankcase breather separator bolt (arrow) . . .

23.12b . . . the crankcase breather separator and the gasket

23.13 Temporarily reinstall the guide plate with its bearing stopper plates, dowels and collars to keep the bearing from falling out of the case

23.15 Remove the right crankcase bolt (arrow)

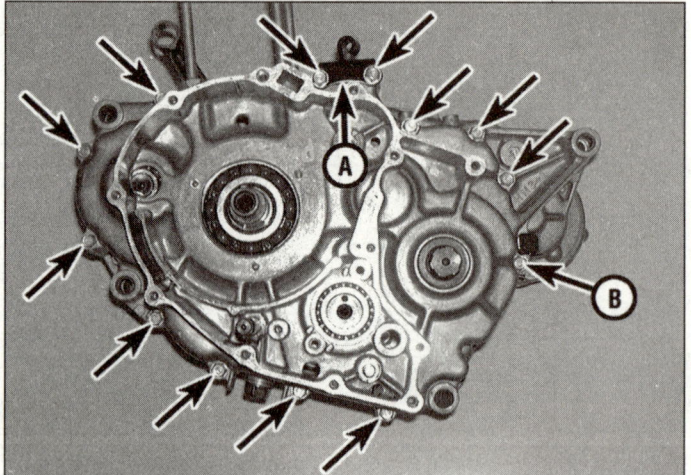

23.16 Loosen the left crankcase bolts evenly in two or three stages and label them with their locations

A   Clutch cable bracket   B   Wiring harness retainer

20 Place the shift collar on the drum shifter **(see illustration 22.7b)**.

21 Place the thrust washer on the gearshift spindle and slide the spindle into the engine **(see illustration 22.6b)**. Slide the spindle all the way in, making sure its upper hole engages the shifter collar and its lower hole is centered on the return spring post **(see illustration 22.6a)**.

22 The remainder of installation is the reverse of removal.

23 Check the engine oil level and add some, if necessary (see Chapter 1).

23 Crankcase - disassembly and reassembly

1 To examine and repair or replace the crankshaft, connecting rod, bearings and transmission components, the crankcase must be split into two parts.

Disassembly

Refer to illustrations 23.12a, 23.12b, 23.13, 23.15, 23.16, 23.17a, 23.17b, 23.18a, 23.18b and 23.19

2 Remove the engine from the vehicle (see Section 5).

3 Remove the right crankcase cover and the clutch (see Section 18).

4 Remove the reverse mechanism (see Section 19).

5 Remove the external shift mechanism (see Section 22).

6 Remove the starter motor, the left crankcase cover and the alternator rotor (see Chapter 8).

7 Remove the cylinder head cover, the cam chain tensioner, the camshaft and sprocket and the cylinder head (see Sections 7, 8, 9 and 10).

8 Remove the front and rear cam chain guides **(see illustration 9.15a)**. Remove the cam chain.

9 Remove the cylinder and the piston (see Sections 13 and 14).

10 Remove the oil pipe and the oil pump (see Section 20).

11 Remove the oil pump gear and the primary drive gear (see Section 21).

12 Remove the crankcase breather separator retaining bolt, remove the crankcase breather separator and remove the gasket **(see illustrations)**.

13 Temporarily install the guide plate for the external shift linkage, along with its bearing stopper plates, dowels and collars **(see illustration)**. Refer to Section 23 for details if necessary.

14 Look over the crankcase carefully to make sure there aren't any remaining components that attach the two halves of the crankcase together.

15 Remove the right crankcase bolt **(see illustration)**. (There's only one bolt on the right side.)

16 Loosen the 13 left crankcase bolts **(see illustration)** in two or three stages, in a criss-cross pattern. Remove the bolts and label them; they are different lengths. Note the location of the clutch cable bracket; it must be installed at this same location when the cases are reassembled.

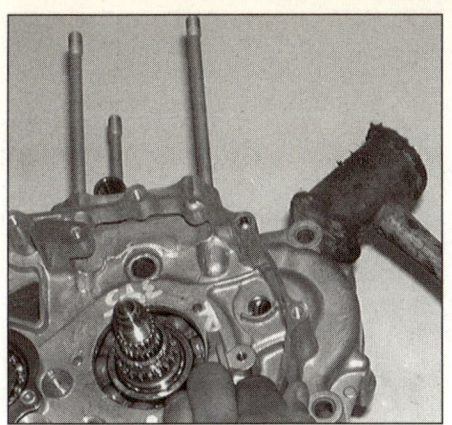

23.17a Place the crankcase assembly on its left side and tap the case halves apart with a plastic mallet

23.17b Separate the right case half from the left half as shown

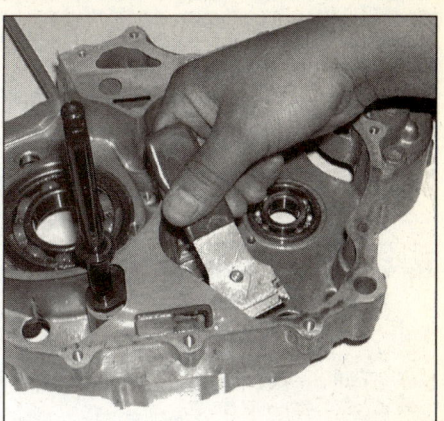

23.18a Remove the crankcase gasket . . .

23.18b . . . and the two crankcase dowels (arrows)

23.19 Remove the reverse gearshift spindle

2A

17   Place the crankcase assembly on its left side and remove the right crankcase half from the left half by tapping them apart with a plastic mallet **(see illustrations)**. If the crankcase halves are stuck together, carefully pry them apart. **Caution:** *Pry only on casting protrusions - don't pry against the mating surfaces, or leaks will develop.*

18   Remove the crankcase gasket and the two crankcase dowels **(see illustrations)**.

19   Remove the reverse gearshift spindle **(see illustration)**.

20   Refer to Sections 24 through 26 for information on the internal components of the crankcase.

Reassembly

Refer to illustrations 23.23 and 23.28

21   Remove all traces of old gasket and sealant from the crankcase mating surfaces with a sharpening stone or similar tool. Be careful not to let any fall into the case as this is done and be careful not to damage the mating surfaces.

22   Install the crankshaft, transmission shafts and any other assemblies that were removed.

23   Install the reverse gearshift spindle **(see illustration 23.19)**. Make sure that the punch marks on the shift drum and the shift fork are aligned **(see illustration)**.

24   Pour some engine oil over the transmission gears, the crankshaft bearing surface and the shift drum. Don't get any oil on the crankcase mating surface.

25   Check to make sure the two dowel pins are in place in their holes in the mating surface of the left crankcase half **(see illustration 23.18b)**.

26   Install a new gasket on the crankcase mating surface.

27   Carefully place the right crankcase half onto the left crankcase

half. While doing this, make sure the transmission shafts, shift drum, crankshaft and balancer fit into their ball bearings in the right crankcase half.

28   Install the left crankcase half bolts in the correct holes and tighten them so they are just snug. Don't forget to install the clutch cable bracket **(see illustration 23.16)**; be sure to install it with the OUT mark **(see illustration)** facing out. Tighten the left crankcase bolts in two or three stages, in a criss-cross pattern, to the torque listed in this

23.23 When installing the reverse gearshift spindle, align the punch marks (circled) on the shift drum and the shift fork

Chapter's Specifications.

29 Tighten the single bolt in the right half of the crankcase to the torque listed in this Chapter's Specifications.

30 Turn the transmission mainshaft to make sure it turns freely. Also make sure the crankshaft turns freely.

31 The remainder of installation is the reverse of removal.

24 Crankcase components - inspection and servicing

Refer to illustrations 24.1, 24.3a, 24.3b, 24.3c, 24.4a and 24.4b

1 Separate the crankcase and remove the following:

a) *Transmission shafts and gears*
b) *Output gear*
c) *Crankshaft and main bearings*
d) *Shift drum and forks*
c) *Mainshaft bearing retainer* **(see illustration)**.

2 Clean the crankcase halves thoroughly with new solvent and dry them with compressed air. All oil passages should be blown out with compressed air and all traces of old gasket sealant should be removed

23.28 Install the clutch cable bracket with the OUT mark facing out

24.1 Right crankcase half (outside)

1 *Right crankshaft bearing (remove from other side)*
2 *Right mainshaft bearing retainer*
3 *Right mainshaft bearing (remove from this side)*
4 *Right countershaft bearing (remove from other side)*

24.3a Right crankcase half (inside)

1 *Right crankshaft bearing (remove from this side)*
2 *Right balancer shaft bearing (remove from this side)*
3 *Right mainshaft bearing (remove from other side)*
4 *Right countershaft bearing (remove from this side)*

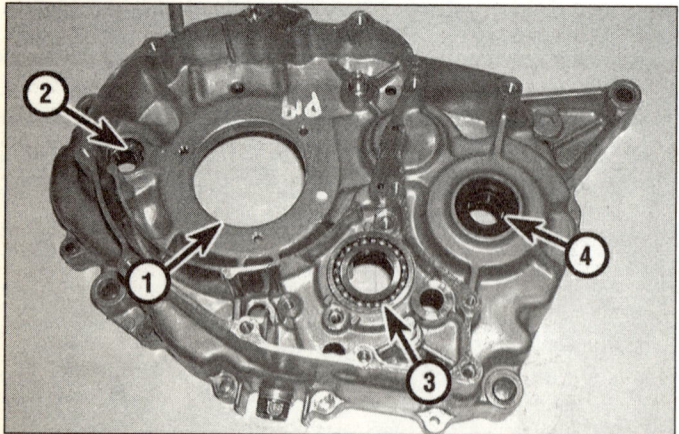

24.3b Left crankcase half (outside)

1 *Left crankshaft bearing (remove from other side)*
2 *Left balancer shaft bearing (remove from other side)*
3 *Shift drum bearing (remove from this side)*
4 *Countershaft bearing seal (remove from this side)*

24.3c Left crankcase half (inside)

1 *Left crankshaft bearing (not shown, came off with crankshaft; but sometimes remains in left case half)*
2 *Left balancer shaft bearing (remove from this side)*
3 *Shift drum bearing (remove from other side)*
4 *Left countershaft bearing (remove from this side)*
5 *Left mainshaft bearing (remove from this side)*

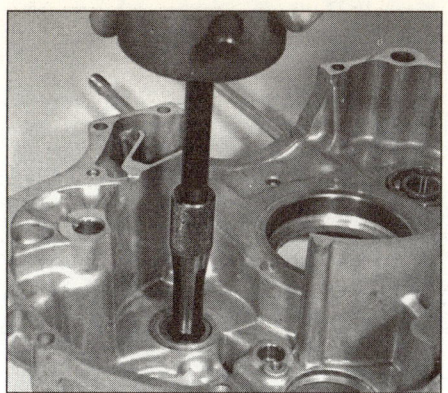

24.4a Bearings that aren't accessible from both sides can be removed with a slide hammer and puller attachment . . .

24.4b . . . and driven in with a bearing driver or socket (ball bearing) or shouldered drift (needle roller bearing)

25.2a Remove the thrust washer and countershaft first gear from the countershaft

25.2b Remove the reverse idler gear

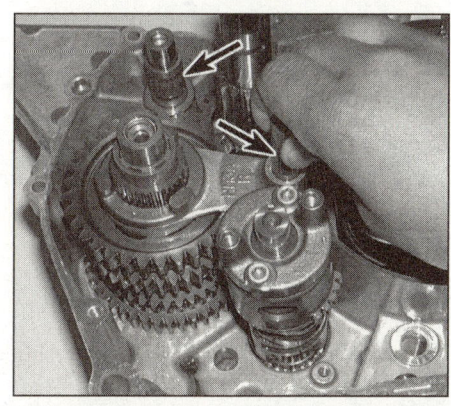

25.2c Remove the shift fork shaft and the reverse idler shaft (arrows)

25.2d Remove the shift drum

2A

from the mating surfaces. **Caution:** *Be very careful not to nick or gouge the crankcase mating surfaces or leaks will result. Check both crankcase sections very carefully for cracks and other damage.*
3   Inspect the bearings in the case halves **(see illustrations)**. If they don't turn smoothly, replace them.
4   You'll need a blind-hole puller to remove bearings that are accessible from only one side, such as the mainshaft needle roller bearing and balancer ball bearing **(see illustration)**. Drive out the other bearings with a bearing driver or a socket having an outside diameter slightly smaller than that of the bearing outer race. Before installing the bearings, allow them to sit in the freezer overnight, and about fifteen-minutes before installation, place the case half in an oven, set to about 200-degrees F, and allow it to heat up. The bearings are an interfer-

ence fit, and this will ease installation. **Warning:** *Before heating the case, wash it thoroughly with soap and water so no explosive fumes are present. Also, don't use a flame to heat the case.* Install ball bearings with a socket or bearing driver that contacts the bearing outer race **(see illustration)**. Install needle roller bearings with a shouldered drift that fits inside the bearing to keep it from collapsing while the shoulder applies pressure to the outer race.
5   If any damage is found that can't be repaired, replace the crankcase halves as a set.
6   Assemble the case halves (see Section 23) and check to make sure the crankshaft and the transmission shafts turn freely.

25 Transmission shafts and shift drum - removal, inspection and installation

Note: *When disassembling the transmission shafts, place the parts on a long rod or thread a wire through them to keep them in order and facing the proper direction.*

Removal and disassembly

Refer to illustrations 25.2a through 25.2v, 25.3a through 25.3g, 25.4a and 25.4b
1   Separate the case halves (see Section 23).
2   The transmission components and shift drum remain in the left case half when the case is separated. To remove the shafts, forks and shift drum, refer to the accompanying photo sequence **(see illustrations)**.
3   To disassemble the shift drum, refer to the accompanying photo sequence **(see illustrations)**.
4   To disassemble the reverse idler shaft, refer to the accompanying photo sequence **(see illustrations)**.

25.2e Remove the right shift fork . . .

25.2f . . . the center shift fork . . .

25.2g . . . and the left shift fork

25.2h Remove the spline bushing and reverse countershaft shifter from the countershaft . . .

25.2i . . . remove the spline collar . . .

25.2j . . . remove the counter reverse gear, spline bushing (upper arrow) and countershaft fifth gear (lower arrow)

25.2k Remove the countershaft third gear (arrow) and lift out the mainshaft

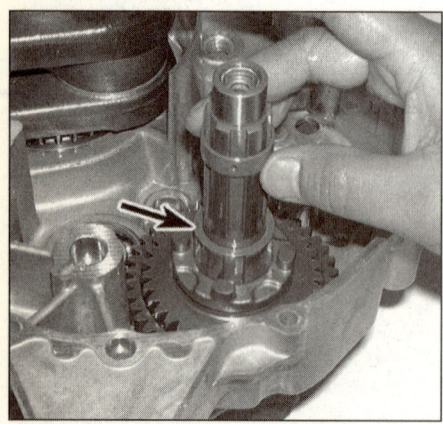

25.2l Remove the countershaft thrust bushing and the 25 mm thrust washer (arrow) . . .

25.2m . . . remove the countershaft fourth shifter gear . . .

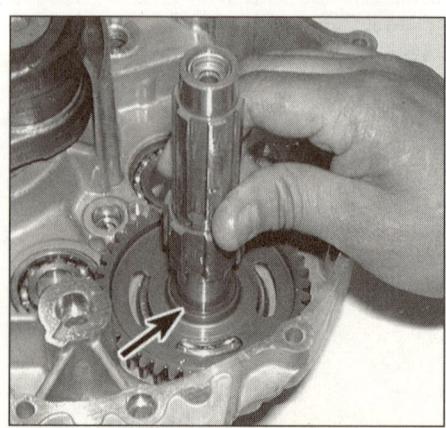

25.2n . . . remove the countershaft and 22 mm thrust washer . . .

25.2o . . . remove the thrust bushing and countershaft second gear . . .

25.2p . . . and finally, remove the 22 mm thrust washer

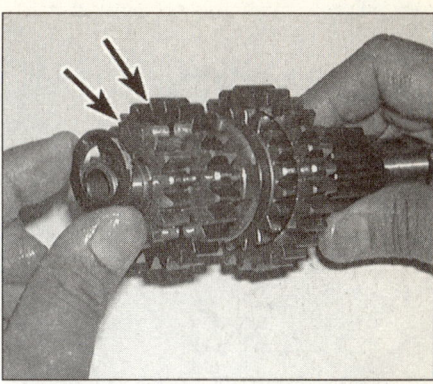

25.2q From the mainshaft, remove the 17.2 mm washer, mainshaft second gear and mainshaft fourth gear (arrows) . . .

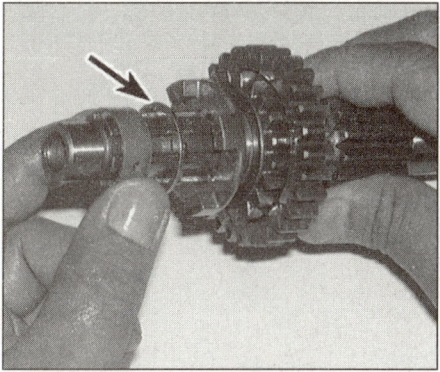

25.2r . . . remove the spline bushing and 22 mm spline washer (arrow)

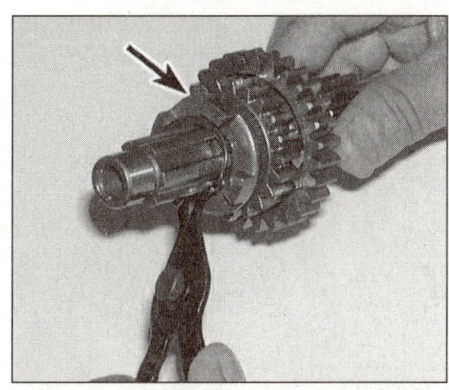

25.2s . . . remove the snap-ring and mainshaft third gear (arrow) . . .

25.2t . . . remove the snap-ring . . .

25.2u . . . remove the 22 mm spline washer and mainshaft fifth gear (arrow) . . .

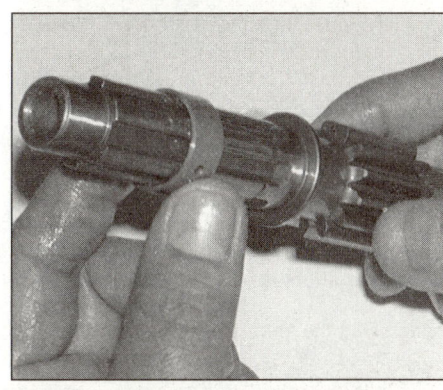

25.2v . . . and, finally, remove the thrust bushing from the mainshaft

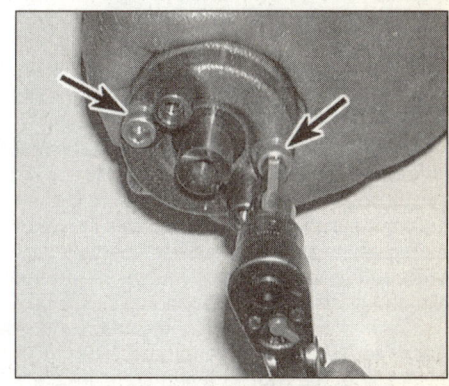

25.3a Remove the two shift drum Allen bolts (arrows) . . .

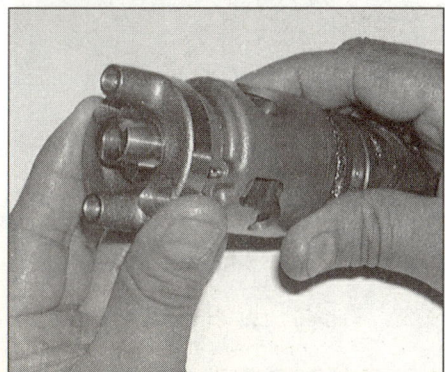

25.3b . . . remove the reverse gearshift drum bearing holder . . .

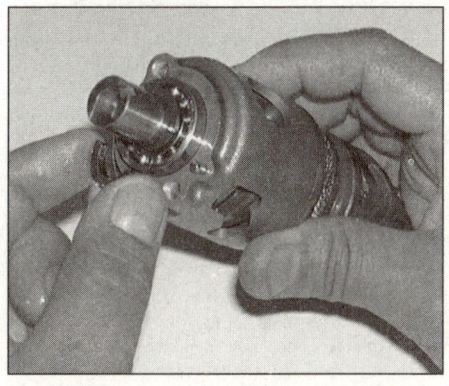

25.3c . . . remove the two bearing retaining half-rings . . .

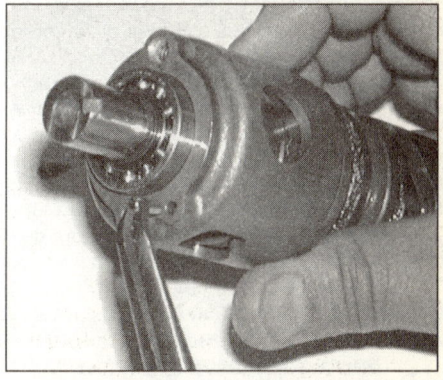

25.3d . . . remove the dowel pin . . .

2A

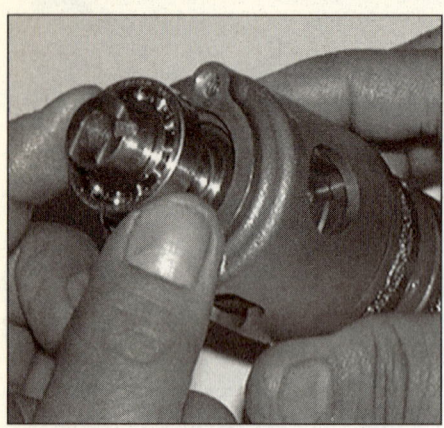

25.3e . . . remove the radial
ball bearing . . .

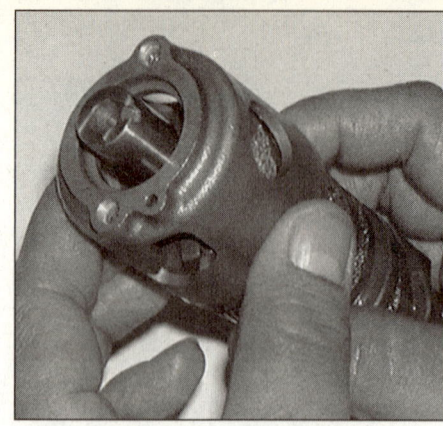

25.3f . . . remove the reverse
gearshift drum . . .

25.3g . . . and remove the two special
needle bearing halves

25.4a Remove the 14 mm thrust washer
and reverse idler gear . . .

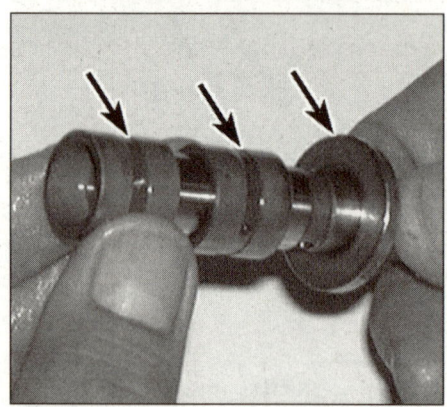

25.4b . . . remove the two reverse idler
gear bushings and the 14 mm thrust
washer (arrows) . . .

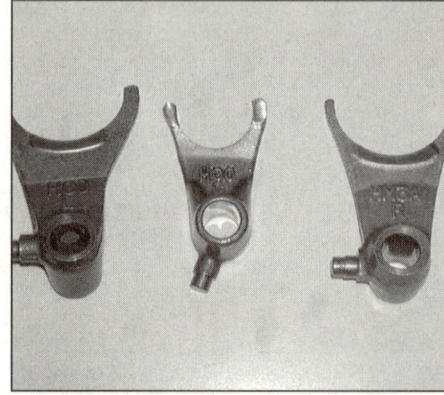

25.14a Make sure that the identification
marks on the three shift forks ("L," "C"
and "R," for left, center and right) face
toward the right crankcase half

Inspection

5   Wash all of the components in clean solvent and dry them off. Rotate the countershaft, feeling for tightness, rough spots, and excessive looseness and listening for noises in the bearing or output gear.

6   Inspect the shift fork grooves in the countershaft first/reverse shifter, mainshaft third gear and countershaft fourth gear between third and fourth gears. If a groove is worn or scored, replace the affected part and inspect its corresponding shift fork.

7   Check the shift forks for distortion and wear, especially at the fork ears **(see illustrations 24.9a and 24.9b in Chapter 2B)**. Measure the thickness of the fork ears and compare your findings with this Chapter's Specifications. If they are discolored or severely worn they are probably bent. Inspect the guide pins for excessive wear and distortion and replace any defective parts with new ones.

8   Measure the inside diameter of the forks and the outside diameter of the fork shaft and compare to the values listed in this Chapter's Specifications. Replace any parts that are worn beyond the limits. Check the shift fork shaft for evidence of wear, galling and other damage. Make sure the shift forks move smoothly on the shafts. If the shafts are worn or bent, replace them with new ones.

9   Check the edges of the grooves in the drum for signs of excessive wear **(see illustration 24.11 in Chapter 2B)**.

10   Hold the inner race of the shift drum bearing with fingers and spin the outer race. Replace the bearing if it's rough, loose or noisy.

11   Check the gear teeth for cracking and other obvious damage. Check the bushing and surface in the inner diameter of the freewheeling gears for scoring or heat discoloration. Replace damaged parts.

12   Inspect the engagement dogs and dog holes on gears so equipped for excessive wear or rounding off **(see illustration 25.1a in Chapter 2b)**. Replace the paired gears as a set if necessary.

13   Check the mainshaft bearings in the crankcase for wear or heat discoloration and replace them if necessary (see Section 24).

Installation

Refer to illustrations 25.14a, 25.14b and 25.14c

14   Installation is the basically the reverse of the removal procedure, but note the following points:

  a) Lubricate each part with engine oil before installing it.
  b) Be sure to align the oil holes in the spline bushing and the mainshaft.

26.14b Place the reverse idler shaft pin (arrow)
in the crankcase notch

26.14c The assembled transmission should look like this when you're done

26.2a Press the crankshaft out of the left crankcase half . . .

c) Use new snap-rings on the mainshaft (worn snap-rings can easily turn in their grooves). Install both snap-rings, and both spline washers, with their chamfered (rolled) edges facing away from the thrust load (the gears they retain on the shaft) The sharp edge of each snap-ring and thrust washer faces toward the thrust load. Be sure to align the gap in each snap-ring with a spline groove (between two splines), so that the ends of the snap-ring are supported by the splines. Make sure each snap-ring is fully seated in its groove by opening it slightly and rotating it in the groove.

d) Make sure that the identification marks on the three shift forks are all facing toward the right crankcase half **(see illustration)**.

e) Place the reverse idler shaft pin in the crankcase notch **(see illustration)**.

f) After assembly, check the gears to make sure they're installed correctly **(see illustration)**.

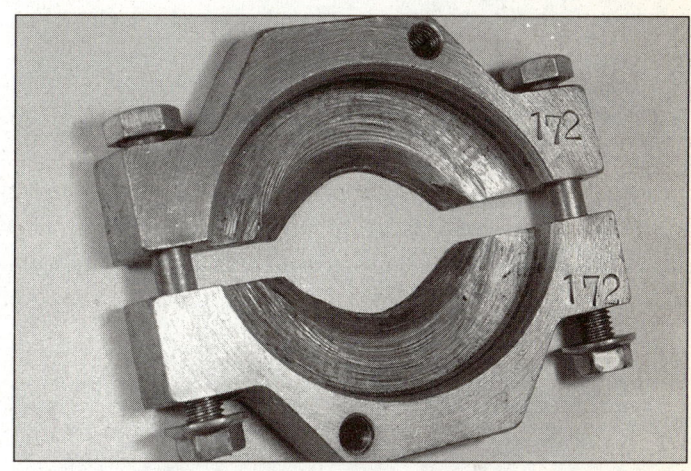

26.2b . . . if the bearing stays on the crankshaft, remove it with the press and a bearing splitter . . .

26 Crankshaft and balancer - removal, inspection and installation

Note: *The following procedure requires a hydraulic press and some special Honda tools (or suitable equivalents). Look over this procedure carefully and decide whether you can do it. If you don't have the right tools, have the crankshaft removed and installed by a Honda dealer.*

Removal

Refer to illustrations 26.2a, 26.2b and 26.2c

1   Remove the engine, separate the crankcase halves and remove the transmission (Sections 5, 23 and 25, respectively).

2   Place the left crankcase half in a press and press out the crankshaft, removing the balancer shaft at the same time **(see illustration)**. The ball bearing may remain in the crankcase or come out with the crankshaft. If it stays on the crankshaft, remove it with the press and a bearing splitter **(see illustration)** or with a suitable puller **(see illustration)**. Discard the bearing, no matter what its apparent condition, and use a new one on installation.

2A

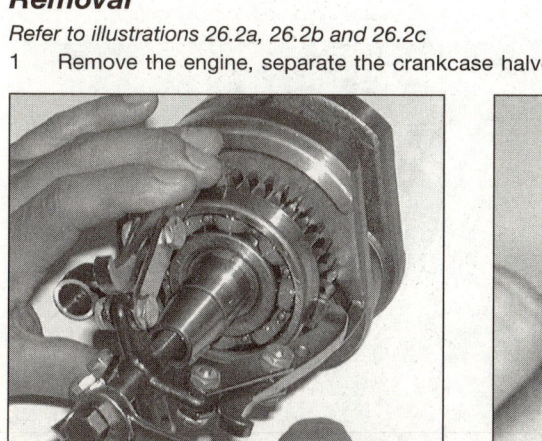

26.2c . . . or with a suitable three-jaw puller

26.3 Check the connecting rod side clearance with a feeler gauge

26.4 Check the connecting rod radial clearance with a dial indicator

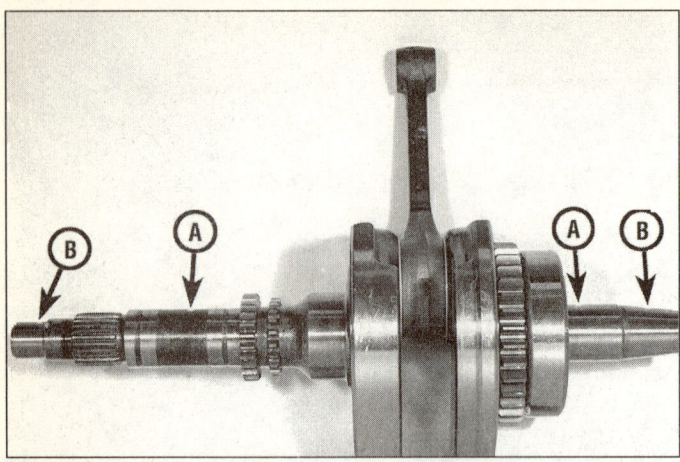

26.6 Place a V-block on each side of the crankshaft (A) and measure runout at the ends (B)

26.7 Check the balancer gear and bearing journals for wear or damage

Inspection

Refer to illustrations 26.3, 26.4, 26.6 and 26.7

3   Measure the side clearance between connecting rod and crankshaft with a feeler gauge **(see illustration)**. If it's more than the limit listed in this Chapter's Specifications, replace the crankshaft and

26.8 Install new crankshaft bearings in the case halves with a suitable bearing driver

26.9a Place the crankshaft and balancer in the right crankcase half . . .

connecting rod as an assembly.

4   Set up the crankshaft in V-blocks with a dial indicator contacting the big end of the connecting rod **(see illustration)**. Move the connecting rod up-and-down against the indicator pointer and compare the reading to the value listed in this Chapter's Specifications. If it's beyond the limit, replace the crankshaft and connecting rod as an assembly.

5   Check the crankshaft gear, sprockets and bearing journals for visible wear or damage, such as chipped teeth or scoring. If any of these conditions are found, replace the crankshaft and connecting rod as an assembly.

6   Set the crankshaft in a pair of V-blocks, with a dial indicator contacting each end **(see illustration)**. Rotate the crankshaft and note the runout. If the runout at either end is beyond the limit listed in this Chapter's Specifications, replace the crankshaft and connecting rod as an assembly.

7   Check the balancer gear and bearing journals for visible wear or damage and replace the balancer if any problems are found **(see illustration)**.

Installation

Refer to illustrations 26.8, 26.9a, 26.9b, 26.10, 26.11, 26.12, 26.13 and 26.14

8   Install a new balancer shaft bearing, if necessary, in the left crankcase half (see Section 24). Install new crankshaft bearings in the crankcase halves with a special driver, adapter and pilot **(see illustration)**.

26.9b . . . making sure to align the timing marks (arrow)

26.10 Thread a special tool adapter into the end of the crankshaft . . .

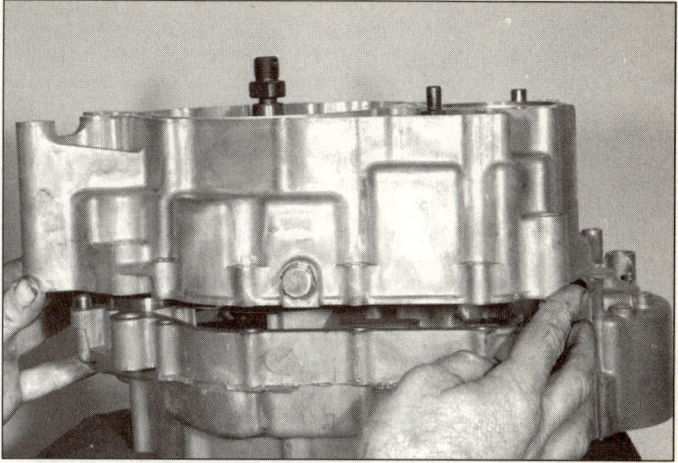

26.11 . . . install the left crankcase half without bolts . . .

9   Install the crankshaft and balancer in the right crankcase half with their timing marks aligned **(see illustrations)**. **Note:** *It's important to align the timing marks exactly throughout this procedure. Severe engine vibration will occur if the crankshaft and balancer are out of time.*

10   Thread a puller adapter into the end of the crankshaft **(see illustration)**.

11   Temporarily install the left crankcase half on the right crankcase **(see illustration)**.

12   Install an assembly collar and crankshaft puller over the crankshaft **(see illustration)**. Hold the puller shaft with one wrench and turn the nut with another wrench to pull the crankshaft into the center race of the ball bearing.

13   Use a suitable aftermarket puller to install the crankshaft **(see illustration)**.

14   Remove the special tools. Lift the right crankcase half off and check to make sure the timing marks on the other side of the crankshaft and balancer are aligned **(see illustration)**.

15   The remainder of installation is the reverse of the removal steps.

26.12 . . . and install the puller on the crankshaft adapter and crankcase half; hold the shaft (upper arrow) with a wrench and turn the nut to pull the crankshaft into the bearing

27   Initial start-up after overhaul

1   Make sure the engine oil level is correct, then remove the spark plug from the engine. Place the engine kill switch in the Off position and unplug the primary (low tension) wires from the coil.

2   Turn on the key switch and crank the engine over with the starter several times to build up oil pressure. Reinstall the spark plug, connect

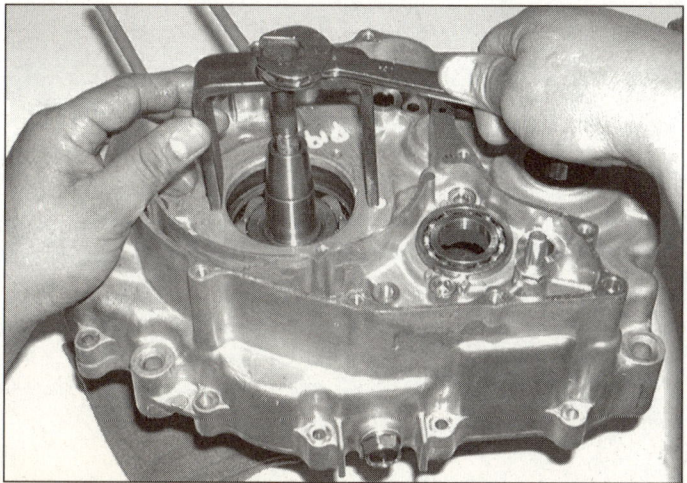

26.13 Use a puller like this one to install the crankshaft

26.14 Remove the right crankcase half and make sure the timing marks are aligned (arrows); if they aren't, remove the crankshaft and balancer and reinstall them correctly

2A

the wires and turn the switch to On.

3   Make sure there is fuel in the tank, then operate the choke.

4   Start the engine and allow it to run at a moderately fast idle until it reaches operating temperature. **Caution:** *If the oil temperature light doesn't go off, or it comes on while the engine is running, stop the engine immediately.*

5   Check carefully for oil leaks and make sure the transmission and controls, especially the brakes, function properly before road testing the machine. Refer to Section 28 for the recommended break-in procedure.

6   Upon completion of the road test, and after the engine has cooled down completely, recheck the valve clearances (see Chapter 1).

28 Recommended break-in procedure

1   Any rebuilt engine needs time to break-in, even if parts have been installed in their original locations. For this reason, treat the machine gently for the first few miles to make sure oil has circulated throughout the engine and any new parts installed have started to seat.

2   Even greater care is necessary if the cylinder has been rebored or a new crankshaft has been installed. In the case of a rebore, the engine will have to be broken in as if the machine were new. This means greater use of the transmission and a restraining hand on the throttle for the first few operating days. There's no point in keeping to any set speed limit - the main idea is to vary the engine speed, keep from lugging (laboring) the engine and to avoid full-throttle operation. These recommendations can be lessened to an extent when only a new crankshaft is installed. Experience is the best guide, since it's easy to tell when an engine is running freely.

3   If a lubrication failure is suspected, stop the engine immediately and try to find the cause. If an engine is run without oil, even for a short period of time, irreparable damage will occur.

Chapter 2 Part B
TRX400EX engine, clutch and transmission

Contents

2B

Specifications

Main rocker arms
Rocker arm inside diameter
- Standard .... 11.500 to 11.518 mm (0.4528 to 0.4535 inch)
- Limit .... 11.53 mm (0.454 inch)

Rocker shaft outside diameter
- Standard .... 11.466 to 11.484 mm (0.4514 to 0.4521 inch)
- Limit .... 11.41 mm (0.449 inch)

Shaft-to-arm clearance
- Standard .... 0.016 to 0.052 mm (0.0006 to 0.0020 inch)
- Limit .... 0.10 mm (0.004 inch)

Sub-rocker arms
Rocker arm inside diameter (intake and exhaust)
- Standard .... 7.000 to 7.015 mm (0.2756 to 0.2762 inch)
- Limit .... 7.05 mm (0.278 inch)

Rocker shaft outside diameter (intake and exhaust)
- Standard .... 6.972 to 6.987 mm (0.2745 to 0.2751 inch)
- Limit .... 6.92 mm (0.272 inch)

Shaft-to-arm clearance
- Standard .... 0.013 to 0.043 mm (0.0005 to 0.0017 inch)
- Limit .... 0.10 mm (0.004 inch)

Camshaft

Lobe height

Intake

Standard .... 30.673 to 30.773 mm (1.2076 to 1.2115 inches)
Limit .... 30.57 mm (1.204 inches)

Exhaust

Standard .... 30.468 to 30.568 mm (1.9957 to 1.2035 inches)
Limit .... 30.37 mm (1.196 inches)
Journal diameter .... Not specified
Bearing journal inside diameter .... Not specified
Journal oil clearance .... Not specified
Camshaft runout limit .... 0.03 mm (0.001 inch)
Camshaft side clearance .... Not specified

Cylinder head, valves and valve springs

Cylinder head warpage limit .... 0.10 mm 0.004 inch)
Valve stem runout .... Not specified
Valve stem diameter

Intake

Standard .... 5.475 to 5.490 mm (0.2156 to 0.2161 inch)
Limit .... 5.46 mm (0.215 inch)

Exhaust

Standard .... 5.455 to 5.470 mm (0.2148 to 0.2154 inch)
Limit .... 5.44 mm (0.214 inch)
Valve guide inside diameter (intake and exhaust)

Standard .... 5.500 to 5.512 mm (0.2165 to 0.2170 inch)
Limit .... 5.52 mm (0.217 inch)
Stem to guide clearance

Intake

Standard .... 0.010 to 0.037 mm (0.0004 to 0.0015 inch)
Limit .... 0.12 mm (0.005 inch)

Exhaust

Standard .... 0.030 to 0.057 mm (0.0012 to 0.0022 inch)
Limit .... 0.14 mm (0.006 inch)
Valve seat width (intake and exhaust)

Standard .... 1.0 to 1.1 mm (0.039 to 0.043 inch)
Limit .... 2.0 mm (0.078 inch)
Valve spring free length

Inner spring

Standard .... 37.19 mm (1.464 inches)
Limit .... 36.3 mm (1.43 inches)

Outer spring

Standard .... 44.20 mm (1.740 inches)
Limit .... 43.1 mm (1.70 inches)
Valve face width .... Not specified

Cylinder

Bore

Standard .... 85.000 to 85.010 mm (3.3465 to 3.3468 inches)
Limit .... 85.10 mm (3.350 inches)
Taper and out-of-round limits .... 0.05 mm (0.002 inch)
Surface warpage limit .... 0.10 mm (0.004 inch)

Piston

Piston diameter

Standard .... 84.960 to 84.985 mm (3.3449 to 3.3459 inches)
Limit .... 84.880 mm (3.3417 inches)
Piston diameter measuring point (above bottom of piston) .... 15 mm (0.6 inch)
Piston-to-cylinder clearance

Standard .... 0.015 to 0.050 mm (0.0006 to 0.0020 inch)
Limit .... 0.10 mm (0.004 inch)
Piston pin bore in piston

Standard .... 20.002 to 20.008 mm (0.7875 to 0.7877 inch)
Limit .... 20.060 mm (0.7898 inch)
Piston pin bore in connecting rod

Standard .... 20.020 to 20.041 mm (0.7882 to 0.7900 inch)
Limit .... 20.067 mm (0.790 inch)
Piston pin outer diameter

Standard .... 19.994 to 20.000 mm (0.7872 to 0.7874 inch)
Limit .... 19.964 mm (0.7860 inch)

Piston pin-to-piston clearance
  Standard... 0.002 to 0.014 mm (0.0001 to 0.0006 inch)
  Limit... 0.096 mm (0.0038 inch)
Piston pin-to-connecting rod clearance
  Standard... 0.020 to 0.047 mm (0.0008 to 0.0019 inch)
  Limit... 0.103 mm (0.0041 inch)
Top ring side clearance
  Standard... 0.030 to 0.065 mm (0.0012 to 0.0026 inch)
  Limit... 0.14 mm (0.006 inch)
Second ring side clearance
  Standard... 0.015 to 0.050 mm (0.0006 to 0.0020 inch)
  Limit... 0.12 mm (0.005 inch)
Oil ring side clearance .... Not specified
Ring end gap
  Top
    Standard .... 0.20 to 0.35 mm (0.008 to 0.014 inch)
    Limit .... 0.50 mm (0.020 inch)
  Second
    Standard .... 0.35 to 0.50 mm (0.014 to 0.020 inch)
    Limit .... 0.65 mm (0.026 inch)
  Oil ring side rails
    Standard .... 0.20 to 0.70 mm (0.008 to 0.028 inch)
    Limit .... 0.90 mm (0.035 inch)

Clutch

Spring free length
  Standard... 52.64 mm (2.072 inches)
  Limit... 50.0 mm (1.97 inches)
Friction plate thickness
  Standard... 2.92 to 3.08 mm (0.115 to 0.121 inch)
  Limit... 2.69 mm (0.106 inch)
Metal plate warpage limit .... 0.15 mm (0.006 inch)
Clutch housing bushing inside diameter
  Standard... 22.010 to 22.035 mm (0.8665 to 0.8675 inch)
  Limit... 22.05 mm (0.868 inch)
Clutch housing bushing outside diameter
  Standard... 27.959 to 27.980 mm (1.1007 to 1.1016 inches)
  Limit... 27.90 mm (1.098 inches)
Clutch housing bushing clearance to mainshaft .... Not specified

Oil pump

Outer rotor-to-body clearance
  Standard... 0.15 to 0.22 mm (0.006 to 0.009 inch)
  Limit... 0.25 mm (0.010 inch)
Inner-to-outer rotor clearance
  Standard... 0.15 mm (0.006 inch) or less
  Limit... 0.20 mm (0.008 inch)
Side clearance (rotors-to-straightedge)
  Standard... 0.02 to 0.09 mm (0.001 to 0.004 inch)
  Limit... 0.12 mm (0.005 inch)

Shift drum and forks

Fork inside diameter
  Standard... 13.000 to 13.021 mm (0.5118 to 0.5126 inch)
  Limit... 13.05 mm (0.514 inch)
Fork shaft outside diameter
  Standard... 12.966 to 12.984 mm (0.5105 to 0.5112 inch)
  Limit... 12.90 mm (0.508 inch)
Fork ear thickness
  Standard... 5.930 to 6.000 mm (0.233 to 0.236 inch)
  Limit... 5.50 mm (0.22 inch)
Shift drum groove width limit.. Not specified

Transmission (1986 through 1995)

Gear inside diameters
  Mainshaft fourth
    Standard .... 25.020 to 25.041 mm (0.9850 to 0.9859 inch)
    Limit .... 25.08 mm (0.987 inch)

2B

Transmission (1986 through 1995)

Gear inside diameters (continued)

Mainshaft fifth

    Standard .... 25.000 to 25.021 mm (0.9843 to 0.9851 inch)

    Limit .... 25.06 mm (0.987 inch)

Countershaft first

    Standard .... 23.000 to 23.021 mm (0.9055 to 0.9063 inch)

    Limit .... 23.07 mm (0.908 inch)

Countershaft second and third

    Standard .... 28.020 to 28.041 mm (1.1031 to 1.1040 inches)

    Limit .... 28.08 mm (1.106 inches)

Bushing outside diameters

Mainshaft fourth

    Standard .... 24.979 to 25.000 mm (0.9834 to 0.9843 inch)

    Limit .... 24.90 mm (0.980 inch)

Mainshaft fifth

    Standard .... 24.959 to 24.980 mm (0.9826 to 0.9843 inch)

    Limit .... 24.90 mm (0.980 inch)

Countershaft first

    Standard .... 22.959 to 22.980 mm (0.9039 to 0.9047 inch)

    Limit .... 20.90 mm (0.902 inch)

Countershaft second and third

    Standard .... 27.979 to 28.000 mm (1.1015 to 1.1024 inch)

    Limit .... 27.94 mm (1.100 inch)

Gear-to-bushing clearance

    Standard .... 0.020 to 0.062 mm (0.0008 to 0.0022 inch)

    Limit .... 0.10 mm (0.004 inch)

Bushing inside diameters

Mainshaft fourth

    Standard .... 22.000 to 22.021 mm (0.8661 to 0.8670 inch)

    Limit .... 22.10 mm (0.870 inch)

Countershaft first

    Standard .... 20.020 to 20.041 mm (0.7882 to 0.7890 inch)

    Limit .... 20.08 mm (0.791 inch)

Countershaft second and third

    Standard .... 25.000 to 25.021 mm (0.9843 to 0.9851 inches)

    Limit .... 25.06 mm (0.987 inches)

Mainshaft diameter (at fourth gear)

    Standard .... 21.959 to 21.980 mm (0.7866 to 0.7874 inch)

    Limit .... 21.92 mm (0.863 inch)

Countershaft diameter

At countershaft first

    Standard .... 19.979 to 20.000 mm (1.1791 to 1.1801 inch)

    Limit .... 19.94 mm (0.785 inch)

At countershaft second and third

    Standard .... 24.959 to 24.980 mm (0.9826 to 0.9835 inch)

    Limit .... 24.92 mm (0.981 inch)

Bushing-to-shaft clearance

    Standard .... 0.020 to 0.062 mm (0.0008 to 0.0062 inch)

    Limit .... 0.10 mm (0.004 inch)

Crankshaft and balancer

Connecting rod side clearance

    Standard .... 0.05 to 0.45 mm (0.0020 to 0.018 inch)

    Limit .... 0.6 mm (0.02 inch)

Connecting rod big end radial clearance

    Standard .... 0.006 to 0.018 mm (0.0002 to 0.0007 inch)

    Limit .... 0.05 mm (0.002 inch)

Runout limit .... 0.12 mm (0.005 inch)

Balancer shaft diameter .... Not specified

Torque specifications

Engine mounting bolts

    Upper engine hanger nut/through-bolt .... 74 Nm (54 ft-lbs)

    Front engine hanger nut/through-bolt (at engine) .... 54 Nm (40 ft-lbs)

    Front engine hanger bolts (at frame) .... 26 Nm (20 ft-lbs)

    Lower engine hanger nut/through-bolt (engine-to-frame) .... 74 Nm (54 ft-lbs)

External oil hose bolts

    End fitting bolts at engine .... Not specified

    Oil cooler flare nuts .... 20 Nm (14 ft-lbs)

Oil pump assembly bolts .... 13 Nm (108 inch-lbs)

Cylinder head cover bolts

    6 mm bolts .... Not specified

    8 mm bolt .... 23 Nm (17 ft-lbs)

Cam sprocket bolts .... 20 Nm (14 ft-lbs) (1)

Cylinder head nuts .... 44 Nm (33 ft-lbs) (2)

Cam chain tensioner

    Plug .... 4 Nm (35 inch-lbs)

    Mounting bolts .... Not specified

Cylinder bolts .... 44 Nm (33 ft-lbs)

Right crankcase cover bolts .... Not specified

Clutch spring bolts .... Not specified

Primary drive gear locknut .... 88 Nm (65 ft-lbs) (3)

Clutch locknut .... 108 Nm (80 ft-lbs) (4)

Shift cam plate to shift drum bolt .... Not specified (1)

Shift drum stopper arm bolt .... 12 Nm (108 inch-lbs)

Shift pedal pinch bolt .... 20 Nm (14 ft-lbs)

Crankcase bolts .... Not specified

Transmission mainshaft bearing set plate (inside crankcase) .... 12 Nm (108 inch-lbs)

1. *Apply non-permanent thread locking agent to the threads.*
2. *Apply engine oil to the threads.*
3. *Apply engine oil to the threads and locknut face.*
4. *Stake after tightening.*

2B

1 General information

The engine/transmission unit is of the air-cooled, single-cylinder four-stroke design. The four valves are operated by an overhead camshaft, which is chain driven off the crankshaft. The valves are arranged in a circle, rather than in the parallel pairs more commonly used in four-valves-per-cylinder engines. To accommodate this arrangement, the camshaft operates four main rocker arms, each of which in turn operates its valve through a sub-rocker arm rather than directly. Honda refers to this design as Radial Four Valve Combustion (RFVC).

The engine/transmission assembly is constructed from aluminum alloy. The crankcase is divided vertically.

The pressure-fed, dry-sump lubrication system uses a gear-driven rotor-type oil pump which pumps the engine oil through the oil cooler, through the oil tank, then back to the engine, where it is pumped to the transmission, crankshaft bearings, the underside of the pistons and the cylinder head. As oil drains back to the crankcase, it's picked up and sucked through an oil strainer screen and a paper element oil filter, then is pumped back to the oil cooler. The oil pump has two sets of rotors, one to circulate oil under pressure to the engine and the other to scavenge oil from the engine.

Power from the crankshaft is routed to the transmission via a wet, multi-plate type clutch. The transmission has five forward gears.

A decompressor reduces the effort required to start the engine.

2 Operations possible with the engine in the frame

The components and assemblies listed below can be removed without having to remove the engine from the frame. If, however, a number of areas require attention at the same time, removal of the engine is recommended.

Cylinder head cover
Camshaft and rocker arms
Cylinder head
Cam chain tensioner
Cylinder and piston
External shift mechanism
Clutch
External oil hoses
Oil pump and pipe

3 Operations requiring engine removal

It is necessary to remove the engine/transmission assembly from the frame and separate the crankcase halves to gain access to the following components:

Crankshaft, balancer and connecting rod
Transmission shafts
Internal shift mechanism (shift shaft, shift drum and forks)
Crankcase bearings

5.4 Back off the wire retainers (arrows) at both ends of the crankcase breather hose and remove the breather hose

5.15a Remove the nut from the other end of the upper engine mounting bolt (arrow), then remove the bolt and its two spacers

4   Major engine repair - general note

1   It is not always easy to determine when or if an engine should be completely overhauled, as a number of factors must be considered.

2   High mileage is not necessarily an indication that an overhaul is needed, while low mileage, on the other hand, does not preclude the need for an overhaul. Frequency of servicing is probably the single most important consideration. An engine that has regular and frequent oil and filter changes, as well as other required maintenance, will most likely give many miles of reliable service. Conversely, a neglected engine, or one that has not been broken in properly, may require an overhaul very early in its life.

3   Exhaust smoke and excessive oil consumption are both indications that piston rings and/or valve guides are in need of attention. Make sure oil leaks are not responsible before deciding that the rings and guides are bad. Refer to Chapter 1 and perform a cylinder compression check to determine for certain the nature and extent of the work required.

4   If the engine is making obvious knocking or rumbling noises, the connecting rod and/or main bearings are probably at fault.

5   Loss of power, rough running, excessive valve train noise and high fuel consumption rates may also point to the need for an overhaul, especially if they are all present at the same time. If a complete tune-up does not remedy the situation, major mechanical work is the only solution.

6   An engine overhaul generally involves restoring the internal parts to the specifications of a new engine. During an overhaul the piston rings are replaced and the cylinder walls are bored and/or honed. If a rebore is done, then a new piston is also required. The crankshaft and connecting rod are permanently assembled, so if one of these components needs to be replaced, both must be. Generally the valves are serviced as well, since they are usually in less than perfect condition at this point. While the engine is being overhauled, other components such as the carburetor can be rebuilt also. The end result should be a like-new engine that will give as many trouble-free miles as the original.

7   Before beginning the engine overhaul, read through all of the related procedures to familiarize yourself with the scope and requirements of the job. Overhauling an engine is not all that difficult, but it is time consuming. Plan on the motorcycle being tied up for a minimum of two weeks. Check on the availability of parts and make sure that any necessary special tools, equipment and supplies are obtained in advance.

8   Most work can be done with typical shop hand tools, although a number of precision measuring tools are required for inspecting parts to determine if they must be replaced. Often a dealer service department or repair shop will handle the inspection of parts and offer advice concerning reconditioning and replacement. As a general rule, time is the primary cost of an overhaul so it doesn't pay to install worn or substandard parts.

9   As a final note, to ensure maximum life and minimum trouble from a rebuilt engine, everything must be assembled with care in a spotlessly clean environment.

5   Engine - removal and installation

Note: *Engine removal and installation should be done with the aid of an assistant to avoid damage or injury that could occur if the engine is dropped. A hydraulic floor jack should be used to support and lower the engine if possible (they can be rented at low cost).*

Removal

Refer to illustrations 5.4, 5.15a, 5.15b, 5.15c and 5.15d

1   Drain the engine oil (see Chapter 1).

2   Remove the seat and the front fender (see Chapter 8).

3   Remove the fuel tank, the heat protector, the carburetor and the exhaust system (see Chapter 3).

4   Disconnect the oil tank breather hose from the cylinder head cover and disconnect the crankcase breather hose from the crankcase **(see illustration)**.

5   Disconnect the spark plug wire (see Chapter 1).

6   Label and disconnect the neutral switch wire, the alternator wires and the starter motor cable (see Chapter 8). Detach all wires from their retainers (most retainers are bendable clamps that can be opened up, then clamped around the wires again after reassembly).

7   Remove the rear starter motor bolt and detach the brake lock cable guide from the crankcase (see "Starter motor - removal and installation" in Chapter 8). It's not necessary to remove the starter motor.

8   Remove the gearshift pedal pinch bolt and remove the gearshift pedal (see Section 20).

9   Remove the sprocket cover, loosen the drive chain and remove the chain and the drive sprocket (see Chapter 5).

10   Loosen the clutch cable locknut, remove the two cable guide bolts and disengage the cable end from the clutch lifter arm (see Section 16).

11   Disconnect the oil cooler lines from the engine (see Section 21).

12   Remove the brake pedal (see Chapter 6).

13   Remove the engine guard (see Chapter 7).

14   Support the engine with a floor jack. Put a block of wood between the jack head and the engine to protect the crankcase.

15   Remove the engine mounting nuts, bolts and brackets at the top, upper front and lower front **(see illustrations)**. You'll also need to remove the swingarm pivot bolt; it passes through the rear of the crankcase to act as an engine support (see Chapter 5).

16   Have an assistant help you lift the engine out of the right side of the frame.

17   Slowly lower the engine to a suitable work surface.

5.15b At the right front of the engine, remove the through-bolt nut (upper arrow) and the right engine mounting bracket bolts (lower arrows) (note the oil cooler hose bracket)

5.15c Remove the lower front mounting bolt and nut (arrow) . . .

5.15d . . . at the left front of the engine, pull out the through-bolt (upper arrow) and remove the mounting bracket bolts (lower arrows)

Installation

Refer to illustration 5.19

18 Have an assistant help lift the engine into the frame so it rests on the jack and block of wood. Use the jack to align the mounting bolt holes, then install the brackets, bolts and nuts. Tighten them to the torques listed in this Chapter's Specifications. Refer to the Chapter 5 Specifications for the swingarm pivot bolt torque.

19 The remainder of installation is the reverse of the removal steps, with the following additions:

a) *The two plates that form the upper front mount are labeled L for left side and R for right side* **(see illustration)**.
b) *Use new gaskets at all exhaust pipe connections.*
c) *Adjust the throttle cable, decompression cable(s) and clutch cable following the procedures in Chapter 1.*
d) *Fill the engine with oil, also following the procedures in Chapter 1.*
d) *Run the engine and check for oil or exhaust leaks.*

6   Engine disassembly and reassembly - general information

Refer to Section of 6 of Chapter 2A for these procedures.

7   Cylinder head cover and rocker arms - removal, inspection and installation

Removal

Refer to illustration 7.6

1   Place the engine at top dead center on its compression stroke (see "Valve clearance – check and adjustment" in Chapter 1).

2   Remove the seat and front fender (see Chapter 7). Remove the fuel tank and heat protector (see Chapter 3).

3   Disconnect the oil tank breather hose from the cylinder head cover.

4   Disconnect the spark plug cap from the spark plug.

5   Remove the upper engine mounting nut, bolt and spacers **(see illustration 5.15a)**.

6   Loosen the 12 6mm cover bolts and the single 8 mm bolt in a criss-cross pattern, in two or three stages **(see illustration)**. If any of the bolts secure cable or hose retainers, note their location. It's a good idea to punch holes in a piece of cardboard in the shape of the cover and place the bolts in the same positions they're in when installed.

7   Lift the cover off the engine. If it's stuck, don't attempt to pry it off - tap around its sides with a plastic hammer to dislodge it.

8   Remove the gasket from the cylinder head cover. Use a new gasket whenever the cover is removed.

2B

5.19 The left engine mounting bracket is marked with an "L" and the right bracket is marked with an "R"

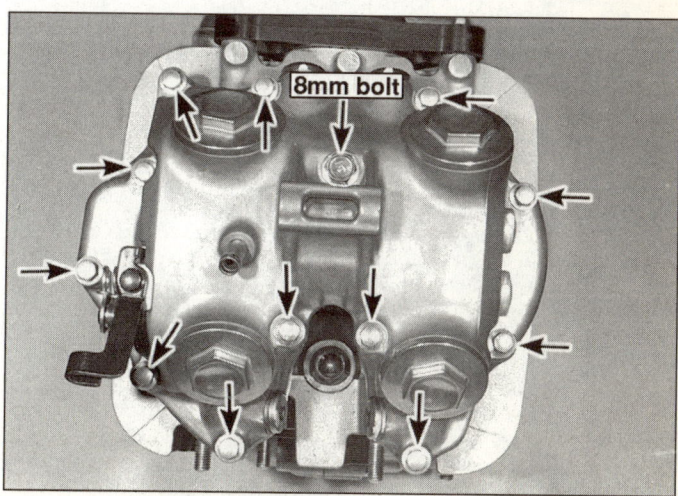

7.6 Loosen the twelve 6 mm cover bolts, and the single 8 mm bolt, evenly in two or three stages

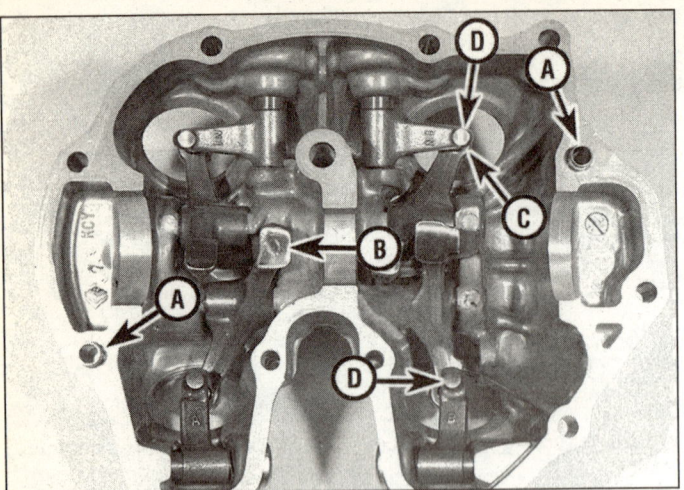

7.10 Lift off the cover, remove the gasket and note the locations of the dowels; they may come off with the cover or stay in the head

A   *Dowels*
B   *Main rocker arm camshaft contact surfaces*
C   *Main rocker arm to sub-rocker arm contact surfaces*
D   *Sub-rocker arm valve stem contact surfaces*

Inspection

Refer to illustrations 7.10 and 7.11

9   Inspect the cam bearing surfaces in the cylinder head and its cover for wear and damage (see Section 9).
10   Inspect the rocker arms for wear at the cam contact surfaces and at the tips of the valve adjusting screws **(see illustration)**. Try to twist the rocker arms from side-to-side on the shafts. If they're loose on the shafts or if there's visible wear, remove them as described below.
11   Unscrew the rocker shafts and sub-rocker shafts and pull them out of the cover **(see illustration**. Remove the sealing washers, wave washers (sub-rocker arms only) and the rocker arms. **Note:** *Some of the rocker arms (all of the rocker arms on some models) have identification marks so they can be returned to their original positions. Look for the marks, and make your own if they aren't clearly visible. It's a good idea to label all of the rocker arms and sub-rocker arms so they can be reinstalled in their original locations. In addition to developing wear patterns with their shafts, some of the rocker arms are shaped differently from the others and won't work in the wrong location.*
12   Measure the outer diameter of each rocker shaft and the inner diameter of the rocker arms with a micrometer and compare the mea-

7.11 Remove the rocker shafts

surements to the values listed in this Chapter's Specifications. If rocker arm-to-shaft clearance is excessive, replace the rocker arm or shaft, whichever is worn.

Installation

Refer to illustrations 7.15a and 7.15b

13   Coat the main rocker shafts and rocker arm bores with moly-based grease containing 40-percent or more molybdenum disulfide. Apply non-hardening gasket sealant to the threads of the main rocker shafts. Install the rocker shafts and rocker arms in the cylinder head cover, using new copper sealing washers on the shafts. Depending on model, the main rocker arms will be marked as follows:

a)   *No marks*
b)   *A and B*
c)   *A, B, C and D*

14   If the rocker arms are unmarked, you'll need to refer to the labels made on disassembly.
15   If the rocker arms have factory markings, install them in the specified locations **(see illustrations)**.
16   Coat new sub-rocker shaft sealing rings with clean engine oil and install them on the shafts. Coat the shaft threads with non-hardening gasket sealant. Install the sub-rocker arms and their wave washers in the correct locations in the cylinder head cover, then install the shafts and sealing washers.
17   Clean the mating surfaces of the cylinder head and cover with lacquer thinner, acetone or brake system cleaner. Install a new gasket,

7.15a Intake sub-rocker shaft wave washer locations; both intake sub-rocker arms are labeled IN

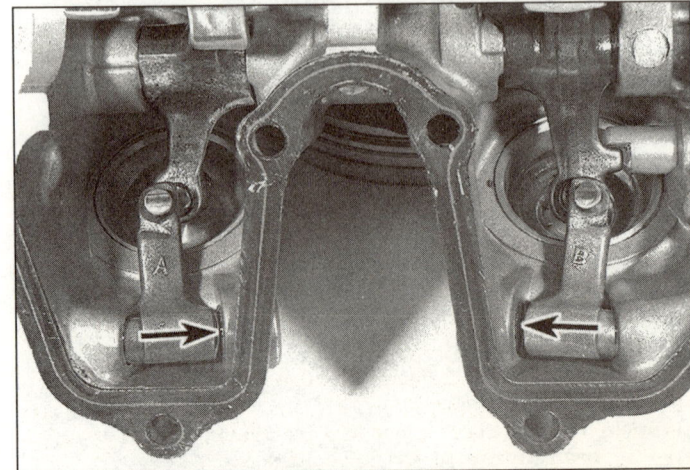

7.15b Exhaust sub-rocker shaft wave washer locations; the exhaust sub-rocker arms are labeled A and B

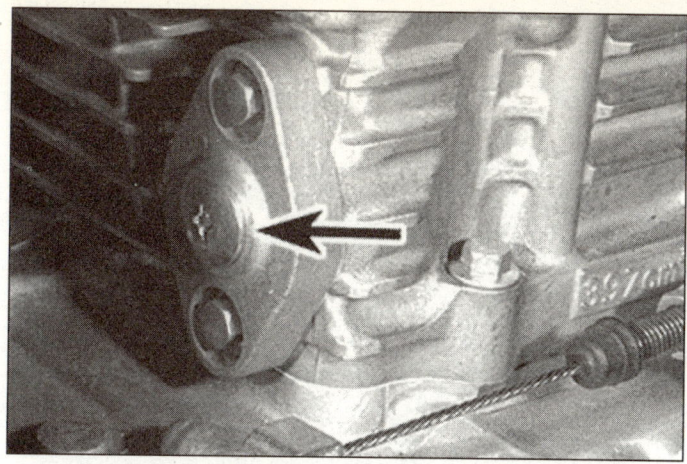

8.1 Loosen the cover screw (arrow) while the tensioner is on the engine

8.2 Pull the tensioner out of the engine and remove the gasket

taking care not to damage its silicone coating.
18   Make sure the piston is still at top dead center on its compression stroke (both cam lobes pointing downward). Fill the oil pockets in the top of the cylinder head with clean engine oil so the oil covers the cam lobes.
19   Loosen the valve adjusting screws all the way, then install the cover on the cylinder head. Install the bolts and tighten them evenly in two or three stages to the torques listed in this Chapter's Specifications. Note

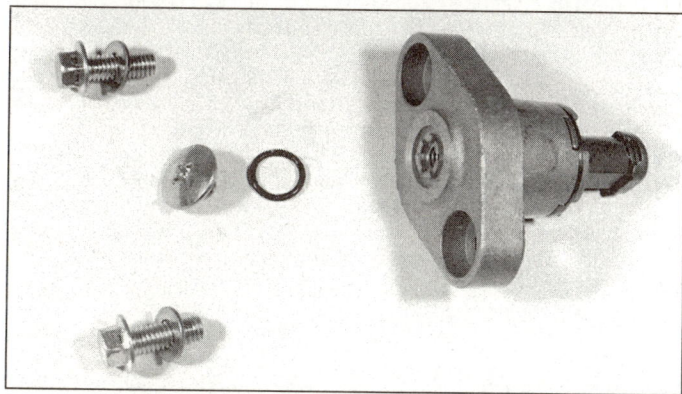

8.3 Use a new O-ring and sealing washers on installation

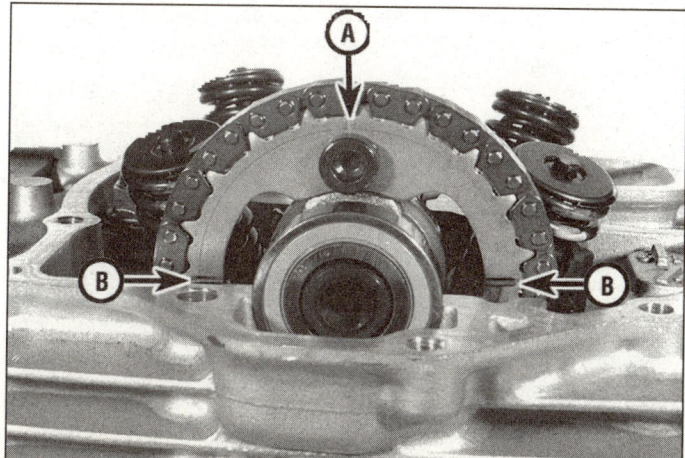

9.3 Rotate the camshaft sprocket so the Top Dead Center mark is straight up and the timing marks are aligned with the cylinder head surface, then remove the exposed bolt . . .

A   Top Dead Center mark
B   Timing marks (align with cylinder head surface)

that there are different types of bolts with different torque settings.
20   Adjust the valve clearances (see Chapter 1).
21   The remainder of installation is the reverse of removal.

8   Cam chain tensioner - removal, inspection and installation

Removal

Refer to illustrations 8.1 and 8.2
1   You'll need to remove the tensioner cover screw to reset the tensioner for installation, so loosen the screw while the tensioner is still on the engine (see illustration).
2   Remove two bolts and take the tensioner and its gasket off the cylinder (see illustration). Clean all traces of old gasket from the tensioner and engine.

Inspection

Refer to illustration 8.3
3   Check the tensioner for visible wear and damage and replace it if any problems are found. Replace the sealing washers and O-ring whenever the tensioner is removed (see illustration).

Installation

4   Place a new gasket on the tensioner. Insert a small screwdriver into the tensioner and turn it clockwise to retract the tensioner piston. **Caution:** *If you try to tighten the tensioner bolts without retracting the piston, the tensioner body or cam chain will be damaged.*
5   Hold the tensioner piston in the retracted position and install the tensioner on the engine. Install the bolts, using new sealing washers. Tighten the bolts securely, but don't overtighten them and strip the threads. Release the screwdriver.
6   Install the tensioner cover screw, using a new O-ring. Tighten it to the torque listed in this Chapter's Specifications.

9   Camshaft, guides and chain – removal, inspection and installation

Removal

Refer to illustrations 9.3, 9.4 and 9.5
1   Remove the cylinder head cover and the cam chain tensioner (see Sections 7 and 8).
2   Stuff rags into the cam chain opening so the camshaft bolts won't fall into the crankcase.
3   Position the engine at TDC on the compression stroke (see illustration) and remove the sprocket bolt. Now rotate the crankshaft one

2B

9.4 . . . rotate the sprocket, remove the other bolt and disengage the sprocket from the chain (its OUT mark (arrow) faces outward on installation)

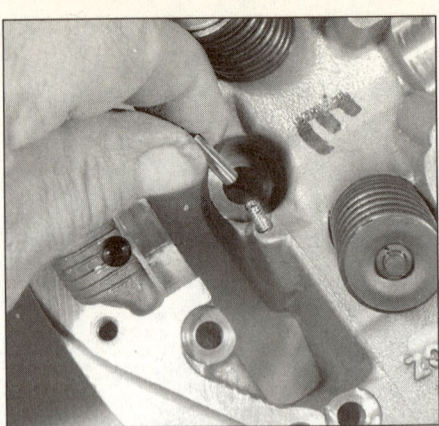

9.5 Pull the decompressor pin and spring out of the cylinder head

9.6 Check the cam bearing surfaces (arrows) for scoring or wear

revolution (clockwise), which will turn the camshaft one-half turn, and remove the other cam sprocket bolt.

4   Disengage the sprocket from the cam chain **(see illustration)**. Support the cam chain so it won't drop into the engine and remove the sprocket.

5   Lift the camshaft out of the cylinder head, then remove the decompressor pin and spring **(see illustration)**.

Inspection

Refer to illustrations 9.6, 9.8a, 9.8b, 9.8c and 9.11

Note: *Before replacing the camshaft or the cylinder head cover and cylinder head because of damage, check with local machine shops specializing in motorcycle engine work. In the case of the camshaft, it may be possible for cam lobes to be welded, reground and hardened, at a cost far lower than that of a new camshaft. If the bearing surfaces in the center of the cylinder head or cover are damaged, it may be possible for them to be bored out to accept bearing inserts. Due to the cost of a new cylinder head it is recommended that all options be explored before condemning it as trash!*

6   Inspect the cam bearing surfaces of the cylinder head and cover **(see illustration 7.10 and the accompanying illustration)**. Look for score marks, deep scratches and evidence of spalling (a pitted appearance). The bearing surfaces that support the ends of the camshaft contain ball bearings, so the surfaces shouldn't be scored or worn. If they are, the bearings may have spun in their bores due to seizure or a loose cylinder head cover.

7   Refer to Chapter 2A for camshaft inspection procedures. Also,

be sure to check the condition of the rocker arms, as described in Section 7.

8   Check the chain guides for wear or damage. If they are worn or damaged, replace them. To remove the exhaust side (front) chain guide, you'll need to remove the cylinder head (see Section 10), then lift the guide out of its notches **(see illustration)**. To remove the intake side guide, you'll need to remove the right crankcase cover and clutch (see Section 17), then unscrew the bolt and take off the chain guide, collar and washer **(see illustrations)**.

9   Except in cases of oil starvation, the camshaft chain wears very little. If the chain has stretched excessively, which makes it difficult to maintain proper tension, replace it with a new one. To remove the chain from the crankshaft sprocket, it's necessary to remove the chain guides as described above, as well as the oil pump (see Section 18). Once this is done, the chain can be lowered away from the sprocket.

10   Inspect the sprocket for wear, cracks and other damage, replacing it if necessary. If the sprocket is worn, the chain is also worn, and possibly the sprocket on the crankshaft. If wear this severe is apparent, the entire engine should be disassembled for inspection.

11   Slip the ball bearings off the ends of the camshaft **(see illustration)**. Hold the center race of each bearing with fingers and turn the outer race. The bearing should spin freely without roughness, looseness or noise. If it has obvious problems, or if you aren't sure it's in good condition, replace the bearing. Spin the outer race of the decompressor one-way clutch and make sure it turns in only one direction. If not, have it pressed off and a new one pressed on by a Honda dealer or motorcycle repair shop.

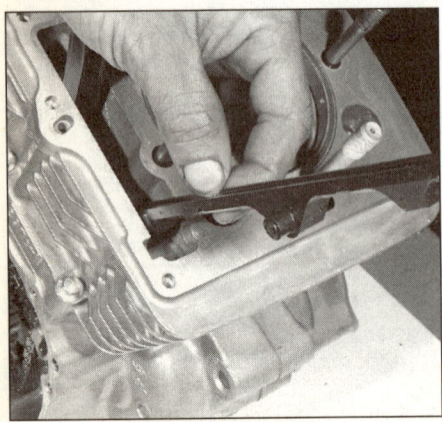

9.8a The front chain guide fits in notches in the cylinder head

9.8b The rear chain guide is secured by a bolt and pivot collar; there's a washer between the chain guide and crankcase

9.8c The finger on the front chain guide fits in a pocket in the crankcase (arrow)

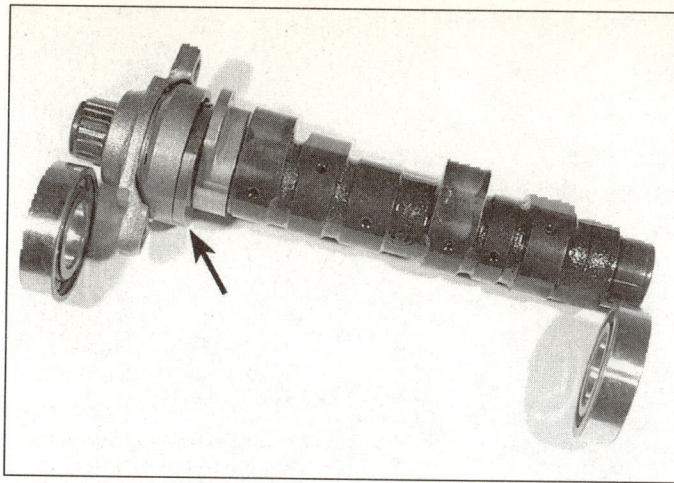

9.11 There's a ball bearing on each end of the camshaft; the sealed side of the bearing at the sprocket end faces away from the camshaft; the one-way decompressor clutch (arrow) in located at the sprocket end of the camshaft

9.14a Coat the cam lobes with moly-based grease

Installation

Refer to illustrations 9.14a and 9.14b

12   Install the cam chain guides, cylinder head, oil pump, clutch and right engine cover if they were removed.

13   Make sure the bearing surfaces in the cylinder head and cylinder head cover are clean. Install the ball bearings on the ends of the camshaft **(see illustration 9.11)**. If there's only one sealed bearing, install it on the sprocket end of the camshaft with its sealed side facing away from the center of the engine. If both bearings are sealed, install both with their sealed sides facing away from the center of the engine.

14   Lubricate the cam bearing journals with molybdenum disulfide grease. Lay the camshaft in the cylinder head with the lobes downward **(see illustration)**. Install the bearing retaining dowels in the cylinder head at each end of the camshaft **(see illustration)**.

15   Engage the sprocket with the chain so its timing mark will be straight up and its OUT mark faces away from the center of the engine **(see illustration 9.4)**. Place the sprocket on the camshaft so its bolt holes align with the camshaft bolt holes. Rotate the sprocket as needed for access, install the bolts and tighten them to the torque listed in this Chapter's Specifications.

16   Turn the sprocket so its timing mark is straight up and the marks are aligned with the cylinder head gasket surface **(see illustration 9.3)**.

17   Recheck the crankshaft timing mark on the alternator rotor to make sure it's still at the TDC position (see "Valve clearance - check

and adjustment" in Chapter 1). If it's out of position and the camshaft sprocket is aligned as described in Step 3, you'll need to remove the chain from the sprocket and reposition it. Don't run the engine with the marks out of alignment or severe engine damage could occur.

18   Remove the holder from the cam chain tensioner.

19   Adjust the valve clearances (see Chapter 1).

20   The remainder of installation is the reverse of removal.

10   Cylinder head - removal and installation

Caution: *The engine must be completely cool before beginning this procedure, or the cylinder head may become warped.*

Removal

Refer to illustrations 10.2a, 10.2b and 10.4

1   Remove the cylinder head cover, the cam chain tensioner and the camshaft (see Sections 7, 8 and 9).

2   Loosen the main cylinder head nuts in two or three stages, in a criss-cross pattern **(see illustration)**. Remove the bolts or nuts and their washers. The washer nearest the spark plug hole is thicker than the others **(see illustration)**.

3   Lift the cylinder head off the cylinder. If the head is stuck, use wooden dowels inserted into the intake or exhaust ports to lever the head off. Don't attempt to pry the head off by inserting a screwdriver between the head and the cylinder - you'll damage the sealing surfaces. Don't hammer against the side of the head or the cooling fins may be broken.

9.14b Position the bearing in its saddle and make sure the dowel is in position, then do the same at the other end of the camshaft

10.2a The cylinder head is secured by four nuts (arrows)

10.2b The washer near the spark plug hole is thicker than the others

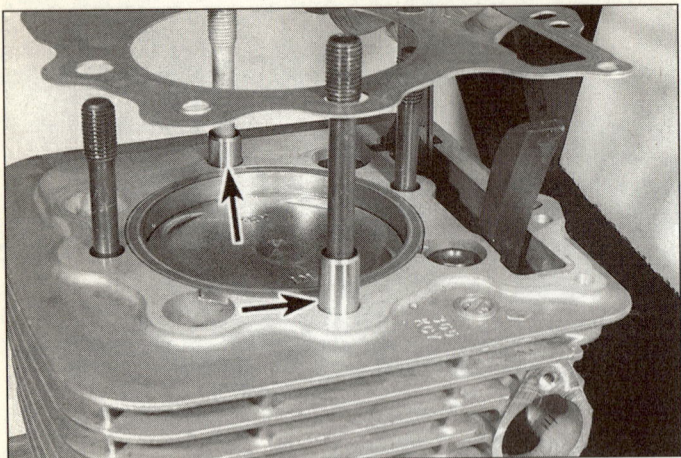

10.4 Lift the head off, remove the gasket and locate the dowels (arrows); they may come off with the head or stay in the cylinder

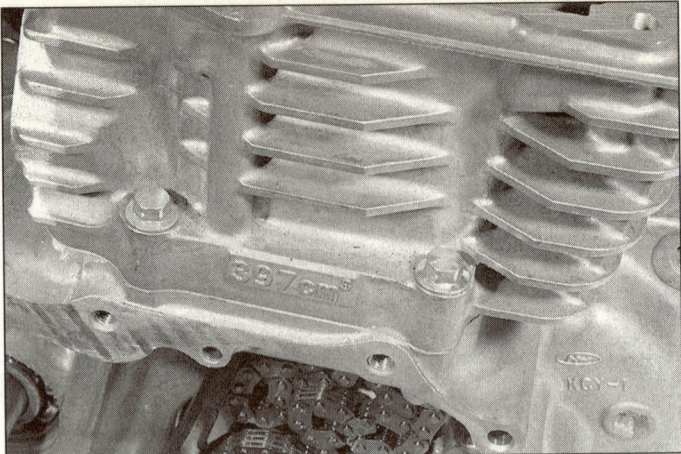

13.2 Remove the two small bolts that secure the cylinder to the crankcase . . .

4   Support the cam chain so it won't drop into the cam chain tunnel, and stuff a clean rag into the tunnel to prevent the entry of debris. Once this is done, remove the gasket and two dowel pins from the cylinder **(see illustration)**.
5   If the front (exhaust) side chain guide is worn, lift it out of its notches **(see illustration 9.8a)**.
6   Inspect the cylinder head gasket and the mating surfaces on the cylinder head and cylinder for leakage, which could indicate warpage. Check the flatness of the cylinder head (see Section 12).
7   Clean all traces of old gasket material from the cylinder head and cylinder. Be careful not to let any of the gasket material fall into the crankcase, the cylinder bore or the bolt holes.

Installation

8   Install the two dowel pins, then lay the new gasket in place on the cylinder **(see illustration 10.4)**. Never reuse the old gasket and don't use any type of gasket sealant.
9   Make sure the cam chain front guide fits in its notches **(see illustration 9.8a)**.
10   Carefully lower the cylinder head over the dowels. It's helpful to have an assistant support the camshaft chain with a piece of wire so it doesn't fall and become kinked or detached from the crankshaft. When the head is resting on the cylinder, wire the cam chain to another component to keep tension on it.
11   Install the cylinder head nuts and tighten them securely, but don't overtighten them and strip the threads.
12   The remainder of installation is the reverse of the removal steps.
13   Change the engine oil (see Chapter 1).

11   Valves/valve seats/valve guides - servicing

1   Because of the complex nature of this job and the special tools and equipment required, servicing of the valves, the valve seats and the valve guides (commonly known as a valve job) is best left to a professional.
2   The home mechanic can, however, remove and disassemble the head, do the initial cleaning and inspection, then reassemble and deliver the head to a dealer service department or properly equipped motorcycle repair shop for the actual valve servicing. Refer to Section 12 for those procedures.
3   The service department will remove the valves and springs, recondition or replace the valves and valve seats, replace the valve guides, check and replace the valve springs, spring retainers and keepers (as necessary), replace the valve seals with new ones and reassemble the valve components.
4   After the valve job has been performed, the head will be in like-new condition. When the head is returned, be sure to clean it again very thoroughly before installation on the engine to remove any metal

particles or abrasive grit that may still be present from the valve service operations. Use compressed air, if available, to blow out all the holes and passages.

12   Cylinder head and valves - disassembly, inspection and reassembly

The procedures are the same as for the TRX300EX engine (refer to Chapter 2A).

13   Cylinder - removal, inspection and installation

Removal

Refer to illustrations 13.2, 13.3, 13.4 and 13.5
1   Remove the cylinder head cover, camshaft and cylinder head (see Sections 7, 9 and 10). Make sure the crankshaft is positioned at Top Dead Center (TDC).
2   Remove two small bolts securing the cylinder to the crankcase **(see illustration)**.
3   If the cylinder head dowels stayed in the cylinder, remove them. Loosen the main cylinder attaching bolts in two or three stages in a criss-cross pattern **(see illustration)**.
4   Lift the cylinder straight up, off the piston and the front cam chain guide **(see illustration)**. If it's stuck, tap around its perimeter with a

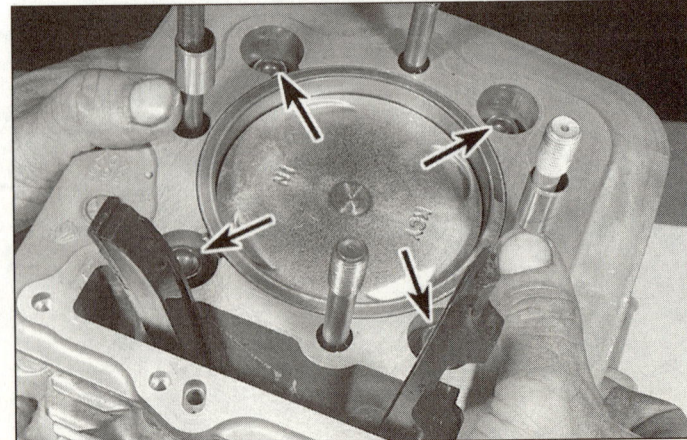

13.3 . . . then lift off the dowels if they're still in the cylinder and loosen the main bolts (arrows) evenly, in a criss-cross pattern - note the location of the IN mark on the piston; it must face the same way on installation

13.4 Lift the cylinder off the crankcase; if you're experienced and very careful, the cylinder can be installed over the rings without a ring compressor, but a compressor is recommended

13.5 Locate the dowels and remove the base gasket

soft-faced hammer (but don't tap on the cooling fins or they may break). Don't attempt to pry between the cylinder and the crankcase, as you'll ruin the sealing surfaces.

5   Locate the dowel pins (they may have come off with the cylinder or still be in the crankcase) **(see illustration)**. Be careful not to let these drop into the engine. Stuff rags around the piston and remove the gasket and all traces of old gasket material from the surfaces of the cylinder and the crankcase.

Inspection

Refer to illustration 13.6

Caution: *Don't attempt to separate the liner from the cylinder.*

6   Check the top surface of the cylinder for warpage, using the same method as for the cylinder head (see Section 12). Measure along the sides and diagonally across the stud holes **(see illustration)**.

7   To inspect and recondition the cylinder bore, refer to Chapter 2A.

Installation

8   Lubricate the cylinder bore with plenty of clean engine oil. Apply a thin film of moly-based grease to the piston skirt .

9   Install the dowel pins, then lower a new cylinder base gasket over them **(see illustration 13.5)**.

10   Attach a piston ring compressor to the piston and compress the piston rings. A large hose clamp can be used instead - just make sure it doesn't scratch the piston, and don't tighten it too much.

11   Install the cylinder over the studs and carefully lower it down until the piston crown fits into the cylinder liner **(see illustration 13.4)**. While

doing this, pull the camshaft chain up, using a hooked tool or a piece of stiff wire. Push down on the cylinder, making sure the piston doesn't get cocked sideways, until the bottom of the cylinder liner slides down past the piston rings. A wood or plastic hammer handle can be used to gently tap the cylinder down, but don't use too much force or the piston will be damaged.

12   Remove the piston ring compressor or hose clamp, being careful not to scratch the piston.

13   The remainder of installation is the reverse of the removal steps.

14   Piston - removal, inspection and installation

Removal, inspection and installation procedures for the piston, pin and rings are the same as for the TRX300EX (see Chapter 2A).

15   Piston rings - installation

Installation procedures for the piston rings are the same as for the TRX300EX (see Chapter 2A).

2B

16   Clutch cable – removal and installation

Refer to illustrations 16.1 and 16.3

1   At the handlebar, peel back the rubber dust cover, back off the locknut and unscrew the clutch cable adjuster **(see illustration)**.

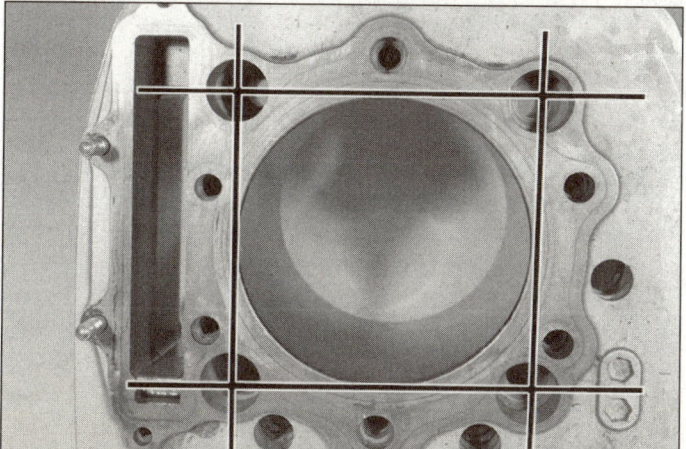

13.6 Check the cylinder top surface for warpage in the directions shown

16.1 Line up the slots in the adjuster and locknut (left arrows), pivot the cable out and lower it from the lever slot (right arrow)

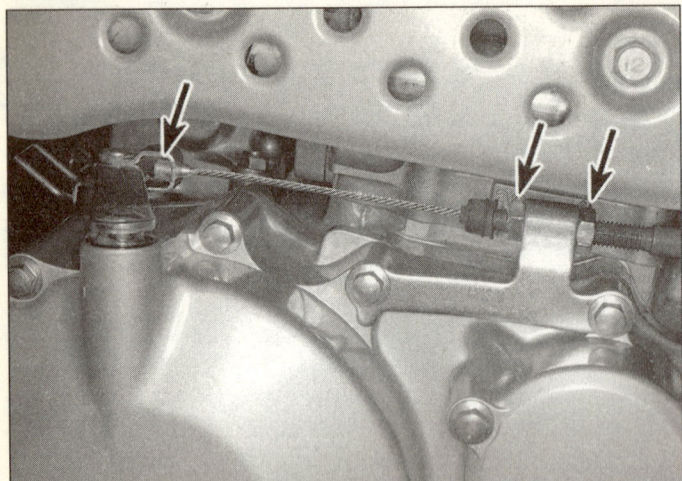

16.3 Loosen the locknut and adjusting nut (right arrows), slip the cable out of the bracket, then disengage the cable end plug from the lifter arm (left arrow)

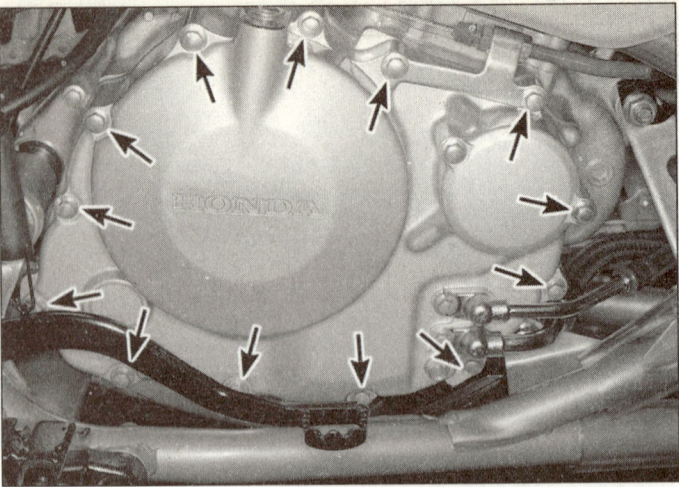

17.5a Remove the right crankcase cover bolts (arrows), then pull off the cover and simultaneously . . .

2   To disengage the upper end of the cable from the clutch lever, rotate the cable adjuster and locknut so that the slots in both are aligned with the cable, then slip the cable out of the adjuster and locknut **(see illustration 16.1)**. Pivot the cable so it aligns with the slot in the lower side of the clutch lever, then lower the cable end plug from the lever.

3   Trace the clutch cable down to the cable bracket on top of the right crankcase cover, loosen the locknut and back off the adjuster (see Section 18 in Chapter 1, if necessary), and unbolt the cable bracket from the engine **(see illustration)**.

4   Disengage the lower end of the cable from the clutch lifter arm **(see illustration 16.3)**. Note the routing of the clutch cable, then remove the cable.

5   Slide the cable back and forth in the housing and make sure it moves freely. If it doesn't, try lubricating it as described in Chapter 1. If that doesn't help, replace the cable.

6   To replace the clutch lever, simply remove the pivot bolt nut and pull out the pivot bolt.

7   To remove the clutch lever assembly, disconnect the parking brake cable (see Chapter 6), unplug the electrical connector from the clutch switch (see Chapter 8), remove the two clamp bolts **(see illustration 17.7 in Chapter 2A)** and remove the lever assembly.

8   Installation is the reverse of removal. If you removed the clutch/parking brake lever bracket assembly, make sure that the UP arrow on the bracket clamp faces up **(see illustration 17.7 in Chapter**

2A), and be sure to adjust the parking brake cable (see "Parking brake – adjustment" in Chapter 1). Make sure that the parking brake cable is correctly routed and adjust the clutch cable when you're done (see "Clutch cable - check and adjustment" in Chapter 1).

17   Clutch - removal, inspection and installation

Right crankcase cover and clutch release mechanism

Removal

Refer to illustrations 17.5a, 17.5b, 17.5c and 17.5d

1   Drain the engine oil (see Chapter 1).

2   Remove the rear brake pedal (see Chapter 6).

3   Unbolt the clutch cable bracket **(see illustration 16.3)**.

4   Disconnect the oil hoses (see Section 21).

5   Remove the right crankcase cover bolts **(see illustration)** and pull off the cover while simultaneously turning the clutch lifter arm counterclockwise to disengage the lifter arm spindle from the lifter piece **(see illustrations)**. Don't pry against the mating surfaces of the cover and crankcase. Once the cover is off, locate the dowels **(see illustration)**; they may have stayed in the crankcase or come off with the cover.

6   Note how the spring is installed, then pull the lifter arm shaft out of the cover **(see illustration 17.5b)**.

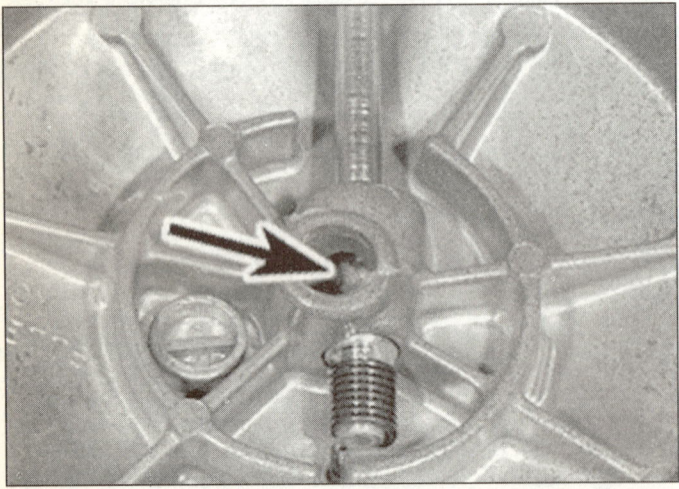

17.5b . . . turn the lifter arm shaft to disengage its center portion (arrow) . . .

17.5c . . . from the lifter piece (A); the pressure plate is secured by five bolts (B)

17.5d Note the locations of the dowels (arrows) (if they're not in the crankcase, look in the cover)

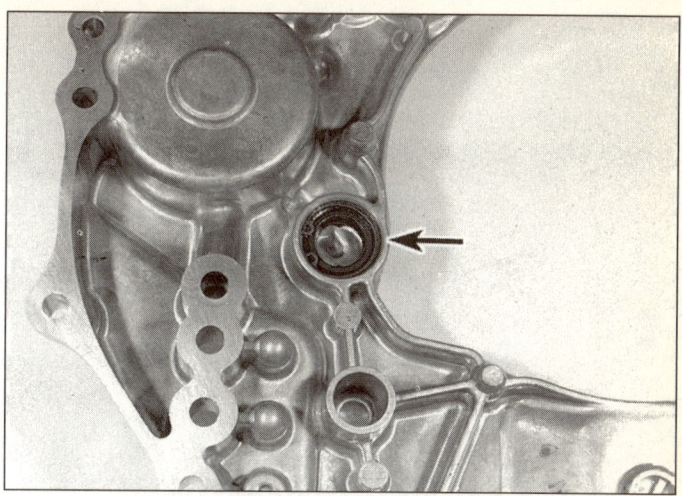

17.7 The crankshaft end seal is held in the right crankcase cover by a snap-ring

17.10a The tabs of the outermost friction plate fit in the clutch housing's short grooves (arrow); all other friction and metal plate tabs fit in the long grooves

Inspection

Refer to illustration 17.7

7   Inspect the clutch lifter arm assembly for obvious wear and damage at the contact points of the lifter arm shaft and the lifter piece. Replace any parts that show problems. Replace the lifter shaft seal in the engine cover whenever it's removed. There's also a crankshaft seal, held in place by a snap-ring **(see illustration)**, in the cover. Replace this seal if it's damaged or worn.

Installation

8   Installation is the reverse of the removal steps. Refill the engine oil and adjust the clutch (see Chapter 1).

Clutch

Removal

Refer to illustrations 17.10a through 17.10g

9   Remove the clutch cover as described above.

10   Refer to the accompanying illustrations to remove the clutch components **(see illustrations)**. **Note:** *When you loosen the clutch housing locknut, wedge the primary drive gear and the driven gear on the clutch housing so they won't turn* **(see illustration 19.2a)**.

2B

17.10b With the bolts and springs removed, pull off the pressure plate together with the metal and friction plates

17.10c With the primary and driven gears wedged, bend back the staked portion of the locknut (arrow) and unscrew it (don't forget to stake the new locknut on installation)

17.10d Pull off the lockwasher (its OUT SIDE mark faces away from the engine on installation)

17.10e Remove the outer thrust washer, then pull off the clutch center and remove the inner thrust washer

17.10f Pull the clutch housing off the bushing . . .

17.10g . . . and remove the bushing from the crankshaft

17.11 Check the ball bearing is in the pressure plate for roughness, looseness or noise; check the friction surface (arrow) for scoring

Inspection

Refer to illustrations 17.11, 17.13 and 17.14

11   Rotate the release bearing **(see illustration)** and check it for rough, loose or noisy operation. If the bearing's condition is in doubt, push it out of the pressure plate and push in a new one.

12   Inspect the friction surface on the pressure plate for scoring or wear. Replace the pressure plate if any defects are found.

13   Inspect the edges of the slots in the clutch housing for indentations made by the friction plate tabs **(see illustration)**. If the indentations are deep they can prevent clutch release, so the housing should be replaced with a new one. If the indentations can be removed easily with a file, the life of the housing can be prolonged to an extent. Also, check the driven gear teeth for cracks, chips and excessive wear and the springs on the back side (if equipped) for breakage. If the gear is worn or damaged or the springs are broken, the clutch housing must be replaced with a new one.

14   Inspect the bearing surface in the center of the clutch housing for score marks, scratches and excessive wear **(see illustration)**. Measure the inside diameter of the bearing surface, the inside and outside diameters of the clutch housing bushing and the bushing's mounting surface on the transmission mainshaft. Compare these to the values listed in this Chapter's Specifications. Replace any parts worn beyond the service limits. If the bushing mounting surface on the mainshaft is worn excessively, the mainshaft will have to be replaced.

15   Inspect the clutch center's friction surface and slots for scoring, wear and indentations **(see illustration 17.13)**. Also inspect the splines in the middle of the clutch center. Replace the clutch center if problems are found.

16   Measure the free length of the clutch springs and compare the

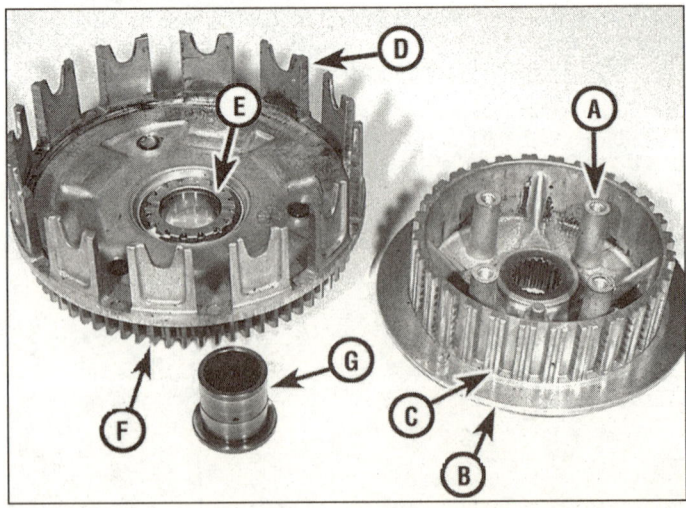

17.13 Clutch inspection points

A	Clutch center posts	E	Clutch housing bushing
B	Clutch center friction		surface
	surface	F	Driven gear
C	Clutch center splines	G	Clutch housing bushing
D	Clutch housing slots		

17.14 Check the clutch housing bushing surface (arrow) for wear

17.20 There's a metal plate between each pair of friction plates; friction plates go on first and last

18.2 Remove the oil pump driven gear; the flat on the shaft aligns with the flat in the gear (arrow)

18.3a Remove the crankcase oil screen bolt, oil pump bolts and O-ring (arrows)

results to this Chapter's Specifications. If the springs have sagged, or if cracks are noted, replace them with new ones as a set.

17   If the lining material of the friction plates smells burnt or if it is glazed, new parts are required. If the metal clutch plates are scored or discolored, they must be replaced with new ones. Measure the thickness of the friction plates (see Chapter 2A) and replace with new parts any friction plates that are worn.

18   Check the metal plates for warpage (see Chapter 2A).

19   Inspect the tabs on the friction plates for excessive wear and mushroomed edges. They can be cleaned up with a file if the deforma-

18.3b The oil pump shim surrounds an oil passage

tion is not severe. Check the friction plates for warpage as described in Step 18.

Installation

Refer to illustration 17.20

20   Installation is the reverse of the removal steps, with the following additions:

a) *Install the lockwasher with its OUT SIDE mark facing away from the engine, then install the clutch nut and tighten it to the torque listed in this Chapter's Specifications.*

b) *Stake the nut into the notch on the mainshaft.*

c) *Coat the friction plates with clean engine oil before you install them.*

d) *Install a friction plate, then alternate the remaining metal and friction plates until they're all installed. Friction plates go on first and last, so the friction material contacts the metal surfaces of the clutch center and the pressure plate* **(see illustration).**

---

18   Oil pump - removal, inspection and installation

Note: *The oil pump can be removed with the engine in the frame.*

Removal

Refer to illustrations 18.2, 18.3a and 18.3b

1   Remove the right crankcase cover and the clutch (see Section 17).

2   Pull the driven gear off the oil pump shaft **(see illustration)**.

3   Remove the oil pump mounting bolts and take the pump off the engine, together with the mounting shim **(see illustrations)**. Note the locations of the two dowels.

2B

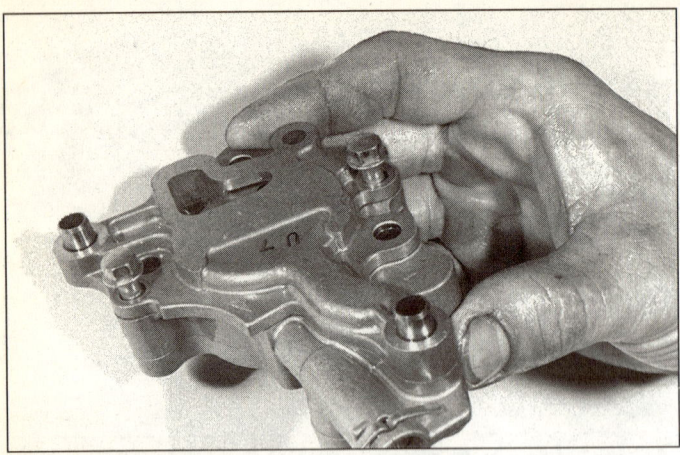

18.5a Remove the oil pump cover screws or bolts . . .

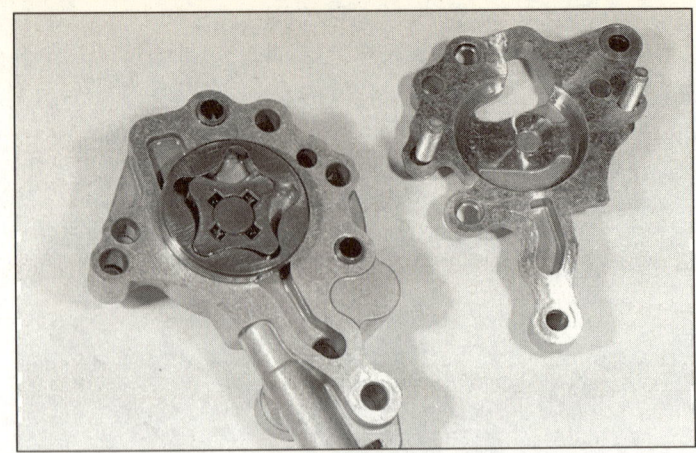

18.5b . . . then lift off the housing to expose one set of rotors

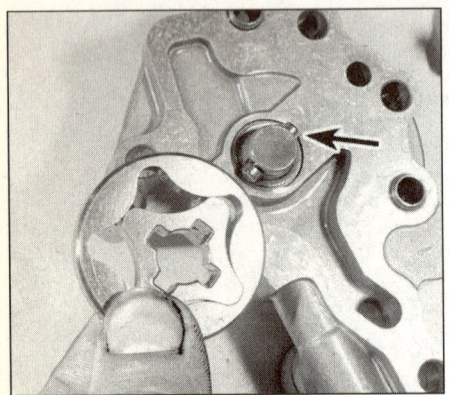

18.6a Push out the drive pin (arrow) . . .

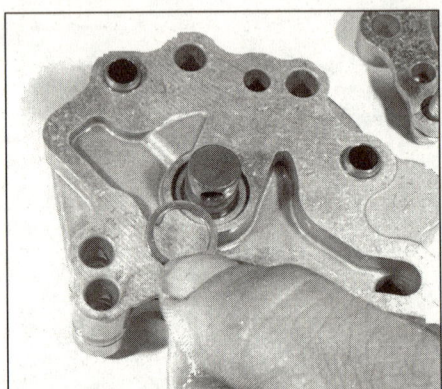

18.6b . . . and remove the washer . . .

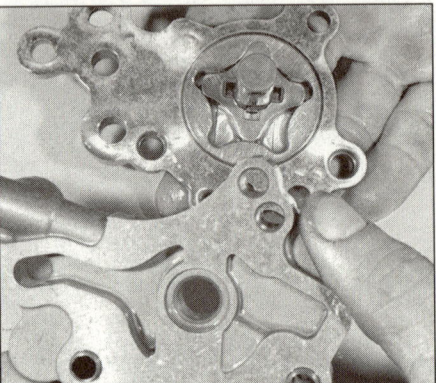

18.6c . . . then remove the remaining housing and rotors

Inspection

Refer to illustrations 18.5a, 18.5b, 18.6a, 18.6b, 18.6c, 18.6d, 18.8a, 18.8b, 18.9 and 18.11

4   The oil pump has an outer housing, an outer pair of rotors, a central spacer, an inner pair of rotors and an inner housing. A drive pin that passes through a hole in the shaft and fits in notches in the inner rotors drives the rotors. There's a separate drive pin for each set of rotors. The pump is held together by bolts.

5   Remove the screws or bolts and lift off the pump cover with its rotors **(see illustrations)**.

6   Pull the drive pins and remove the washer. Pull out the pump shaft, then remove the remaining set of rotors **(see illustrations)**.

7   Wash all the components in solvent, then dry them off. Check the pump body, the rotors, the drive gear and the covers for scoring and wear. If any damage or uneven or excessive wear is evident, replace the pump. If you are rebuilding the engine, it's a good idea to install a new oil pump.

8   Place the rotors in the pump body. Measure the clearance between the outer rotor and body, and between the inner and outer rotors, with a feeler gauge **(see illustrations)**. If any of the clearances are beyond the limits listed in this Chapter's Specifications, replace the pump.

9   Place a straightedge across the outer pump body and rotors and measure the gap with a feeler gauge **(see illustration)**. If the clearance is beyond the limits listed in this Chapter's Specifications, replace the pump.

10   To check the end clearance of the inner rotors and pump, you'll need some Plastigage. Place the rotors in the pump. Cut a strip of Plastigage, lay it across the rotors, then install the pump cover and tighten the screws or bolts. Remove the screws or bolts, lift off the cover, and measure the width of the crushed Plastigage with the scale on the envelope it comes in. If the end clearance is beyond the limit

listed in this Chapter's Specifications, replace the pump.

11   Remove the cotter pin, washer, spring and relief valve **(see illustration)**. If any of the parts are damaged or if the relief valve or its bore are scored, replace the pump. If you're planning to reuse the pump, use a new cotter pin when you install the relief valve.

12   Reassemble the pump by reversing the disassembly steps, with the following additions:

a) *Before installing the covers, pack the cavities between the rotors with petroleum jelly - this will ensure the pump develops suction quickly and begins oil circulation as soon as the engine is started.*

b) *Tighten the cover screws or bolts securely.*

18.6d Oil pump details

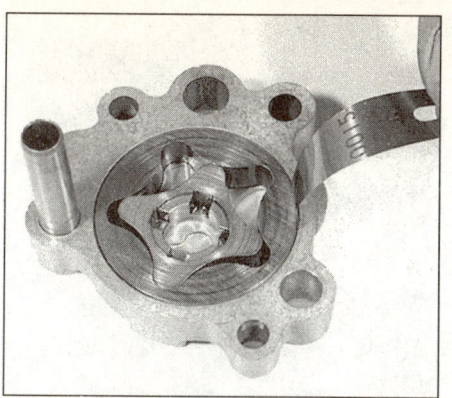

18.8a Measure the clearance between the outer rotor and body . . .

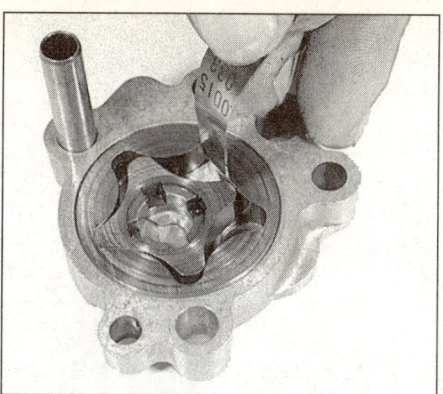

18.8b . . . between the inner and outer rotors . . .

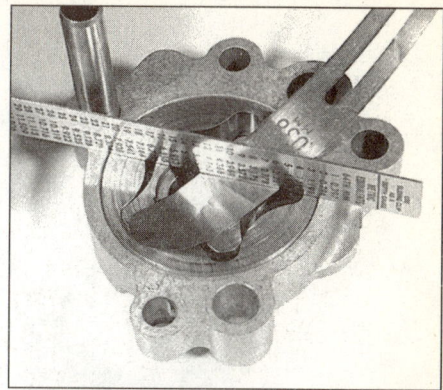

18.9 . . . and between the thin rotors and a straightedge laid across the pump body

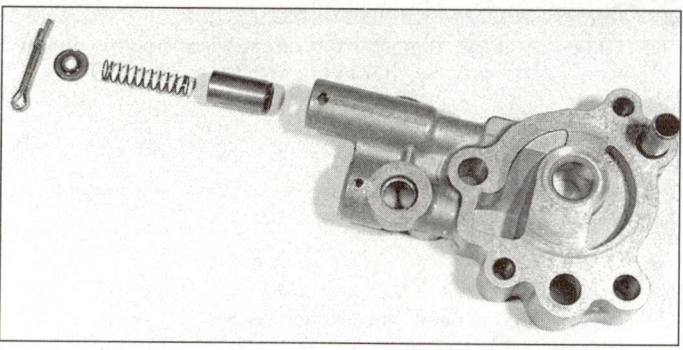

18.11 Remove the cotter pin, washer, spring and check valve

Installation

13   Installation is the reverse of removal, with the following additions:
  a) *Be sure you don't forget the shim and pump dowels.*
  b) *Install new O-rings on the outer body.*
  c) *Tighten the oil pump mounting screws or bolts securely, but don't overtighten them.*

---

19 Primary drive gear and camshaft sprocket - removal, inspection and installation

Removal

Refer to illustrations 19.2a, 19.2b and 19.2c
1   Remove the oil pump (Section 18).

19.2a Wedge a rag between the primary drive and driven gears (arrow) and loosen the locknut, then remove the clutch . . .

2   Wedge a rag between the teeth of the primary drive gear and the primary driven gear on the clutch housing. Remove the clutch (Section 17). Unscrew the primary drive gear locknut, then remove the lockwasher and the oil pump drive gear (**see illustrations**).
3   Slide the primary drive gear off the crankshaft.
4   If necessary, slide the camshaft sprocket off as well.

19.2b . . . unscrew the nut and slide off the lockwasher and the oil pump drive gear; on installation, the OUT SIDE mark on the lockwasher faces away from the engine

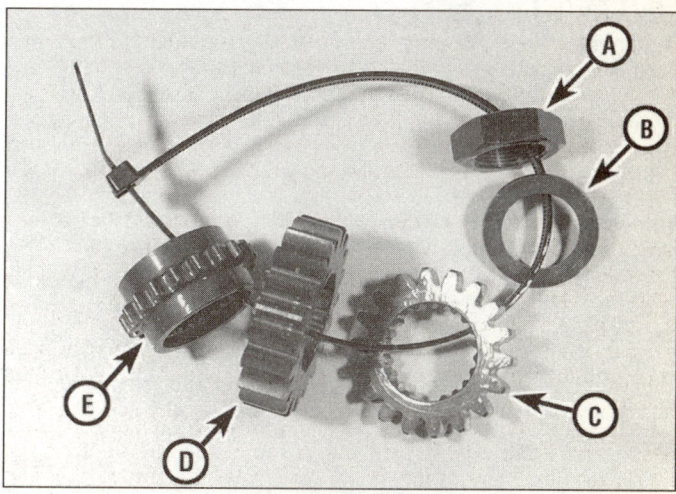

19.2c A tie wrap is a handy way to keep the parts together and in order

2B

19.6 The wide grooves (arrows) in the cam sprocket, primary drive gear and oil pump drive gear align with a wide spline on the crankshaft

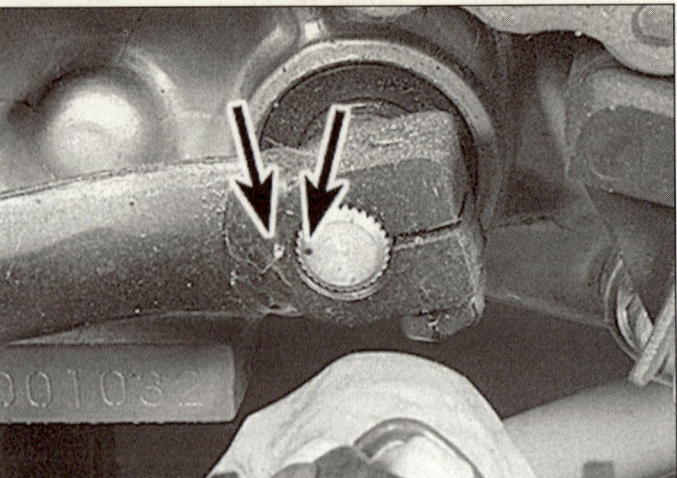

20.1 If you don't see a punch mark on the pedal and the end of the spindle, make your own (arrows)

Inspection

5   Check the primary drive gear and camshaft sprocket for obvious damage such as chipped or broken teeth. Replace the drive gear or cam sprocket if either is damaged or worn.

Installation

Refer to illustration 19.6

6   Installation is the reverse of the removal steps, with the following additions:

a) *The wide spline on the crankshaft aligns with a wide groove on the camshaft sprocket, primary drive gear, pulse generator rotor (if equipped) and oil pump drive gear so they can only be installed one way* **(see illustration)**.
b) *Install the pulse generator rotor (if equipped) and lockwasher with its OUT SIDE mark away from the engine.*
c) *Tighten the locknut to the torque listed in this Chapter's Specifications.*

---

20   External shift mechanism - removal, inspection and installation

Shift pedal

Removal

Refer to illustration 20.1

1   Look for alignment marks on the end of the shift pedal and shift shaft **(see illustration)**. If they aren't visible, make your own marks with a sharp punch.
2   Remove the shift pedal pinch bolt and slide the pedal off the shaft.

Inspection

Refer to illustration 20.4

3   Check the shift pedal for wear or damage such as bending. Check the splines on the shift pedal and shaft for stripping or step wear. Replace the pedal or shaft if these problems are found.
4   Check the shift shaft seal for signs of oil leakage **(see illustration)**. If it has been leaking, refer to Chapter 4 and remove the left engine cover. Pry the seal out of the cover and install a new one. You may be able to push the seal in with your thumbs; if not, tap it in with a hammer and block of wood or a socket the same diameter as the seal.

Installation

5   Line up the punch marks, install the shift pedal and tighten the pinch bolt.

External shift linkage

6   Shift linkage components accessible without splitting the crankcase include the stopper plate and cam plate.

Removal

Refer to illustrations 20.9a, 20.9b and 20.10

7   Remove the shift pedal as described above.
8   Remove the clutch (see Section 17).
9   Note how the stopper arm spring presses against the case and hooks around the stopper arm **(see illustration)**. Pull the stopper arm away from the shift drum cam, then loosen the stopper arm bolt and release the spring tension **(see illustration)**. Remove the bolt and take the stopper arm and spring off the crankcase.
10   Remove the bolt from the shift drum cam and take the cam off the drum **(see illustration)**.

Inspection

11   Check all parts for visible wear or damage and replace any parts that show problems.

Installation

12   Position the shift drum cam on the shift drum, aligning the hole in the back of the cam with the pin on the shift drum. Apply non-perma-

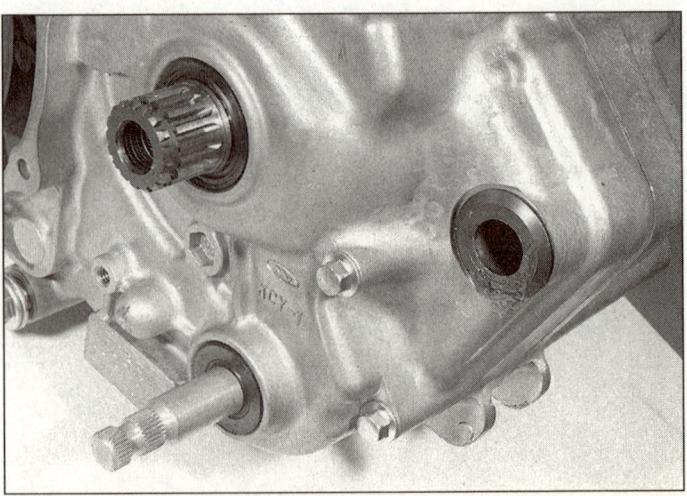

20.4 The smaller seal in the left engine cover is for the shift shaft; the other is for the transmission countershaft

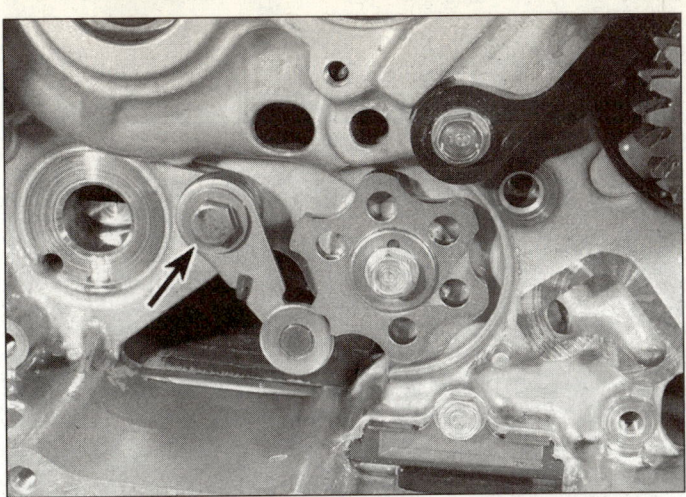

20.9a Note how the ends of the spring are positioned, then loosen the stopper arm bolt (the stopper arm pivots upward from below) . . .

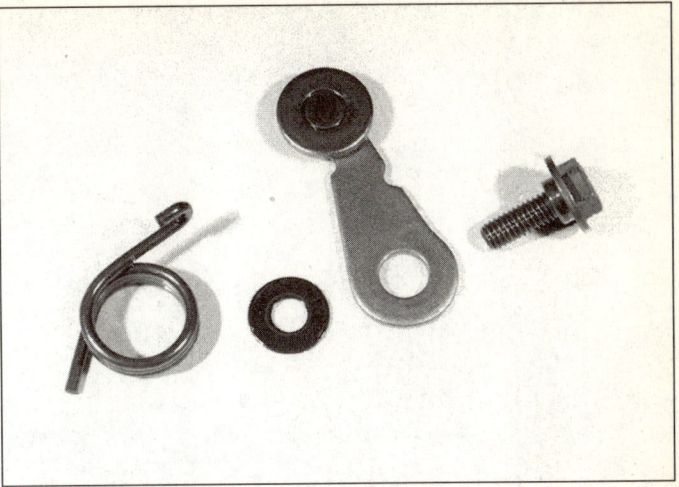

20.9b . . . release the spring tension and remove the bolt, stopper arm, washer and spring

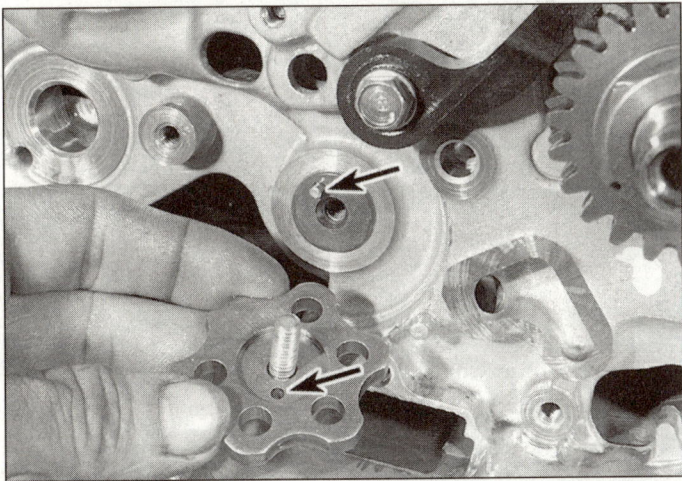

20.10 The hole in the shift drum cam (lower arrow) aligns with the pin in the shift drum (upper arrow) on installation

in the shift drum cam (see illustration 20.9a). Tighten the bolt to the torque listed in this Chapter's Specifications.

14   The remainder of installation is the reverse of the removal steps.

15   Check the engine oil level and add some, if necessary (see Chapter 1).

21   Oil cooler, tank and lines - removal and installation

1   Drain the engine oil (see Chapter 1).

Oil lines

Refer to illustrations 21.3a, 21.3b, 21.3c, 21.3d, 21.4a and 21.4b

2   There are two oil lines connected to the engine at the lower right front corner of the right crankcase cover. The shorter (lower) line is the inlet line from the oil tank to the engine. The longer line returns oil from the engine to the oil cooler. A line routes oil from the oil cooler back to the oil tank.

3   Unbolt the hose mounting plates at the crankcase (see illustration). Pull out the two hoses and their O-rings (see illustration). At the oil tank, disconnect the oil inlet hose from the tank (see illustration). At the right end of the oil cooler, unbolt the return hose from the cooler (see illustration).

4   To remove the third oil line, unbolt it from the oil cooler and unscrew it from the oil tank (see illustrations).

5   Installation is the reverse of removal. Be sure to use new O-rings at the fittings and tighten all bolts and nuts securely.

nent thread locking agent to the threads of the bolt, then tighten it to the torque listed in this Chapter's Specifications.

13   Position the spring on the stopper arm, then install the stopper arm on the engine and tighten its bolt loosely (see illustration 20.9b). Pull up the stopper arm and engage its roller end with the neutral notch

2B

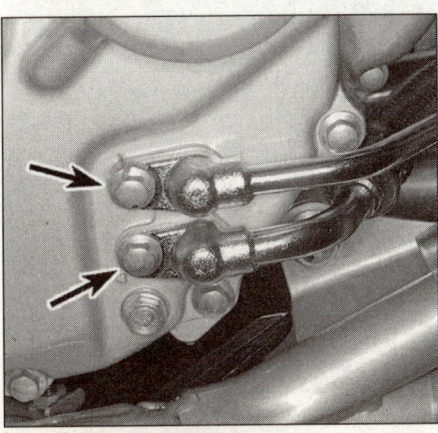

21.3a Remove the oil line mounting plate bolts (arrows) . . .

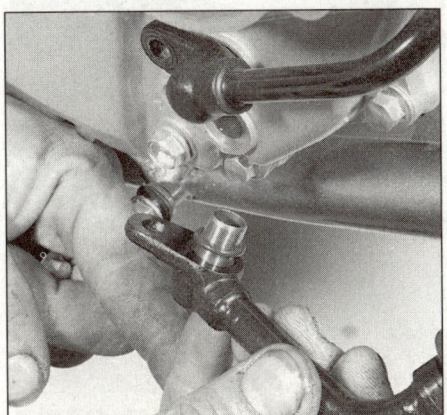

21.3b . . . and pull the fittings and their O-rings out of the engine

21.3c Place a pan beneath the fitting (arrow) for the inlet line, then unscrew the fitting

21.3d Remove the bolts (arrows) to detach the return line from the right end of the oil cooler; discard the old O-ring and replace it with a new one

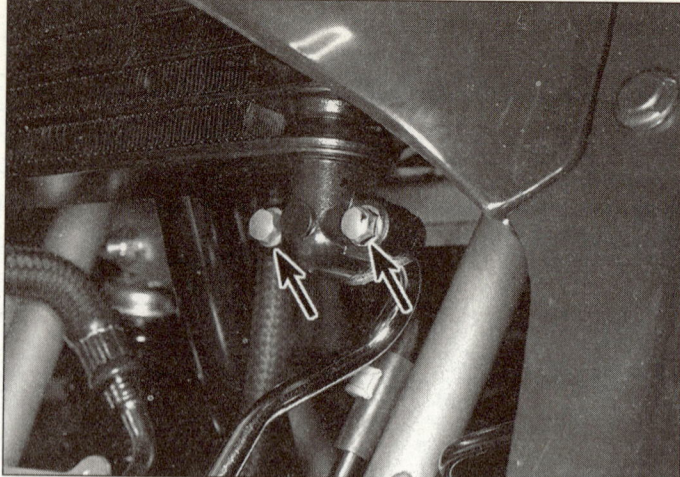

21.4a To detach the line between the oil cooler and the oil tank, remove these two bolts (arrows), discard the old O-ring . . .

Oil tank

Refer to illustration 21.7

6   Disconnect the two oil lines and the oil tank breather hose from the oil tank **(see illustrations 21.3c and 21.4b)**.
7   To detach the oil tank from the frame, remove the two mounting bolts, one on each side **(see illustration)**.
8   Installation is the reverse of removal.

Oil cooler

Refer to illustration 21.10

9   Disconnect both oil lines from the oil cooler **(see illustrations 21.3d and 21.4a)**.
10   Remove the mounting bolt and grommet **(see illustration)** from each end of the oil cooler and remove the cooler.
11   Inspect the grommets for wear and deterioration and replace as needed.
12   Installation is the reverse of removal. Tighten the bolts securely, but don't overtighten them and strip the threads.

22 Crankcase - disassembly and reassembly

1   The crankcase must be split into two parts in order to examine and repair or replace the crankshaft, connecting rod, balancer, bearings and transmission components.

Disassembly

Refer to illustrations 22.11a, 22.11b, 22.11c and 22.12

2   Remove the carburetor and exhaust system (see Chapter 3).
3   Remove the engine (see Section 5).
4   Remove the cylinder head cover, the cam chain tensioner, the camshaft, the cylinder head, the cylinder, the cam chain, the chain guides and the piston (see Sections 7, 8, 9, 10, 13 and 14, respectively).
5   Remove the alternator rotor (see Chapter 8).
6   Remove the clutch (see Section 17).
7   Remove the oil pump (see Section 18).
8   Remove the primary drive gear and camshaft sprocket (see Section 19).
9   Remove the external shift mechanism (see Section 20).
10   Carefully inspect the crankcase to make sure there are no other components attaching the halves of the crankcase together.
11   Loosen the crankcase bolts in the left side of the crankcase evenly in two or three stages, then remove them **(see illustration)**. Remove the bolt(s) from the other side of the crankcase **(see illustrations)**.
12   Place the crankcase with its left side down on a pair of wood blocks so the shift pedal shaft and transmission countershaft shaft can extend downward. Carefully tap the crankcase apart and lift the right half off the left half **(see illustration)**. Don't pry against the mating surfaces or they'll develop leaks.

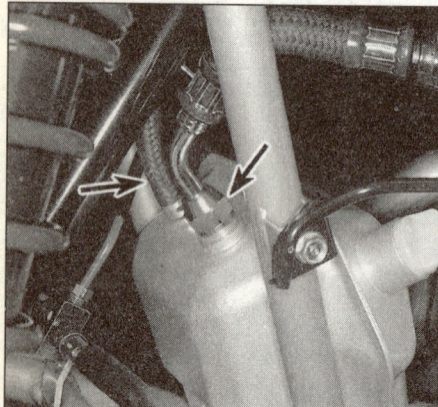

21.4b . . . and unscrew this fitting (right arrow) from the oil tank; if you plan to remove the oil tank, also disconnect the breather hose (left arrow) from the tank

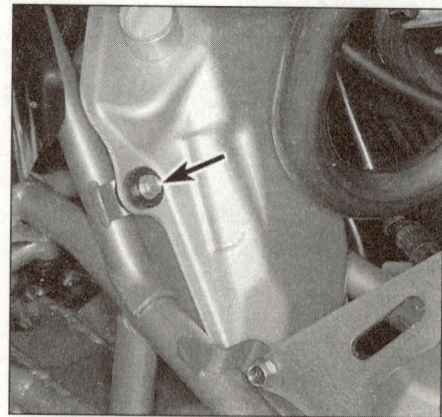

21.7 To detach the oil tank from the frame, remove the left mounting bolt (arrow) and the right bolt (other side of the tank)

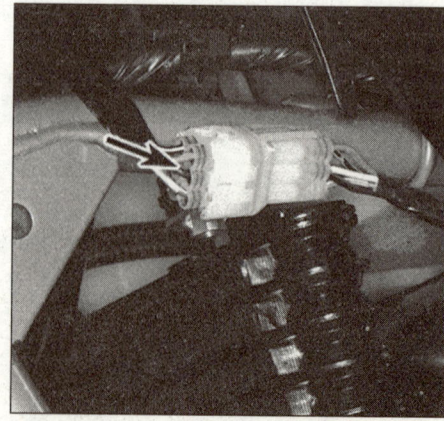

21.10 To detach the oil cooler, remove the left mounting bolt (arrow) and the right bolt (other end of cooler) (the white connector is for the handlebar switches)

22.11a Remove the bolts (arrows) from the left side of the crankcase ...

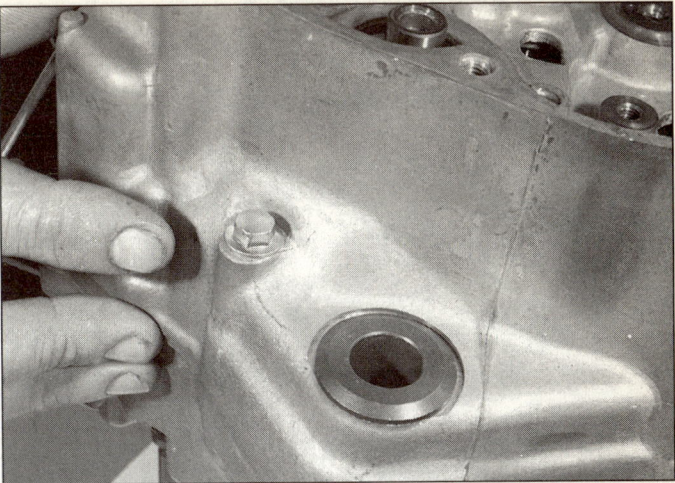

22.11b ... remove the bolt from the right side of the case near the rear of the engine ...

13 Locate the two crankcase dowels (see illustration 22.12).
14 Refer to Sections 24 through 27 for information on the internal components of the crankcase.

Reassembly

15 Remove all traces of old gasket and sealant from the crankcase mating surfaces with a sharpening stone or similar tool. Be careful not to let any fall into the case as this is done and be careful not to damage the mating surfaces.
16 Check to make sure the two dowel pins are in place in their holes in the mating surface of the left crankcase half (see illustration 22.12).
17 Pour some engine oil over the transmission and balancer gears, the right crankshaft bearing and the shift drum. Don't get any oil on the crankcase mating surface.
18 Coat the crankcase mating surface with Gaskacinch or equivalent sealant, then install a new gasket on the crankcase mating surface (see illustration 22.12). Cut out the portion of the gasket that crosses the cylinder opening.
19 Carefully place the right crankcase half onto the left crankcase half. While doing this, make sure the transmission shafts, shift drum, crankshaft and balancer fit into their bearings in the right crankcase half.
20 Install the crankcase bolts and tighten them so they are just snug. Then tighten them evenly in two or three stages to the torque listed in

this Chapter's Specifications.
21 Turn the transmission mainshaft to make sure it turns freely. Also make sure the crankshaft turns freely.
22 The remainder of assembly is the reverse of disassembly.

23 Crankcase components - inspection and servicing

Refer to illustrations 23.3a, 23.3b and 23.3c
1 Separate the crankcase and remove the following:
a) *Balancer*
b) *Transmission shafts and gears*
c) *Crankshaft*
d) *Shift drum and forks*
2 Clean the crankcase halves thoroughly with new solvent and dry them with compressed air. All oil passages should be blown out with compressed air and all traces of old gasket should be removed from the mating surfaces. **Caution:** *Be very careful not to nick or gouge the crankcase mating surfaces or leaks will result.* Check both crankcase halves very carefully for cracks and other damage.
3 Inspect the bearings in the case halves (see illustrations). For details of the transmission and balancer bearings in the left side of the crankcase, see Sections 25 and 26. If the bearings don't turn smoothly, replace them. For bearings that aren't accessible from the

2B

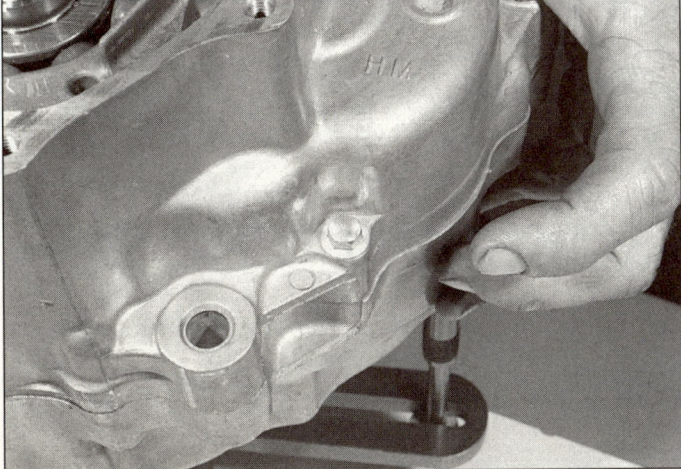

22.11c ... and remove the bolt from the right side near the front

22.12 Support the crankcase on blocks so it isn't resting on the shift or transmission shafts, then lift the right case half off the left half; the two dowels may stay in either side of the case (arrows)

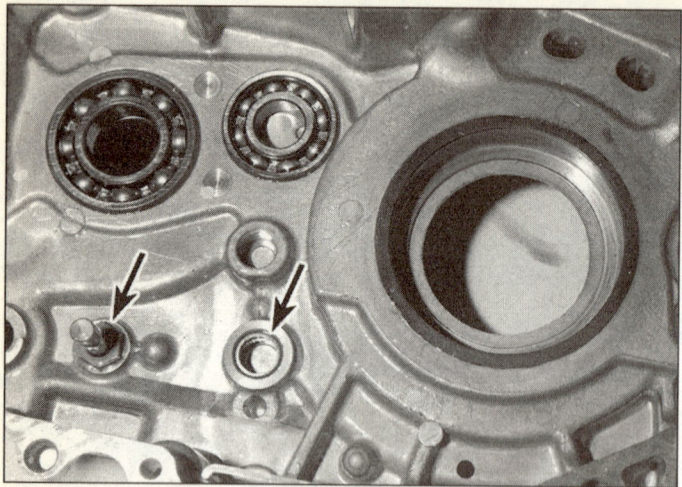

23.3a Check the case bearings for roughness, looseness or noise; don't forget the shift drum needle bearing (right arrow); the return spring pin (left arrow) should be tight

23.3b A retainer secures one transmission bearing in the right case half

23.3c A blind hole puller like this one is needed to remove bearings which are only accessible from one side

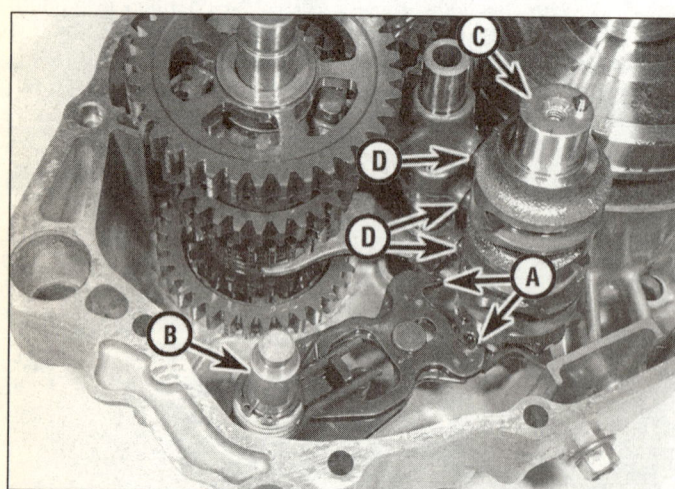

24.2 Pull the shift shaft pawls clear of the shift drum and pull the shift shaft out of the crankcase

A Shift pawls	C Shift drum
B Shift shaft	D Shift fork pins

outside, a blind hole puller will be needed for removal **(see illustration)**. Drive the remaining bearing out with a bearing driver or a socket having an outside diameter slightly smaller than that of the bearing outer race. Before installing the bearings, allow them to sit in the freezer overnight, and about fifteen-minutes before installation, place the case half in an oven, set to about 200-degrees F, and allow it to heat up. The bearings are an interference fit, and this will ease installation. **Warning:** *Before heating the case, wash it thoroughly with soap and water so no explosive fumes are present. Also, don't use a flame to heat the case.* Install the ball bearings with a socket or bearing driver that contacts the bearing outer race.

4   If any damage is found that can't be repaired, replace the crankcase halves as a set.

5   Assemble the case halves (see Section 22) and check to make sure the crankshaft and the transmission shafts turn freely.

24   Internal shift mechanism - removal, inspection and installation

1   Refer to Section 23 and separate the crankcase halves.

Removal

Refer to illustrations 24.2 and 24.4

2   Pull the pawls of the gearshift plate back against spring pressure until they clear the shift drum, then lift the shift shaft out of the crankcase **(see illustration)**.

3   Lift the shift drum out of the case, disengaging it from the pins on the shift forks.

4   Pull up on the shift rod until it clears the case, then move the rod and forks away from the gears **(see illustration)**.

Inspection

Refer to illustrations 24.6a, 24.6b, 24.9a, 24.9b and 24.11

5   Wash all of the components in clean solvent and dry them off.

6   Check the shift shaft for bends and damage to the splines **(see illustration)**. If the shaft is bent, you can attempt to straighten it, but if the splines are damaged it will have to be replaced. Check the condition of the gearshift plate and the pawl spring **(see illustration)**. Replace them if they're worn, cracked or distorted.

7   Make sure the return spring pin isn't loose **(see illustration 23.3a)**. If it is, unscrew it, apply a non-hardening locking compound to the threads, reinstall it and tighten it securely.

8   Inspect the shift fork grooves in the gears. If a groove is worn or scored, replace the affected gear (see Section 25) and inspect its corresponding shift fork.

9   Check the shift forks for distortion and wear, especially at the fork

24.4 The shift forks are labeled R, C and L (right, center and left) - the letters face the right side of the crankcase when the forks are installed

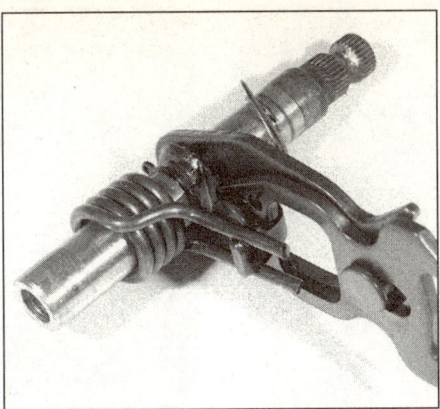

24.6a The return spring is installed like this . . .

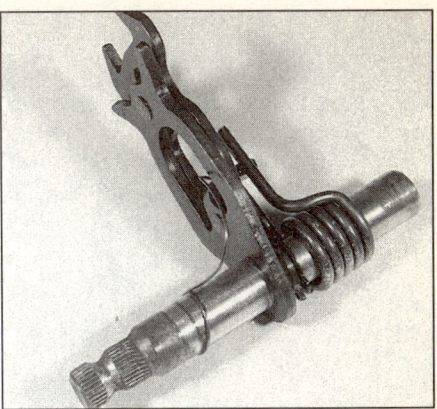

24.6b . . . and the pawl spring is installed like this

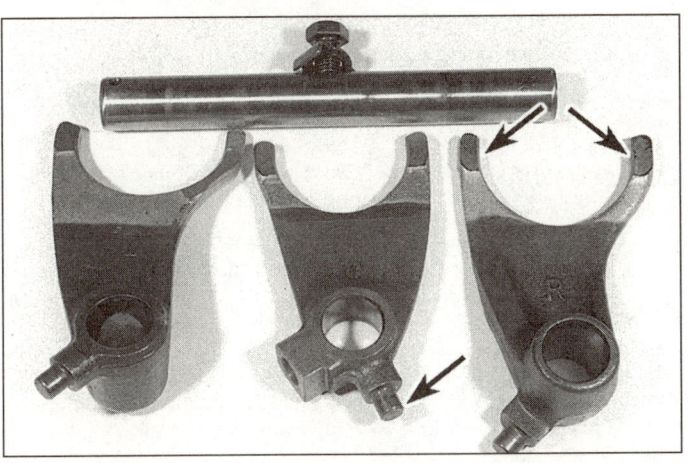

24.9a The fork ears and pins (arrows) are common wear points

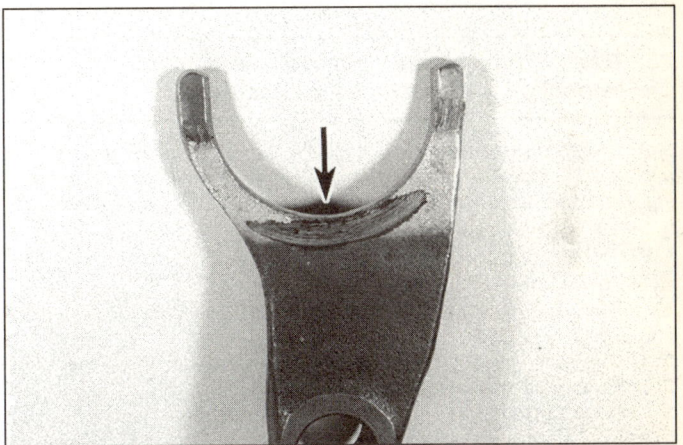

24.9b An arc-shaped burn mark like this means the fork was rubbing against a gear, probably due to bending or worn fork ears

2B

ears **(see illustrations)**. Measure the thickness of the fork ears and compare your findings with this Chapter's Specifications. If they are discolored or severely worn they are probably bent. Inspect the guide pins for excessive wear and distortion and replace any defective parts with new ones.

10 Measure the inside diameter of the forks and the outside diameter of the fork shaft and compare to the values listed in this Chapter's Specifications. Replace any parts that are worn beyond the limits. Check the shift fork shaft for evidence of wear, galling and other damage. Make sure the shift forks move smoothly on the shaft. If the shaft is worn or bent, replace it with a new one.

11 Check the edges of the grooves in the drum for signs of excessive wear **(see illustration)**.

12 Spin the shift drum bearing with fingers and replace it if it's rough, loose or noisy.

Installation

Refer to illustration 24.13

13 Pull back the pawls of the gearshift plate, slide the shaft into its bore and release the pawls. Make sure the return spring fits over the post and the pawls engage the shift drum pins **(see illustration)**.

14 Installation is the reverse of the removal steps. Refer to the identifying letters on the forks and make sure they're installed in the correct positions.

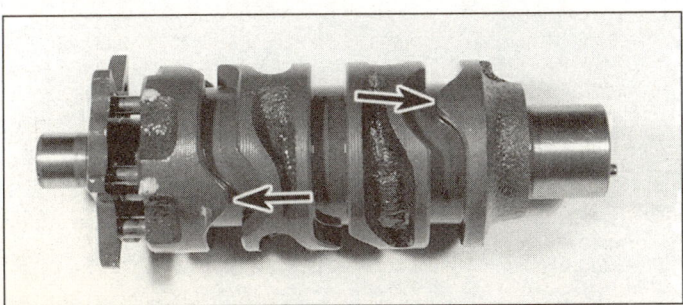

24.11 Check the shift drum grooves for wear, especially at the points

24.13 One end of the return spring should be on each side of the post and the pawls should engage the shift drum

25.4a On the right side, the countershaft has a
thrust washer (arrow) . . .

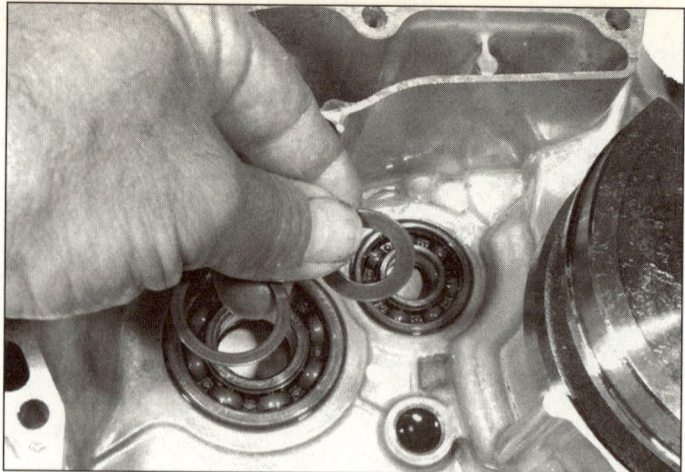

25.4b . . . and both shafts have a thrust washer on the left side

25 Transmission shafts - removal, disassembly, inspection, assembly and installation

Note: *When disassembling the transmission shafts, place the parts on a long rod or thread a wire through them to keep them in order and facing the proper direction.*

Removal

Refer to illustrations 25.4a and 25.4b

1   Remove the engine, then separate the case halves (see Sections 5 and 23).
2   The transmission components remain in the left case half when the case is separated.
3   Refer to Section 25 and remove the shift shaft, shift drum and forks.
4   Take the thrust washer off the countershaft (**see illustration**). Lift the transmission shafts out of the case together, then remove the thrust washers from the case (**see illustration**).

5   Separate the shafts once they're lifted out. If you're not planning to disassemble them right away, reinstall the removed components and place a large rubber band over both ends of each shaft so the gears won't slide off.

Disassembly

Refer to illustrations 25.6a and 25.6b

6   To disassemble the shafts, remove the snap-rings and slide the gears, bushings and thrust washers off (**see illustrations**).

Inspection

Refer to illustration 25.10

7   Wash all of the components in clean solvent and dry them off.
8   Inspect the shift fork grooves in gears so equipped. If a groove is worn or scored, replace the affected gear and inspect its corresponding shift fork.
9   Check the gear teeth for cracking and other obvious damage. Check the bushing or surface in the inner diameter of the freewheeling gears for scoring or heat discoloration. Measure the inside diameters of

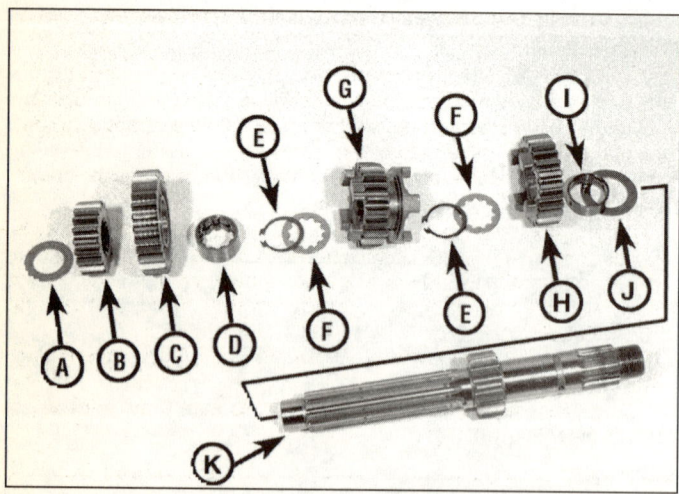

25.6a Mainshaft details

A	Thrust washer	F	Splined washer
B	Second gear	G	Third gear
C	Fifth gear	H	Fourth gear
D	Splined bushing	I	Bushing
E	Snap-ring	J	Thrust washer

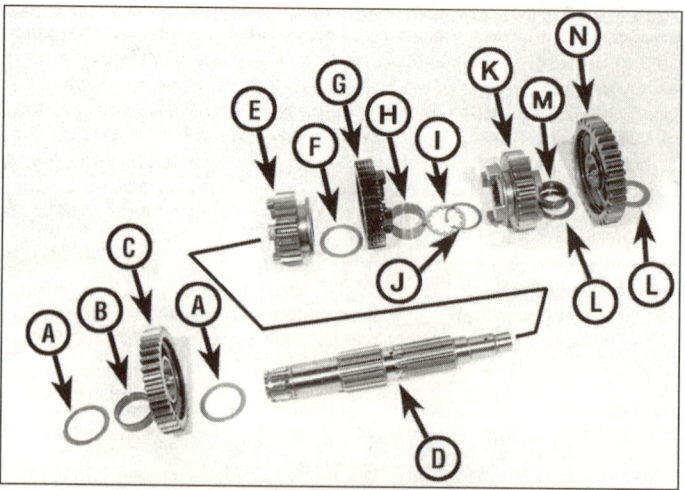

25.6b Countershaft details

A	Thrust washer	H	Bushing
B	Bushing	I	Splined washer
C	Second gear	J	Snap-ring
D	Countershaft	K	Fourth gear
E	Fifth gear	L	Thrust washer
F	Thrust washer	M	Bushing
G	Third gear	N	First gear

25.10 Check the slots (left arrow) and dogs (right arrow) for wear, especially at the edges; rounded corners cause the transmission to jump out of gear - new gears (bottom) have sharp corners

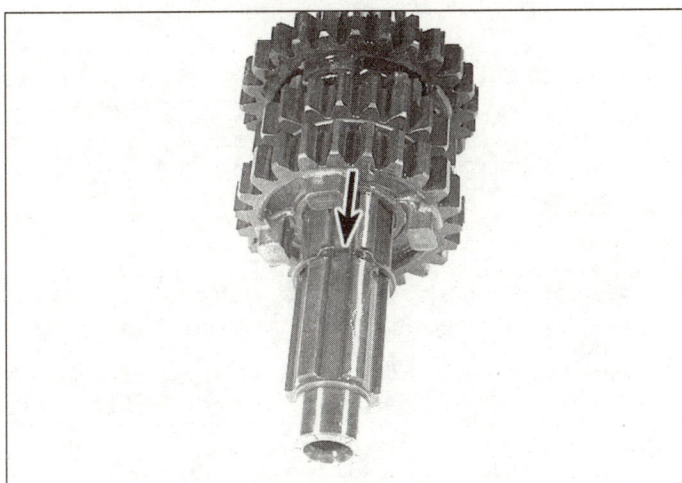

25.18 Install the snap-rings with their rounded sides away from the gears they retain - after installation, rotate them to make sure they're secure in their grooves, then center the snap-ring gap on a spline groove (arrow)

25.20 The assembled shafts and gears should look like this

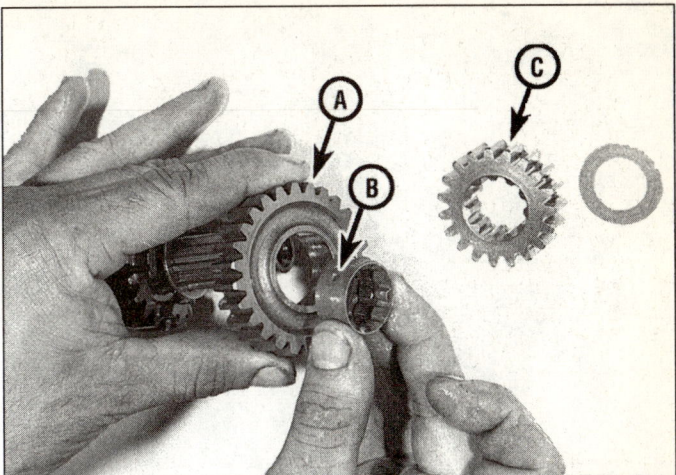

25.17 Align the oil hole in the mainshaft fifth gear splined bushing with the oil hole in the shaft

A   Mainshaft fifth gear   C   Mainshaft second gear
B   Bushing

the gears and compare them to the values listed in this Chapter's Specifications. Replace parts that are damaged or worn beyond the limits.

10   Inspect the engagement dogs and dog holes on gears so equipped for excessive wear or rounding off **(see illustration)**. Replace the paired gears as a set if necessary.

11   Measure the transmission shaft diameters at the points listed in this Chapter's Specifications. If they're worn beyond the limits, replace the shaft(s).

12   Measure the inner and outer diameters of the gear bushings and replace any that are worn beyond the limit listed in this Chapter's Specifications.

13   Inspect the thrust washers. Honda doesn't specify wear limits, but they should be replaced if they show any visible wear or scoring. It's a good idea to replace them whenever the transmission is disassembled.

14   Check the transmission shaft bearings in the crankcase for roughness, looseness or noise and replace them if necessary.

15   Discard the snap-rings and use new ones on reassembly.

Assembly and installation

Refer to illustrations 25.17, 25.18 and 25.20

16   Assembly and installation are the reverse of the removal procedure, but take note of the following points:

17   Align the oil hole in the mainshaft fifth gear splined bushing with the oil hole in the shaft **(see illustration)**.

18   Make sure the snap-rings are securely seated in their grooves, with their sharp sides facing the direction of thrust (toward the gears they hold on the shafts). The ends of the snap-rings must fit in raised splines, so the gap in the snap-ring aligns with a spline groove **(see illustration)**.

19   Lubricate the components with engine oil before assembling them.

20   After assembly, check the gears to make sure they're installed correctly **(see illustration)**.

26   Crankshaft and balancer - removal, inspection and installation

Balancer

Removal

Refer to illustration 26.2

1   Remove the engine and separate the crankcase halves (Sections 5 and 22).

2B

26.2 The timing marks (arrows) on the balancer and crankshaft must be aligned to prevent severe engine vibration

26.5a Press the crankshaft out of the crankcase . . .

26.5b . . . you can also use a puller if you have the correct adapters

26.5c If the bearing stays on the crankshaft, remove it with a puller attached to a bearing splitter

2   Turn the crankshaft so the punch marks on the balancer and crankshaft are aligned **(see illustration)**. The marks must be aligned like this on installation to prevent severe engine vibration.
3   Lift the balancer shaft out of its bearing.

Inspection

4   Inspect the balancer gear teeth and its bearing surface for wear or damage. Replace the balancer if problems can be seen in these areas.

Crankshaft

Removal

Refer to illustrations 26.5a, 26.5b and 26.5c

Note: *Removal and installation of the crankshaft require a press and some special tools. If you don't have the necessary equipment or suitable substitutes, have the crankshaft removed and installed by a Honda dealer.*

5   Place the left crankcase half in a press and push the crankshaft out **(see illustration)**, or remove it with a three-legged puller **(see illustration)**. The ball bearing may remain in the crankcase or come out with the crankshaft. If it stays on the crankshaft, remove it with a bearing splitter **(see illustration)**. Discard the bearing, no matter what its apparent condition, and use a new one on installation.

Inspection

6   Refer to Chapter 2A for the crankshaft inspection procedures.

Installation

Refer to illustrations 26.7, 26.9 and 26.10

7   Align the index marks on the left end of the balancer and crankshaft, then position them simultaneously in the left crankcase half **(see illustration)**.

26.7 Align the crankshaft and balancer marks on the left end before installing them

26.9 Thread the adapter into the end of the crankshaft . . .

26.10 . . . and attach the puller to the adapter

8   Verify that the alignment marks on the right end of the balancer and crankshaft are still aligned **(see illustration 26.2)**.
9   Thread a puller adapter into the end of the crankshaft **(see illustration)**.
10   Install the crankshaft puller and collar on the end of the crankshaft **(see illustration)**.
11   Hold the puller shaft with one wrench and turn the nut with another wrench to pull the crankshaft into the center race of the ball bearing.
12   Remove the special tools from the crankshaft.
13   Make sure the crankshaft and balancer timing marks are still aligned **(see illustration 26.2)**. **Note:** *It's important to align the timing marks exactly throughout this procedure. Severe engine vibration will occur if the crankshaft and balancer are out of time.*
17   Installation is the reverse of the removal steps.

27   Initial start-up after overhaul

1   Make sure the engine oil level is correct, then remove the spark plug from the engine. Unplug the primary wires from the coil.
2   Crank the engine over several times to build up oil pressure. Reinstall the spark plug and connect the wires.
3   Make sure there is fuel in the tank, then operate the choke.
4   Start the engine and allow it to run at a moderately fast idle. Let the engine continue running until it reaches operating temperature.
5   Check carefully for oil leaks and make sure the transmission and controls, especially the brakes, function properly before road testing the machine. Refer to Section 28 for the recommended break-in procedure.
6   Upon completion of the road test, and after the engine has cooled down completely, recheck the oil level and valve clearances (see Chapter 1).

28   Recommended break-in procedure

1   Any rebuilt engine needs time to break-in, even if parts have been installed in their original locations. For this reason, treat the machine gently for the first few miles to make sure oil has circulated throughout the engine and any new parts installed have started to seat.
2   Even greater care is necessary if the cylinder has been rebored or a new crankshaft has been installed. In the case of a rebore, the engine will have to be broken in as if the machine were new. This means greater use of the transmission and a restraining hand on the throttle for the first few operating days. There's no point in keeping to any set speed limit - the main idea is to vary the engine speed, keep from lugging the engine and to avoid full-throttle operation. These recommendations can be lessened to an extent when only a new crankshaft is installed. Experience is the best guide, since it's easy to tell when an engine is running freely.
3   If a lubrication failure is suspected, stop the engine immediately and try to find the cause. If an engine is run without oil, even for a short period of time, irreparable damage will occur.

2B

Notes

Chapter 3
Fuel and exhaust systems

Contents

Specifications

General

Fuel type	Unleaded gasoline; minimum 87 pump octane (91 RON octane)
Idle speed	1300 to 1500 rpm
Throttle lever freeplay	3 to 8 mm (1/8 to 5/16 inch)

Carburetor

Identification number

TRX300EX	QB02A
TRX400EX	QB10A

Jet sizes and settings

Main jet

TRX300EX	
Standard	122
High altitude	120
TRX400EX	
Standard	148
High altitude	142
Jet needle clip position	Third groove from top

Slow jet

TRX300EX	38
TRX400EX	38

Pilot screw adjustment

Initial setting	2-1/4 turns out from lightly seated position
High altitude adjustment	
TRX300EX	1/ 8-turn in from initial opening
TRX400EX	1 turn in from initial opening
Float level	18.5 mm (47/64-inch)

3

Torque settings

Exhaust system

  TRX300EX

    Heat shield bolts .... 6 Nm (144 in-lbs)

    Muffler-to-exhaust pipe clamp bolt .... 23 Nm (17 ft-lbs)

    Muffler mounting bolts .... 55 Nm (40 ft-lbs)

  TRX400EX

    Heat shield bolts .... 20 Nm (168 in-lbs)

    Muffler-to-exhaust pipe clamp bolt .... 23 Nm (17 ft-lbs)

    Muffler mounting bolts .... 32 Nm (24 ft-lbs)

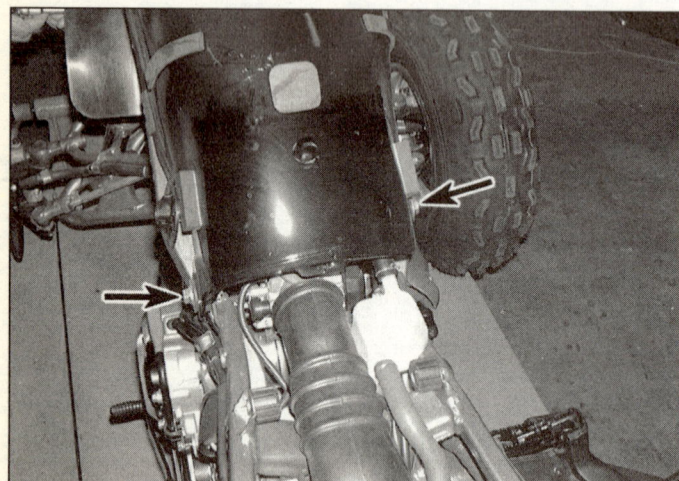

2.2a To detach the rear of a TRX300EX fuel tank from the frame, remove the bolt (arrows) from each side

2.2b To detach the front of a TRX400EX fuel tank from the frame, remove these two bolts (arrows)

1   General information

The fuel system consists of the air cleaner assembly, the fuel tank, the fuel tap and filter screen, the carburetor and the connecting lines, hoses and control cables.

The air cleaner assembly, which is located under the seat, is a plastic housing with a removable, reusable air filter element inside.

The carburetor used on these vehicles is a piston-valve type (or slide type). With this design, there is no separate butterfly-type throttle valve; the piston is the throttle valve. For cold starting, a choke lever on the carburetor activates a butterfly-type choke plate inside the mouth (the air cleaner side) of the carburetor.

The exhaust system consists of the exhaust pipe and a muffler equipped with a removable spark arrester.

Some of the fuel system service procedures are considered routine maintenance items and, for that reason, are in Chapter 1 instead of this Chapter.

2   Fuel tank - removal and installation

Refer to illustrations 2.2a, 2.2b, 2.3 and 2.4

Warning: *Gasoline is extremely flammable, so take extra precautions when you work on any part of the fuel system. Don't smoke or allow open flames or bare light bulbs near the work area, and don't work in a garage where a natural gas-type appliance (such as a water heater or clothes dryer) with a pilot light is present. Since gasoline is carcino-genic, wear latex gloves when there's a possibility of being exposed to fuel, and, if you spill any fuel on your skin, rinse it off immediately with soap and water. Mop up any spills immediately and do not store fuel-soaked rags where they could ignite. When you perform any kind of work on the fuel system, wear safety glasses and have a Class B type fire extinguisher on hand.*

1   Remove the seat and rear fenders and the front fender (see Chapter 7).

2   Remove the fuel tank mounting bolts **(see illustrations)**.

3   On TRX400EX models, remove the rubber retaining straps that secure the rear of the tank to the frame **(see illustration)**.

4   Raise the rear of the tank and either disconnect the fuel line from the fuel tap **(see illustration 2.3)** or disconnect the line from the carburetor **(see illustration)**.

5   Pull the fuel tank backward and lift it off the vehicle.

6   Installation is the reverse of removal.

3   Fuel tank - cleaning and repair

1   All repairs to the fuel tank should be done by a professional with experience in this critical and potentially dangerous work. Even after cleaning and flushing of the fuel system, explosive fumes can remain and ignite during repair of the tank.

2   If the fuel tank is removed from the vehicle, it should not be placed in an area where sparks or open flames could ignite the fumes coming out of the tank. Be especially careful inside garages where a natural gas-type appliance is located, because the pilot light could cause an explosion.

2.3 To detach the rear of a TRX400EX fuel tank from the frame, remove the rubber retaining strap (arrow) from each side (left side shown), then lift the tank and disconnect the fuel line

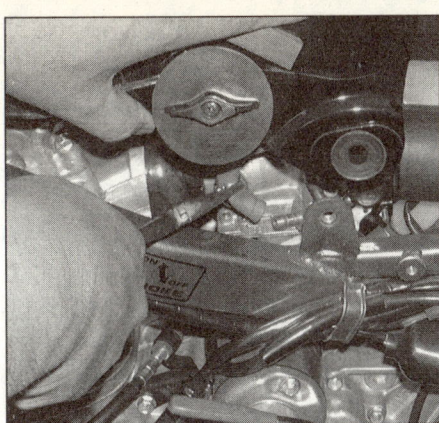

2.4 On TRX300EX models, it's a little easier to disconnect the fuel line from the carburetor instead of the fuel tap

4.3 Turning the pilot screw (arrow) in or out adjusts the idle air/fuel mixture (TRX300EX shown; TRX400EX similar)

4 Idle air/fuel mixture adjustment

Idle drop procedure (normal adjustment)

Refer to illustration 4.3

1   Idle fuel/air mixture on these vehicles is preset at the factory and should not need adjustment unless the carburetor is overhauled or the pilot screw, which controls the mixture adjustment, is replaced.

2   The engine must be in a good state of tune before adjusting the idle air/fuel mixture (valve clearances set to specifications, spark plug in good condition and correctly gapped).

3   To make the initial adjustment, turn the pilot screw **(see illustration)** in until it seats lightly, then back it out the number of turns listed in this Chapter's Specifications. **Caution:** *Turn the screw just far enough to seat it lightly. Do not bottom the screw tightly, or you will damage the screw or its seat, either of which will make accurate mixture adjustment impossible.*

4   Warm up the engine to its normal operating temperature. Shut it off and connect a tachometer, following the tachometer manufacturer's instructions.

5   Restart the engine and compare the idle speed to the value listed in the Chapter 1 Specifications. Adjust it if necessary, using the throttle stop screw (see Section 19 in Chapter 1).

6   Slowly turn the pilot screw in or out to achieve the highest idle speed.

7   Lightly blip the throttle two or three times, then adjust the idle speed with the throttle stop screw again.

8   Gradually turn the pilot screw in until engine speed drops by 100 rpm (TRX300EX) or 50 rpm (TRX400EX).

9   On TRX400EX models only, back out the pilot screw one turn from its position in the previous step.

10   Readjust the idle speed with the throttle stop screw.

High altitude adjustment

11   If the vehicle is normally used at altitudes from sea level to 5000 feet (1500 meters), use the normal main jet size and pilot screw setting. If it's used at altitudes between 3000 and 8000 feet (1000 and 2500 meters), the main jet and pilot screw setting must be changed to compensate for the thinner air. **Caution:** *Don't use the vehicle for sustained operation below 5000 feet (15 meters) with the main jet and pilot screw at the high altitude settings or the engine may overheat and be damaged.*

12   Change the standard main jet (see Section 7) to the high altitude main jet listed in this Chapter's Specifications.

13   Set the pilot screw to the standard setting, then turn it in the additional amount listed in this Chapter's Specifications.

14   Once the vehicle is at high altitude, adjust the idle speed with the throttle stop screw (see Section 19 in Chapter 1).

5 Carburetor overhaul - general information

1   Poor engine performance, hesitation, hard starting, stalling, flooding and backfiring are all signs that major carburetor maintenance may be required.

2   Keep in mind that many so-called carburetor problems are really not carburetor problems at all, but are ignition system malfunctions or mechanical problems within the engine. Try to establish for certain that the carburetor is in need of maintenance before beginning a major overhaul.

3   Before assuming that a carburetor overhaul necessary, inspect the fuel tap and strainer, the fuel lines, the insulator clamp, the O-ring between the insulator and the cylinder head, the vacuum hoses, the air filter element, the cylinder compression, the spark plug and the ignition timing. If the vehicle has been unused for more than a month, refer to Chapter 1, drain the float chamber and refill the tank with fresh fuel.

4   Most carburetor problems are caused by dirt particles, varnish and other deposits that build up in and block the fuel and air passages. Also, in time, gaskets and O-rings shrink or deteriorate and cause fuel and air leaks which lead to poor performance.

5   When the carburetor is overhauled, it is generally disassembled completely and the parts are cleaned thoroughly with a carburetor cleaning solvent and dried with filtered, unlubricated, compressed air. The fuel and air passages are also blown through with compressed air to force out any dirt that may have been loosened but not removed by the solvent. Once the cleaning process is complete, the carburetor is reassembled using new gaskets, O-rings and, generally, a new inlet needle valve and seat.

6   Before disassembling the carburetors, make sure you have a carburetor rebuild kit (which will include all necessary O-rings and other parts), some carburetor cleaner, a supply of rags, some means of blowing out the carburetor passages and a clean place to work.

6 Carburetor - removal and installation

Warning: *Gasoline is extremely flammable, so take extra precautions when you work on any part of the fuel system. Don't smoke or allow open flames or bare light bulbs near the work area, and don't work in a garage where a natural gas-type appliance (such as a water heater or clothes dryer) with a pilot light is present. Since gasoline is carcinogenic, wear latex gloves when there's a possibility of being exposed to fuel, and, if you spill any fuel on your skin, rinse it off immediately with soap and water. Mop up any spills immediately and do not store fuel-soaked rags where they could ignite. When you perform any kind of work on the fuel system, wear safety glasses and have a Class B type fire extinguisher on hand.*

3

Removal

Refer to illustrations 6.2a, 6.2b, 6.3, 6.5, 6.6a and 6.6b

1   Remove the fuel tank (see Section 2) and, on TRX300EX models, the fuel line.

2   On TRX400EX models, disconnect the air vent tube from the plastic heat protector **(see illustration)**. Disengage the fuel line from the rubber heat protector **(see illustration),** then disconnect it from the carburetor.

3   Place a drain pan under the float bowl, open the drain screw **(see illustration)** and drain any residual fuel from the float bowl.

4   Remove the throttle linkage cover and disconnect the throttle cable from the carburetor (see Section 10). (If you have difficulty disconnecting the throttle cable from the throttle linkage with the carburetor installed, wait until you have detached the carburetor from the air inlet tube and the insulator, and then disconnect the cable.)

5   On TRX400EX models, detach the vacuum tube from the air cutoff valve **(see illustration)**.

6   Loosen the screws on the clamps that secure the air inlet tube and the insulator to the carburetor **(see illustrations)**. Work the carburetor free of the inlet tube and insulator and lift it up. If you had trouble disconnecting the throttle cable from the carburetor in Step 4, disconnect it now (see Section 10).

7   Inspect the air inlet tube and the insulator for cracks, deterioration or other damage. If either part is damaged, replace it. There's an O-ring between the insulator and the cylinder head; if you suspect an air leak at the insulator, remove the insulator and inspect the O-ring **(see illustrations 12.4a and 12.4b in Chapter 2A)**.

8   After the carburetor has been removed, stuff clean rags into the intake tube (or the intake port in the cylinder head, if the tube has been removed) to prevent the entry of dirt or other objects.

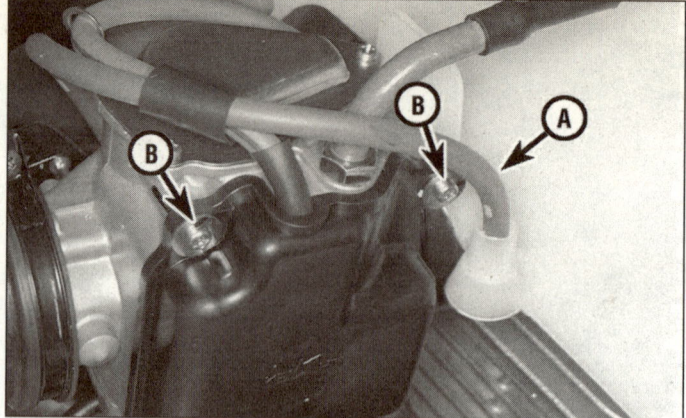

6.2a On TRX400EX models, disengage the air vent tube from the plastic heat protector . . .

A   Air vent tube   *B   Carburetor cover screws*

Installation

9   Connect the throttle cable to the carburetor (see Section 10).

10   Connect the fuel line to the carburetor (if it was disconnected) and connect any vent tubes that were disconnected.

11   Slip the clamping bands onto the air inlet tube and the insulator. If either of the bands is pinned, position the pin on the band in the groove on the air inlet tube or insulator.

12   Adjust the throttle freeplay (see Chapter 1).

13   The remainder of installation is the reverse of the removal steps.

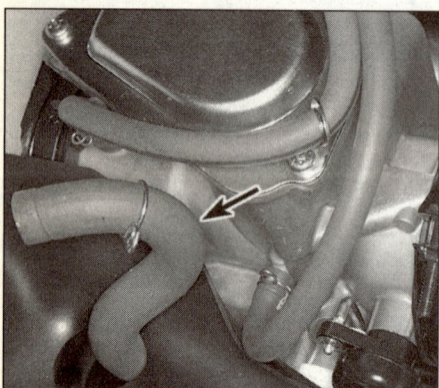

6.2b . . . and the fuel line (arrow) from the rubber heat protector, then disconnect the fuel line from the carburetor

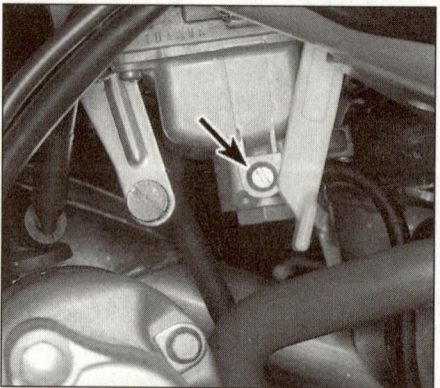

6.3 Open the drain screw (arrow) and drain any fuel in the float bowl (TRX400EX shown, TRX300EX similar)

6.5 Disconnect the vacuum tube (right arrow) from the air cut-off valve; to remove the valve, remove the screw (left arrow) and pull off the valve

6.6a Loosen the clamping band screws on the air inlet tube . . .

6.6b . . . and on the insulator (arrow) (TRX300EX shown, TRX400EX similar)

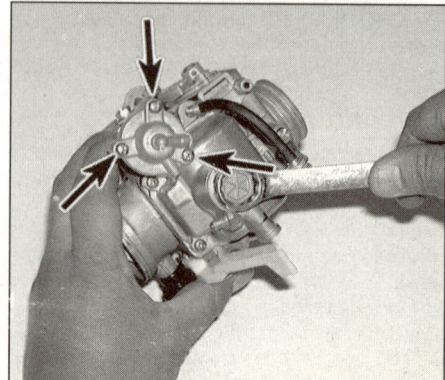

7.2a Remove the float chamber drain plug, O-ring and accelerator pump cover screws (arrows) . . .

7.2b . . . remove the accelerator pump cover, diaphragm spring and diaphragm

7.2c Remove the float chamber screws (arrows), the float chamber and its O-ring

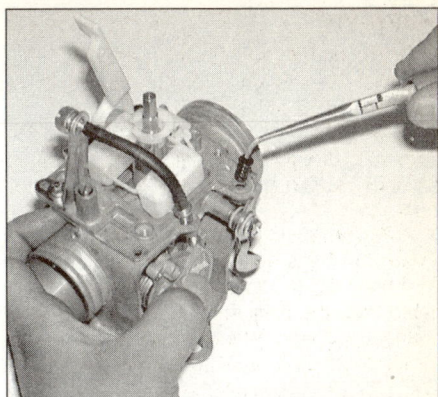

7.2d Remove the accelerator pump diaphragm shaft dust boot

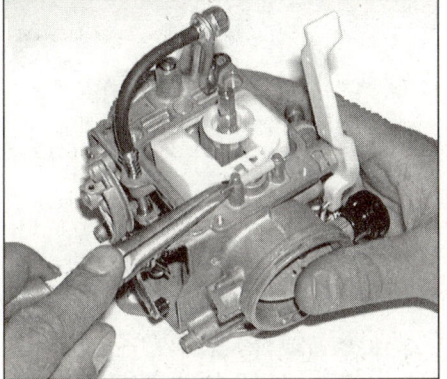

7.2e Remove the float pin . . .

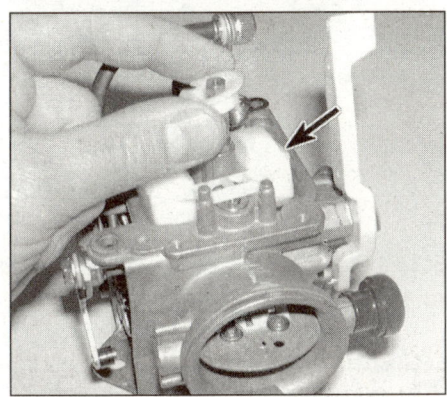

7.2f . . . remove the baffle plate and float (arrow)

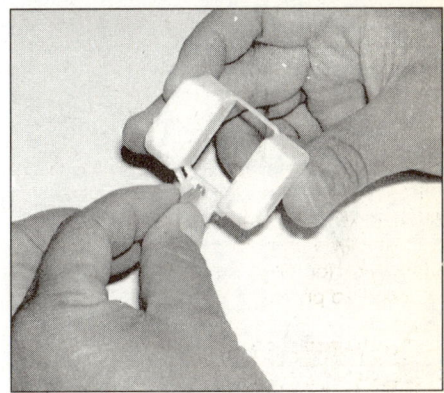

7.2g Remove the needle valve

7   Carburetor - disassembly, cleaning and inspection

Warning: *Gasoline (petrol) is extremely flammable, so take extra precautions when you work on any part of the fuel system. Don't smoke or allow open flames or bare light bulbs near the work area, and don't work in a garage where a natural gas-type appliance (such as a water heater or clothes dryer) with a pilot light is present. Since gasoline is carcinogenic, wear latex gloves when there's a possibility of being exposed to fuel, and, if you spill any fuel on your skin, rinse it off immediately with soap and water. Mop up any spills immediately and do not store fuel-soaked rags where they could ignite. When you perform any kind of work on the fuel system, wear safety glasses and have a Class B type fire extinguisher on hand.*

Disassembly

Refer to illustrations 7.2a through 7.2y

1   Remove the carburetor (see Section 6). Place it on a clean working surface.

2   Refer to the accompanying photographs to disassemble the carburetor **(see illustrations)**. The accompanying photos depict a TRX300EX carburetor, which is virtually identical to the TRX400EX carburetor except that the TRX400EX uses an air cut-off valve **(see illustration 6.5)**. To remove the air cut-off valve, remove the retaining screw, pull off the valve, pull the joint pipe out of the valve and remove and discard the old joint pipe O-rings.

3

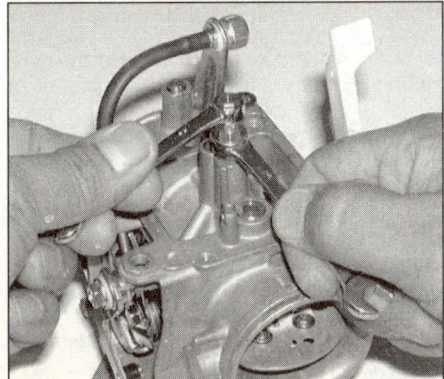

7.2h Using a backup wrench on the needle jet holder, unscrew the main jet, then the needle jet holder and nut . . .

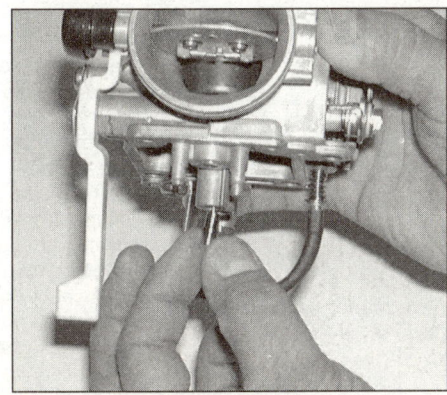

7.2i . . . and remove the needle jet

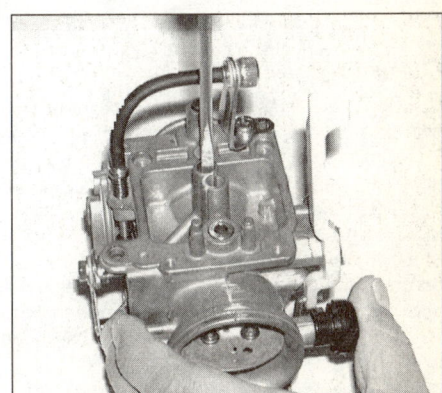

7.2j Unscrew the slow jet

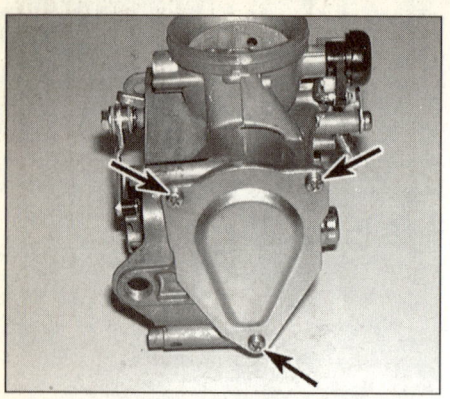

7.2k Remove the throttle valve cover screws (arrows), the cover and its O-ring

7.2l Remove the throttle valve arm screw and washer . . .

7.2m . . . remove the throttle link arm . . .

7.2n . . . remove the return spring . . .

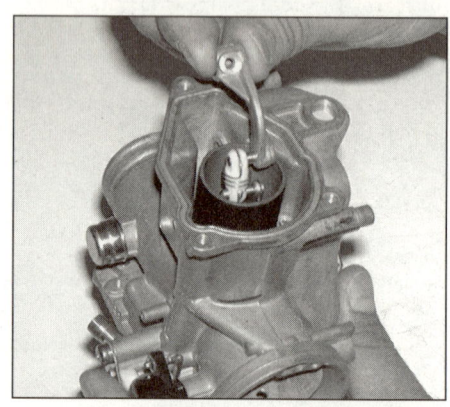

7.2o . . . remove the throttle valve arm . . .

7.2p . . . and remove the spacer

7.2q Remove the throttle valve . . .

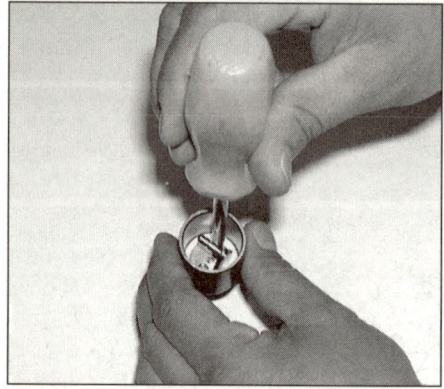

7.2r . . . remove the two jet needle holder screws . . .

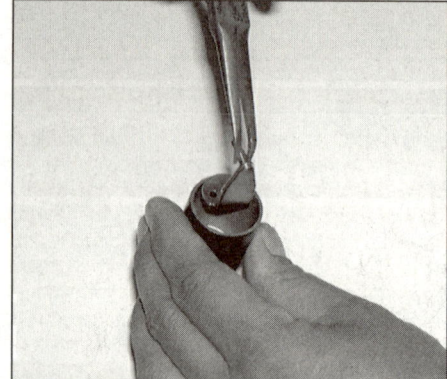

7.2s . . . remove the jet needle holder . . .

Cleaning

Caution: *Use only a carburetor cleaning solution that is safe for use with plastic parts (be sure to read the label on the container).*

3   Submerge the metal components in the carburetor cleaner for approximately thirty minutes (or longer, if the directions recommend it).
4   After the carburetor has soaked long enough for the cleaner to loosen and dissolve most of the varnish and other deposits, use a brush to remove the stubborn deposits. Rinse it again, then dry it with compressed air. Blow out all of the fuel and air passages in the main and upper body. **Caution:** *Never clean the jets or passages with a piece of wire or a drill bit, as they will be enlarged, causing the fuel and air metering rates to be upset.*

Inspection

5   Inspect the float assembly for cracks; note whether there is fuel inside one of the floats. If the float assembly is damaged, replace it.

6   Inspect the accelerator pump diaphragm for tears, holes and general deterioration. Holding it up to a light will help to reveal problems of this nature.
7   Insert the throttle valve in the carburetor body and verify that it travels up and down smoothly. If it doesn't, inspect the walls of the throttle valve piston and the throttle valve chamber for rough areas. Remove minor imperfections with crocus or emery cloth. If the throttle valve still doesn't operate smoothly, replace the carburetor.
8   Inspect the tip of the jet needle. If it has grooves or scratches in it, replace it. Inspect the jet needle for straightness by rolling it on a flat surface (such as a piece of glass). If it's bent, replace it.
9   Inspect the tapered portion of the pilot screw for wear or damage. Replace the pilot screw if necessary.
10   Inspect the carburetor body and float chamber castings for cracks, distorted sealing surfaces and other damage. If any defects are found, replace the carburetor.

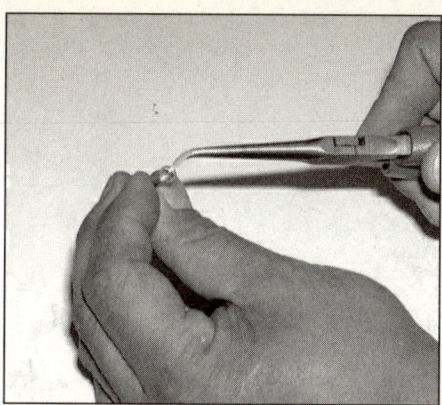

7.2t . . . remove the jet needle . . .

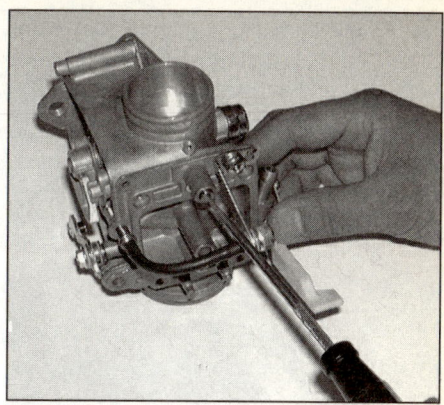

7.2u . . . and remove the jet needle clip

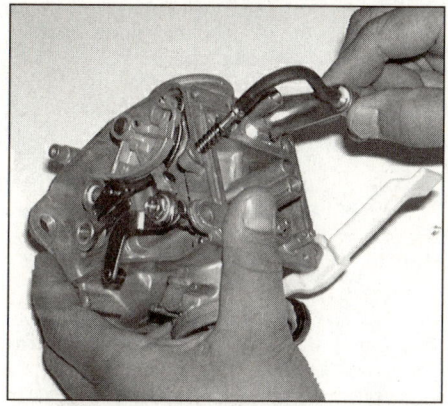

7.2v Remove the pilot screw, its O-ring, spring, washer and gasket

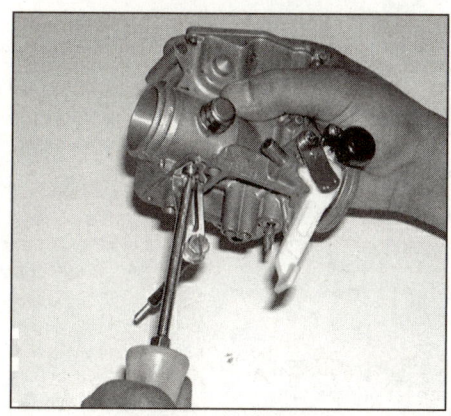

7.2w Remove the retaining screw for the throttle stop screw bracket and remove the throttle stop screw assembly

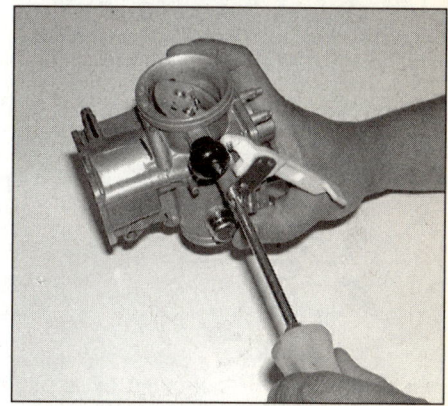

7.2x Remove the stopper plate screw and the stopper plate . . .

7.2y . . . and remove the choke lever

8   Carburetor - reassembly and float height check

Refer to illustrations 8.1, 8.2, 8.3 and 8.7
Caution: *When installing the jets, be careful not to over-tighten them - they're made of soft material and can strip or shear easily.*
Note: *When reassembling the carburetor, be sure to use the new O-rings, gaskets and other parts supplied in the rebuild kit.*

1   Install the choke lever, stopper plate and choke lever screw **(see illustration 7.2y and accompanying illustration)**. Install the throttle stop screw spring, washer and screw **(see illustrations 7.2x and 7.2w)**.

2   Install the pilot screw gasket, washer, spring, O-ring and pilot screw **(see illustration 7.2v and accompanying illustration)**. Turn the screw in (clockwise) until it seats lightly. The pilot screw must be adjusted after the carburetor has been reassembled and installed (see Section 4).

8.1 Carburetor body details

1 Choke lever	6 Throttle stop screw
2 Stopper plate	7 Throttle stop screw
3 Stopper plate screw	retaining screw
4 Throttle stop screw spring	8 Pilot screw
5 Washer	

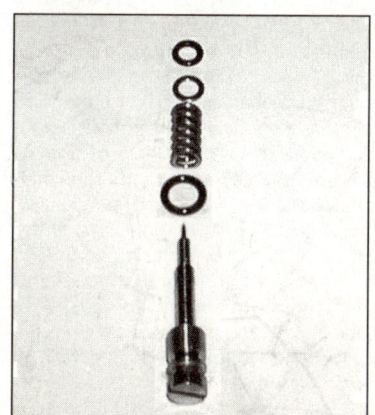

8.2 Pilot screw details

1   O-ring
2   Washer
3   Spring
4   Gasket
5   Pilot screw

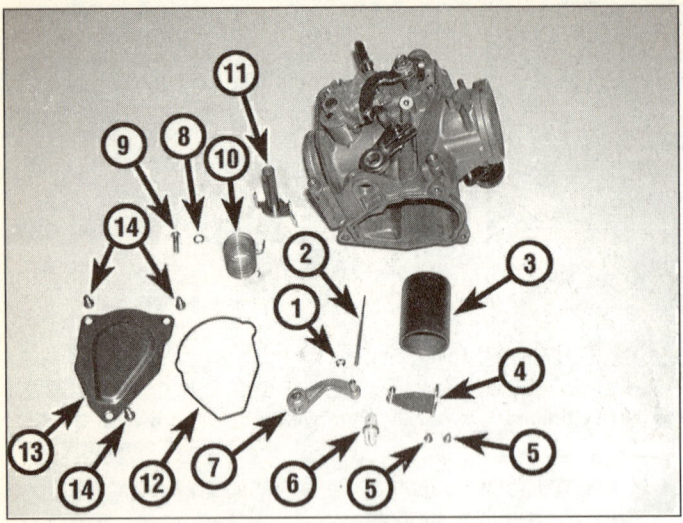

8.3 Throttle valve assembly

1	Jet needle clip	8	Throttle valve arm washer
2	Jet needle	9	Throttle valve arm screw
3	Throttle valve piston	10	Throttle link arm return
4	Jet needle holder		spring
5	Jet needle holder	11	Throttle link arm
	screws (2)	12	Throttle valve cover O-ring
6	Throttle valve arm-to-jet	13	Throttle valve cover
	needle holder link	14	Throttle valve cover
7	Throttle valve arm		screws (3)

3   Install the jet needle clip on the jet needle **(see illustration 7.2u and accompanying illustration)**. Make sure that the jet needle clip is installed in the position listed in this Chapter's Specifications. Install the jet needle in the throttle valve piston **(see illustration 7.2t)**, install the jet needle holder and install the jet needle holder screws **(see illustrations 7.2s and 7.2r)**. Tighten the screws securely, but don't overtighten them and strip the threads.
4   Install the throttle valve piston and jet needle into the carburetor body **(see illustration 7.2q)**.
5   Install the spacer, the throttle valve arm, the return spring, the throttle link arm, the throttle valve arm washer and the throttle valve arm screw **(see illustrations 7.2p, 7.2o, 7.2n, 7.2m and 7.2i)** Tighten the screw securely, but don't overtighten it and strip the threads.
6   Install a new throttle valve cover O-ring, the cover and its screws **(see illustration 7.2k)**. Tighten the screws securely but don't overtighten them and strip the threads.
7   Invert the carburetor. Install the slow jet **(see illustration 7.2j and accompanying illustration)**. Install the jet needle, the needle jet holder and nut and the main jet **(see illustrations 7.2i and 7.2h)**.
8   Install the needle valve, the float, the baffle plate and the float pin **(see illustrations 7.2g, 7.2f and 7.2e)**.
9   To check the float height, hold the carburetor so the float hangs down, then tilt it back until the valve needle is just seated. A special Honda float level gauge (Honda tool no. 07401-0010000) is available from Honda dealers, but you can also use a small ruler for this procedure. Measure the distance from the float chamber gasket surface to the top of the float and compare your measurement to the float height listed in this Chapter's Specifications. There's no means of adjustment; if it isn't as specified, replace the float assembly and recheck. If the float height is still incorrect, replace the float valve and recheck.
10   Install a new O-ring in the float chamber groove and install the diaphragm shaft dust boot, then install the float chamber and its screws **(see illustrations 7.2d and 7.2c)**. Tighten the float chamber screws securely but don't overtighten them and strip the threads.
11   Install the accelerator pump diaphragm, the diaphragm spring, the accelerator pump cover and the cover screws **(see illustrations 7.2b and 7.2a)**. Tighten the pump cover screws securely but don't overtighten them and strip the threads.

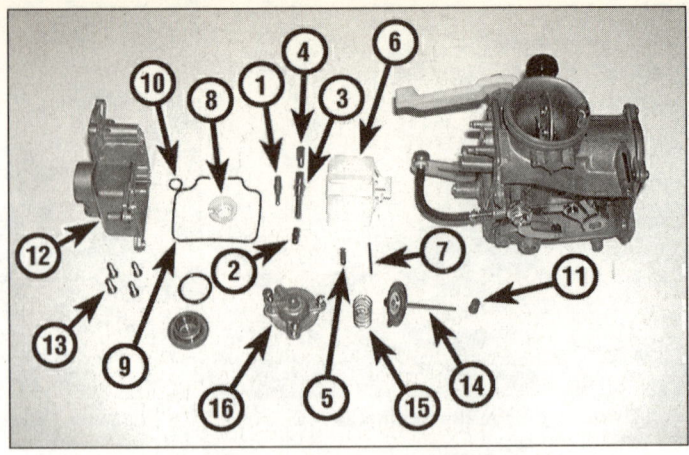

8.7 Float chamber details

1	Slow jet	11	Accelerator pump shaft
2	Needle jet		dust boot
3	Needle jet holder	12	Float chamber
4	Main jet	13	Float chamber retaining
5	Needle valve		screws (4)
6	Float	14	Accelerator pump
7	Float pin		shaft/diaphragm
8	Baffle plate	15	Accelerator pump
9	Float chamber O-ring		diaphragm spring
10	Accelerator pump shaft	16	Accelerator pump cover
	O-ring		and screws (3)

12   Install a new drain plug O-ring and the drain plug **(see illustration 7.2a)**.
13   On TRX400EX models, install new O-rings on the air cut-off valve joint pipe and install the joint pipe in the air cut-off valve with the stepped side facing out (toward the valve). Install the air cut-off valve and the valve retaining screw **(see illustration 6.5)** and tighten the screw securely.
14   Reattach any vacuum, air vent and fuel hoses that were detached.
15   Install the carburetor (see Section 6).

9   Air cleaner housing - removal and installation

Removal

Refer to illustrations 9.3, 9.4a, 9.4b, 9.4c, 9.5 and 9.6
1   Remove the fuel tank (see Section 2).
2   On TRX300EX models, remove the air cleaner housing cover to get to the housing retaining bolt in the bottom of the housing (see "Air

9.3 On TRX300EX models, loosen the clamping bands (arrows) and detach the air inlet tubes from the air cleaner housing

9.4a On TRX400EX models, detach the crankcase breather storage tank (arrow) from the carburetor air inlet tube

9.4b On TRX400EX models, loosen the clamp screw (arrow) and disconnect the air inlet tube from the carburetor . . .

9.4c . . . and loosen this clamp screw (arrow) to disconnect the air inlet tube from the air cleaner housing

9.5 On TRX300EX models, remove this bolt (arrow) from the bottom of the air cleaner housing

9.6 On all models, remove this bolt from the rear edge of the air cleaner housing (TRX400EX shown, TRX300EX similar)

filter element and drain tube – cleaning" in Chapter 1). Even though it's not necessary to remove the filter element to remove the air cleaner housing, you may want to remove the element at this time to wash and inspect it.

3   On TRX300EX models, remove the air inlet tube between the air cleaner housing and the carburetor (see illustration 6.6a) and loosen the clamps that attach the two air inlet tubes to the air cleaner housing (see illustration). (The two air inlet tubes to the air cleaner housing can remain in the frame.)

4   On TRX400EX models, detach the crankcase breather hose storage tank (see illustration) from the air inlet tube between the air cleaner housing and the carburetor by pulling it straight off. A short

pipe on the inner side of the storage tank (not visible in this photo) is attached to the air inlet tube by a wire-type clamp. Then loosen the clamps and detach the air inlet tubes from the carburetor and from the air cleaner housing (see illustrations). (The air inlet tube to the carburetor remains attached to the air cleaner housing; the air inlet tube to the air cleaner housing can remain in the frame.)

5   On TRX300EX models, remove the air cleaner housing retaining bolt from the bottom of the air cleaner housing (see illustration).

6   On all models, remove the air cleaner housing retaining bolt from the rear edge of the housing (see illustration). Lift the air cleaner housing (and, on TRX400EX models, the air inlet tube for the carburetor) out of the frame.

7   Installation is the reverse of removal. When reattaching the crankcase breather hose storage tank to the inlet tube on TRX400EX models, make sure that the tab on the inner side of the tank is inserted between the two tabs on the air inlet tube.

10 Throttle cable - removal, installation and adjustment

Removal

Refer to illustrations 10.3a, 10.3b, 10.3c, 10.3d, 10.4a, 10.4b, 10.5a, 10.5b, 10.5c and 10.6

1   Remove the seat and rear fenders and the front fender (see Chapter 7).

2   Remove the fuel tank (see Section 2).

3   At the handlebar, remove the screws from the throttle housing cover, remove the cover and gasket, and disconnect the cable from the throttle lever arm (see illustrations). Pull back the dust boot,

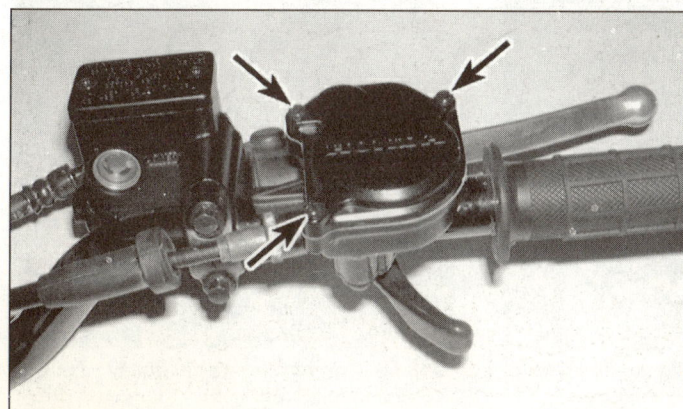

10.3a Remove the throttle housing cover screws and the cover . . .

3

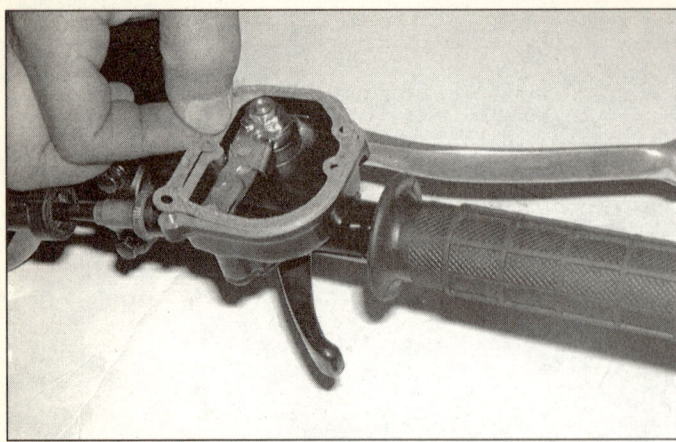

10.3b . . . remove the cover gasket . . .

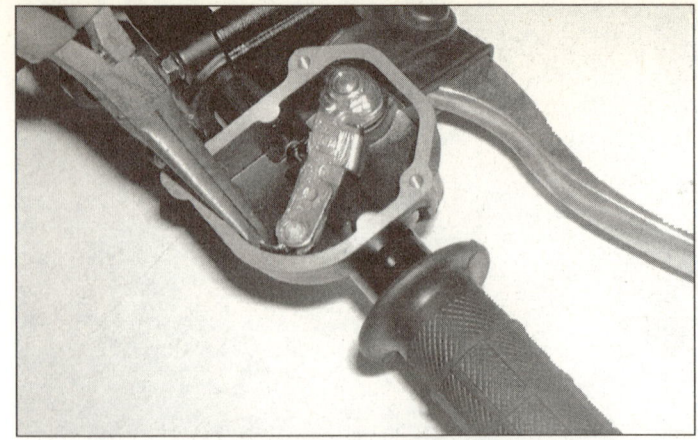

10.3c . . . and disengage the cable end plug from the throttle lever arm

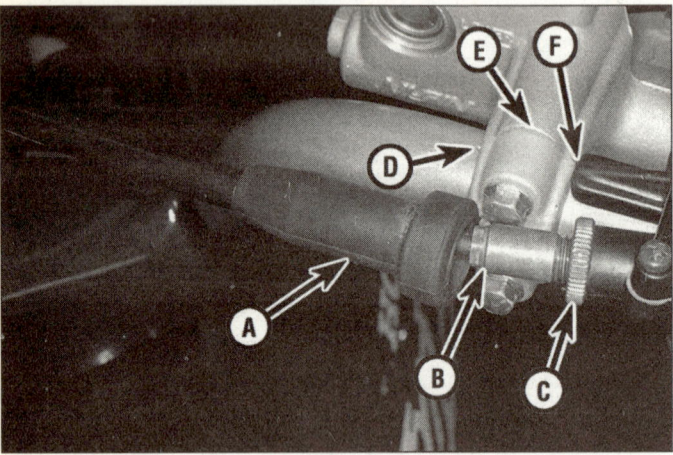

10.3d Throttle housing details

A	Dust boot	E	Split between housing
B	Adjuster		and clamp
C	Locknut	F	Throttle housing
D	Punch mark (for		protrusion
	alignment)		

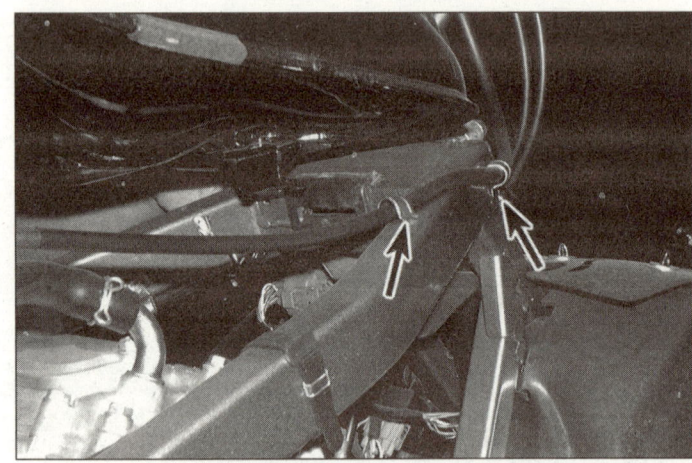

10.4a The throttle cable is clamped (arrows) to the upper right frame member on TRX300EX models

loosen the cable locknut, count the number of threads exposed on the cable adjuster and unscrew the adjuster **(see illustration)**.

4   Carefully note the routing of the cable from the throttle housing down to the carburetor **(see illustrations)** before removing it.

5   Remove the throttle linkage cover from the carburetor **(TRX300EX models, see illustration; TRX400EX models, see illustration 6.2a)**. Rotate the throttle pulley clockwise to put some slack in the end of the cable, align the cable with the slot in the pulley and slip the cable end plug out of the pulley **(see illustration)**. Loosen the cable locknut **(see illustration)**, count (and jot down) the number of threads above the bracket, then unscrew the threaded cable housing and pull the cable through the hole in the cable bracket.

6   If you want to remove or replace the entire throttle housing

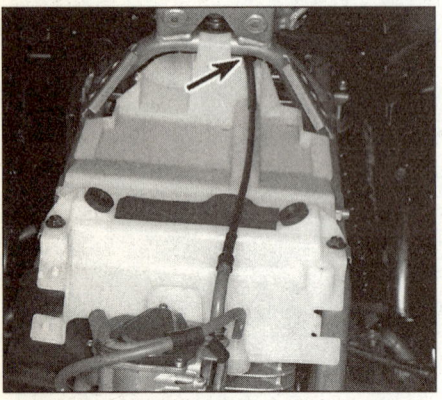

10.4b Note how the throttle cable is routed underneath the crossmember in front of the heat protector on TRX400EX models

10.5a Remove the throttle linkage cover screws (arrows) and the cover

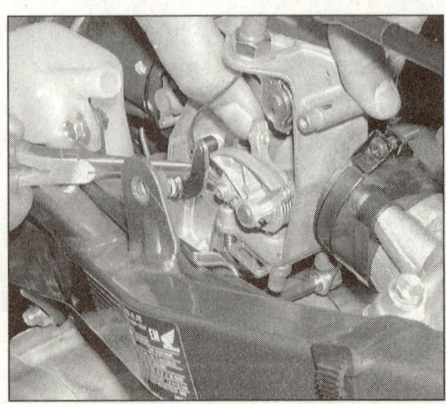

10.5b Rotate the throttle pulley to put some slack in the cable, then align the cable with the slot in the pulley and slip the cable end plug out of the pulley

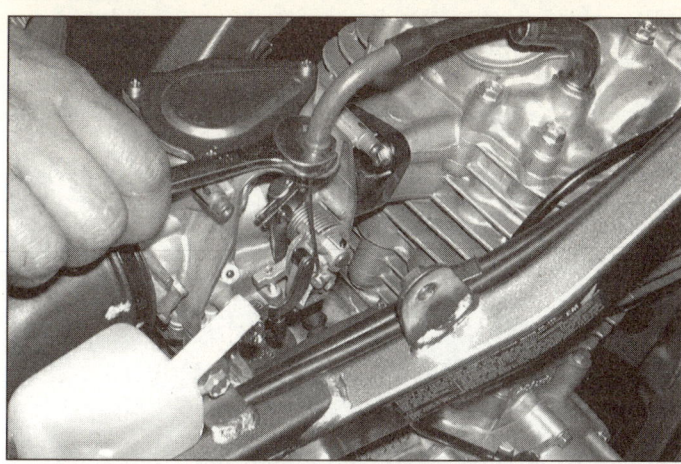

10.5c Loosen the locknut and adjusting nut and slip the cable out of the bracket

10.6 To remove the throttle housing assembly from the handlebar, remove the screws from underneath (arrow)

assembly, remove the throttle housing clamp screws and detach the throttle housing from the handlebar **(see illustration)**.

Installation

Refer to illustration 10.9

7   If the throttle housing was removed, install it on the handlebar and tighten the housing-to-handlebar clamp screws loosely. Position the housing so the projection on the inner end of the throttle housing is aligned with the split between the brake master cylinder housing and its clamp **(see illustration 10.3d)**. Tighten the clamp screws securely.

8   Route the cable exactly as it was routed before. Make sure it does-n't interfere with any other components and isn't kinked or bent sharply.

9   Pass the cable through the bracket on the carburetor, screw the cable housing into the bracket so that the same number of threads are exposed above the bracket as before, then connect the cable end plug to the throttle pulley. Put a dab of multi-purpose grease on the cable end plug and the pulley roller **(see illustration)**. When installing the throttle linkage cover, make sure that the tab on the bottom of the cover fits into the slot on the float chamber.

10   Up at the handlebar, insert the cable through the throttle housing and connect the cable end plug to the throttle lever arm. Again, put a dab of multi-purpose grease on the end plug; put a little grease on the lever arm pivot too. Replace the throttle housing cover gasket if the old gasket is damaged. Installation is otherwise the reverse of removal. Operate the lever and make sure it returns to the idle position by itself under spring pressure. **Warning:** *If the lever doesn't return by itself, find and solve the problem before continuing with installation. A stuck lever can lead to loss of control of the vehicle.*

Adjustment

11   Adjust the cable (see "Throttle cable - check and adjustment" in Chapter 1).

12   Turn the handlebars back and forth to make sure the cables don't cause the steering to bind.

13   When you're sure the cable operates properly, install the covers on the throttle housing and on the carburetor throttle linkage.

14   Install the fuel tank (see Section 2).

15   With the engine idling, turn the handlebars through their full travel (full left lock to full right lock) and note whether idle speed increases. If it does, the cable is routed incorrectly. Correct this dangerous condi-tion before riding the vehicle.

11   Exhaust system - removal and installation

Warning: *Wait until the exhaust system has completely cooled down before working on it.*

Heat shield

Refer to illustration 11.1

1   It's not necessary to remove the heat shield to remove the exhaust pipe, or vice versa. If you want to replace the heat shield, sim-ply detach it **(see illustration)**. Make sure you install any washers, col-lars and spacers in the same order in which they were removed. Tighten the heat shield bolts to the torque listed in this Chapter's Specifications.

3

10.9 Put a dab of grease on the cable end plug and pulley roller (arrows); fit the tab on the cover into the slot (lower arrow) on the carburetor

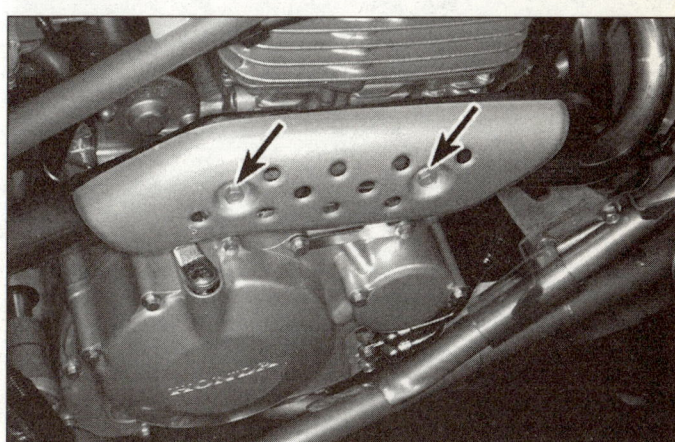

11.1 Unbolt the heat shield (TRX400EX shown, TRX300EX similar); don't forget to install all washers, collars and spacers in the correct order when attaching the heat shield

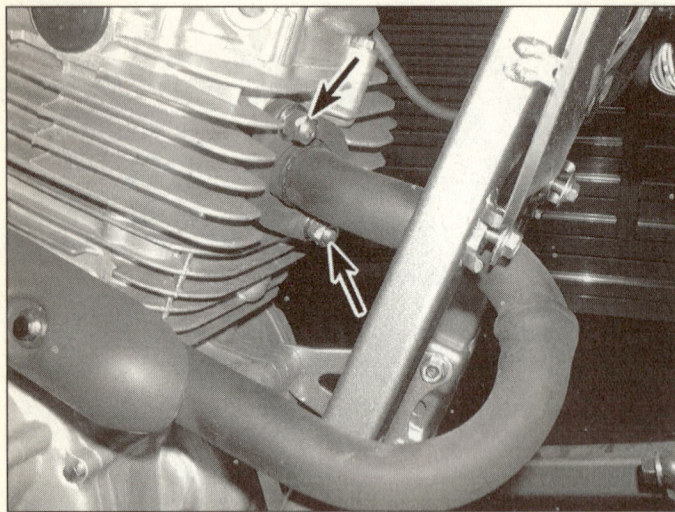

11.3 To detach the exhaust pipe from the cylinder head, remove these two nuts (arrows) (TRX300EX shown)

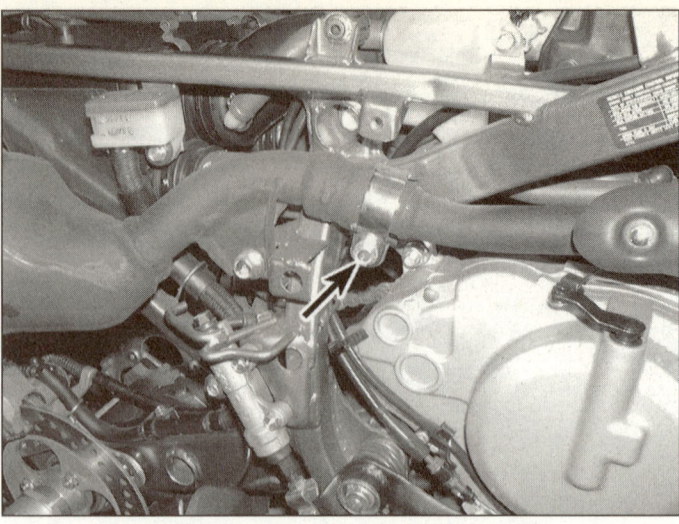

11.4 To disconnect the exhaust pipe from the muffler, loosen this clamp bolt (arrow) (TRX300EX models) . . .

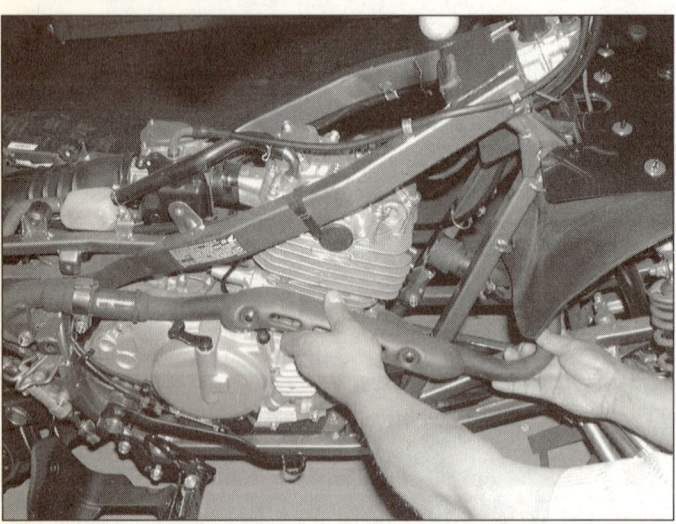

11.5 . . . and pull the pipe forward out of the muffler

11.6 Remove and discard the old exhaust pipe gasket (TRX300EX)

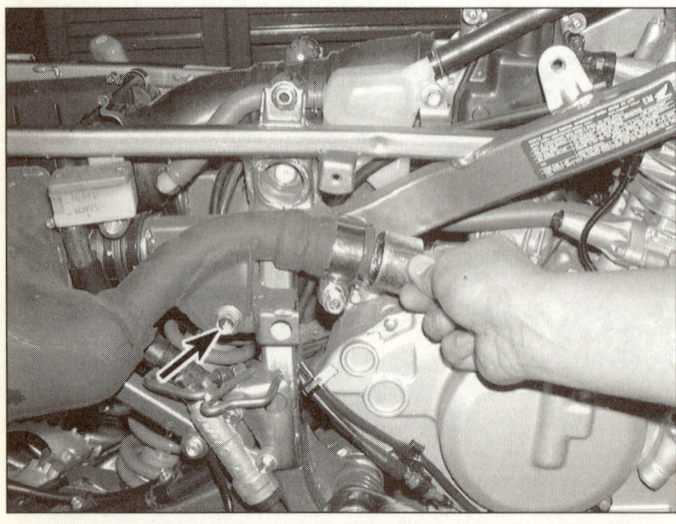

11.7 Pull off the muffler packing; remove the forward muffler mounting bolt (arrow) (TRX300EX) . . .

11.8 . . . and the rear muffler mounting bolt (arrow) (TRX300EX)

11.9 To detach the TRX400EX exhaust pipe from the muffler, loosen these two clamp bolts (arrows)

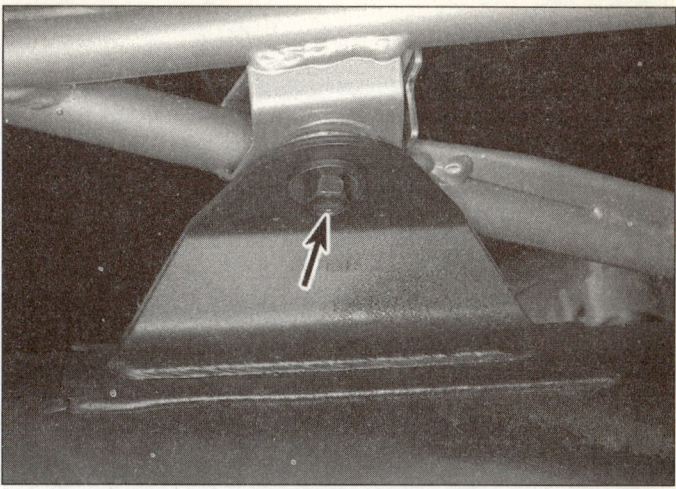

11.10 To detach the TRX400EX muffler from the frame, remove the upper mounting bolt (arrow) and the lower bolt (not shown)

Exhaust pipe and muffler

2   Remove the seat and rear fenders (see Section 7).

Removal

TRX300EX models

Refer to illustrations 11.3, 11.4, 11.5, 11.6, 11.7 and 11.8

3   Remove the exhaust pipe flange nuts **(see illustration)**.
4   Loosen the bolt for the clamp that secures the exhaust pipe to the muffler **(see illustration)**.
5   Pull the exhaust pipe forward to separate it from the cylinder head and from the muffler. Remove the exhaust pipe **(see illustration)**.
6   Remove the old exhaust pipe gasket **(see illustration)** and discard it.
7   Remove the old muffler packing **(see illustration)** and discard it.
8   Remove the forward muffler mounting bolt **(see illustration 11.7)** and the rear muffler mounting bolt **(see illustration)**. Remove the muffler.

TRX400EX models

Refer to illustrations 11.9, 11.10 and 11.15

9   Loosen the muffler band bolts **(see illustration)**.
10   Loosen the upper and lower muffler mounting bolts **(see illustration)** and the washers (between the muffler and the frame).
11   Remove the muffler assembly.
12   Remove the old muffler packing **(see illustration 11.7)**.
13   Remove the four exhaust pipe flange nuts **(see illustration 11.3)**.
14   Remove the exhaust pipe assembly.
15   Remove the old exhaust pipe gaskets **(see illustration)**.

11.15 Be sure to install new gaskets in both of the TRX400EX exhaust ports

Installation

16   Installation is the reverse of removal. Be sure to install a new gasket (TRX300EX models) or gaskets (TRX400EX models) in the cylinder head exhaust port(s) and a new muffler packing between the exhaust pipe and the muffler.
17   Tighten all exhaust pipe, muffler and heat shield bolts to the torque listed in this Chapter's Specifications. (There is no specified torque for the exhaust flange nuts; tighten them securely.)

3

Notes

Chapter 4
Ignition system

Contents

4

Specifications

TRX300EX

Ignition coil resistance (at 20-degrees C/68-degrees F)	
Primary resistance	0.1 to 0.3 ohms
Secondary resistance	
With plug cap attached	7000 to 9000 ohms
With plug cap detached	2500 to 3500 ohms
Pulse generator resistance (at 20-degrees C/68-degrees F)	300 to 500 ohms
Ignition timing	
At 1400 rpm	17 degrees BTDC
At 3000 to 3500 rpm (full advance)	33 degrees BTDC

TRX400EX

Ignition coil primary peak voltage (minimum)	100 volts
Ignition pulse generator peak voltage (minimum)	0.7 volts
Exciter coil peak voltage (minimum)	100 volts
Ignition timing	8 degrees BTDC

Torque specifications

Pulse generator screws or Allen bolts	6 Nm (48 inch-lbs)*

*Apply non-permanent thread locking agent to the bolt threads.

2.3 With the plug wire attached, lay a spark plug on the engine and operate the starter – bright blue sparks should be visible

2.5 Unscrew the spark plug cap from the plug wire and measure its resistance with an ohmmeter

1   General information

This vehicle is equipped with a battery operated, fully transistorized, breakerless ignition system. The system consists of the following components:

> Battery and fuse
> Engine kill (stop) and main (key) switches
> Reverse switch (TRX300EX models)
> Ignition coil
> Ignition pulse generator
> Exciter coil (TRX400EX models)
> Ignition control module (ICM)
> Spark plug
> Primary and secondary circuit wiring

The transistorized ignition system functions on the same principle as a DC ignition system with the pulse generator and ICM performing the tasks previously associated with the breaker points and mechanical advance system. As a result, adjustment and maintenance of ignition components is eliminated (with the exception of spark plug replacement).

Because of their nature, the individual ignition system components can be checked but not repaired. If ignition system troubles occur, and the faulty component can be isolated, the only cure for the problem is to replace the part with a new one. Keep in mind that most electrical parts, once purchased, can't be returned. To avoid unnecessary expense, make very sure the faulty component has been positively identified before buying a replacement part.

2   Ignition system - check

Warning: *Because of the very high voltage generated by the ignition system, extreme care should be taken when these checks are performed.*

1   If the ignition system is the suspected cause of poor engine performance or failure to start, a number of checks can be made to isolate the problem.

2   Make sure the ignition kill (stop) switch is in the Run or On position.

Engine will not start

Refer to illustrations 2.3 and 2.5

3   Disconnect the spark plug wire. Connect the wire to a known good spark plug and lay the plug on the engine with the threads contacting the engine **(see illustration)**. If necessary, hold the spark plug with an insulated tool. Crank the engine over and make sure a well-

defined, blue spark occurs between the spark plug electrodes. **Warning:** *Don't remove the spark plug from the engine to perform this check - atomized fuel being pumped out of the open spark plug hole could ignite, causing severe injury!*

4   If there is a good spark, the old plug is bad. Replace it (see Chapter 1). If no spark occurs, test the ignition system as follows:

5   Unscrew the spark plug cap from the plug wire and check the cap resistance with an ohmmeter **(see illustration)**. If the resistance is infinite, replace the spark plug cap.

6   Make sure all electrical connectors are clean and tight. Inspect all wires for opens, shorts and bad connections.

7   Check the battery voltage with a voltmeter. If the voltage is less than 12-volts, recharge the battery (see Chapter 8).

8   Check the 15-amp ignition fuse and the fuse connections (see Chapter 8). If the fuse is blown, replace it; if the connections are loose or corroded, clean or repair them.

9   Check the ignition coil (see Section 3) and the ignition pulse generator (see Section 5).

10   On TRX400EX models, check the exciter coil (see Section 4).

11   If the preceding checks produce positive results but there is still no spark at the plug, check the ignition control module (ICM) (see Section 6).

Engine starts but misfires

Refer to illustration 2.13

12   If the engine starts but misfires, make the following checks before deciding that the ignition system is at fault.

2.13 A simple spark gap testing fixture can be made from a block of wood, two nails, a large alligator clip, a screw and a piece of wire

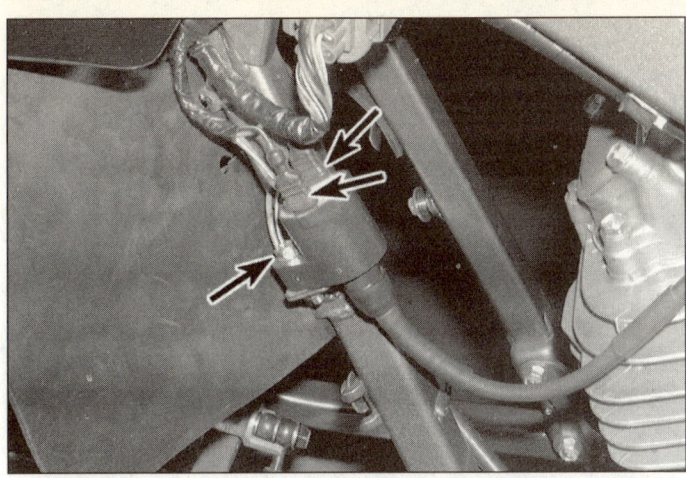

3.3 To check the TRX300EX coil, disconnect the primary connectors (arrows); to remove the coil, remove its retaining bolt (arrow)

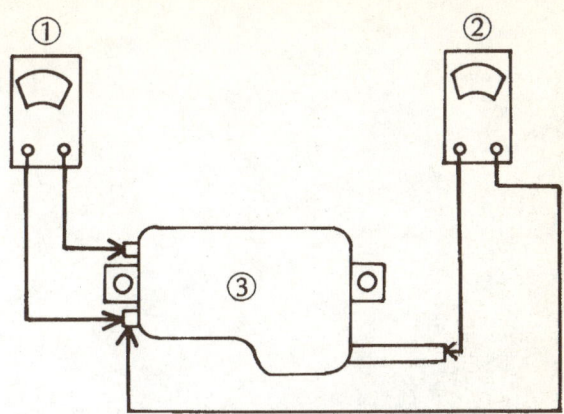

3.4 Ignition coil test (TRX300EX models)

1 Measure primary winding resistance
2 Measure secondary winding resistance
3 Ignition coil

13 The ignition system must be able to produce a spark across a seven millimeter (1/4-inch) gap (minimum). A simple test fixture **(see illustration)** can be constructed to make sure the minimum spark gap can be jumped. Make sure the fixture electrodes are positioned seven millimeters apart.

14 Connect the spark plug wire to the protruding test fixture electrode, then attach the fixture's alligator clip to a good engine ground.

15 Crank over the engine and note whether a well-defined, blue sparks occur between the test fixture electrodes. If the minimum spark gap test is positive, the ignition coil is functioning satisfactorily. If the spark will not jump the gap, or if it is weak (orange colored), refer to Steps 5 through 11 of this Section and perform the component checks described.

3 Ignition coil - check and replacement

TRX300EX models

Check

Refer to illustrations 3.3, 3.4 and 3.6

1 In order to determine conclusively that the ignition coils are defective, they should be tested by an authorized Honda dealer service department which is equipped with the special electrical tester required for this check.

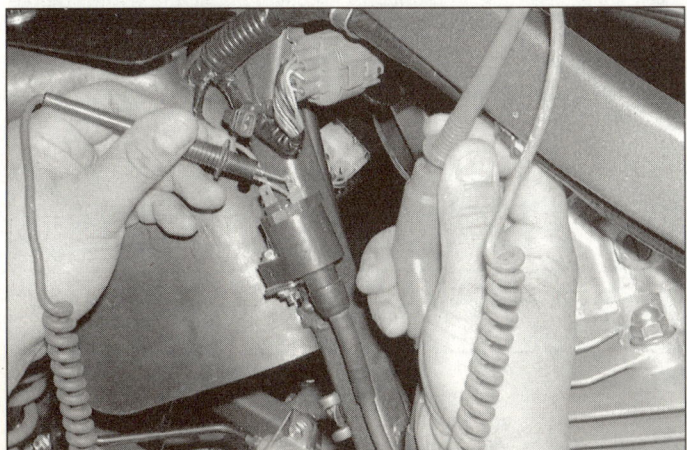

3.6 Measure secondary resistance between the primary terminal for the green wire (ground) and the high tension terminal inside the spark plug cap (TRX300EX models)

2 There are a couple of things you can do at home, however, First, inspect the coil for cracks and other damage. Second, measure the primary and secondary coil resistance with an ohmmeter. If the coil is undamaged, and if the resistance is within the specified range, it's probably okay.

3 To check the coil, unplug the primary circuit electrical connectors **(see illustration)** from the coil and remove the spark plug wire from the spark plug. Mark the locations of all wires before disconnecting them.

4 To check the coil primary resistance, attach one ohmmeter lead to one of the primary terminals and the other ohmmeter lead to the other primary terminal **(see illustration)**.

5 Place the ohmmeter selector switch in the Rx1 position and compare the measured resistance to the value listed in this Chapter's Specifications.

6 If the coil primary resistance is as specified, check the coil secondary resistance by disconnecting the meter leads from the primary terminals and attaching them between the spark plug wire terminal and the green primary terminal **(see illustration)**.

7 Place the ohmmeter selector switch in the Rx100 position and compare the measured resistance to the values listed in this Chapter's Specifications.

8 If the resistances are not as specified, unscrew the spark plug cap from the plug wire and check the resistance between the green primary terminal and the end of the spark plug wire **(see illustration 3.4)**. If it is now within specifications, the spark plug cap is bad. If it's still not as specified, the coil is probably defective and should be replaced with a new one.

Replacement

9 Disconnect the spark plug wire from the plug. Unplug the coil primary circuit electrical connectors **(see illustration 3.3)**. Note which color lead is attached to each primary terminal (there's a green wire and a blue/yellow wire).

10 Remove the coil mounting bolt **(see illustration 3.3)**, then remove the coil.

11 Installation is the reverse of removal. Make sure the primary circuit electrical connectors are attached to the proper terminals. Just in case you forgot to mark the wires, the black/yellow wire connects to the black primary terminal and the green wire attaches to the green terminal.

TRX400EX models

Check

12 Honda doesn't provide resistance specifications for these models. Instead, it recommends testing the coil peak primary voltage with

4

3.40 To replace the TRX400EX ignition coil, remove the spark plug cap from the plug, disconnect the primary connector and remove the coil mounting bolts (arrows)

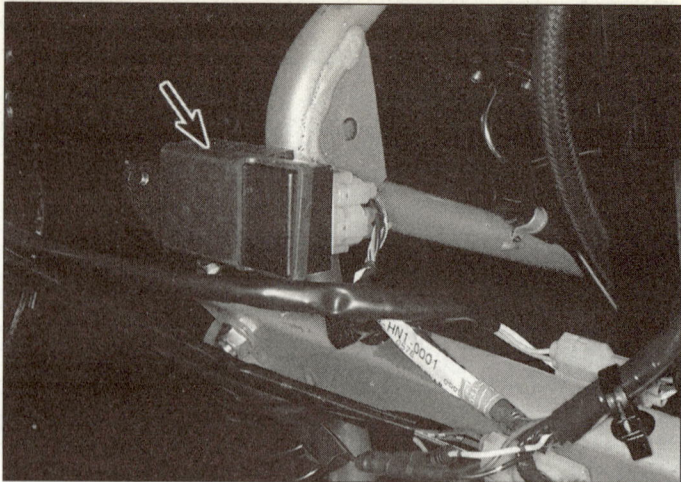

4.3 The TRX400EX ignition control module (arrow) is at the right front corner of the machine; disconnect the connectors and slip the module out of its rubber mount

one of the following:

a) *Imrie diagnostic tester model 625*

b) *Digital multimeter (minimum impedance 10 meg-ohms per DC volt) and peak voltage adapter (Honda part no. 07HGJ-0020100 or equivalent)*

If you don't have one of these special tools, or access to one of them, have the coil tested by a Honda dealer. If you do have the right equipment, the procedure is as follows.

13 Remove the left side cover (see Chapter 7).

14 Connect the positive lead of the peak voltage tester or adapter to the ignition coil primary terminal (black/yellow wire) and the negative lead to ground (bare metal on the engine). The meter must be on its "10 M-ohms/DCV" setting. If your multimeter doesn't have this setting, it won't produce an accurate or meaningful measurement.

15 Turn the ignition switch to ON, crank the engine, note the primary coil voltage and compare the indicated voltage to the coil peak voltage listed in this Chapter's Specifications.

16 There are four possible outcomes: peak voltage is adequate, peak voltage is too low, there is no voltage, or peak voltage is satisfactory but there is still no spark at the spark plug.

a) *If peak voltage is adequate, the ignition coil is operating satisfactorily.*

b) *If peak voltage is low, go to Step 17.*

c) *If there is no voltage, go to Step 25.*

d) *If peak voltage is satisfactory but there is no spark at the plug, go to Step 34.*

Peak voltage is low

17 First, check the peak voltage adapter connections as follows. Check the coil peak voltage again, this time with the connections reversed. If the indicated voltage now exceeds the minimum specified voltage, the coil is okay (the adapter connections were incorrect the first time you measured coil peak voltage).

18 Check your multimeter's impedance setting. It should be at "10 M-ohms/DCV." If it isn't, change the setting and recheck.

19 If the battery is undercharged, the cranking speed might be too low. Charge the battery (see Chapter 8).

20 The sampling timing of the tester and measured pulse are not synchronized. The ignition system is operating satisfactorily if the measured peak voltage is higher than the specified minimum peak voltage at least once (this only applies to the peak voltage tester, not to a multimeter used with the peak voltage adapter).

21 There is a bad connection or an open circuit in the ignition system (see wiring diagrams at the end of Chapter 8).

22 The exciter coil is defective (see Section 4).

23 The ignition coil is defective (substitute a known good coil and recheck).

24 If everything in Steps 17 through 23 is satisfactory, the ignition control module is defective (see Section 6).

No peak voltage

25 First, check the peak voltage adapter connections as follows. Check the coil peak voltage again, this time with the connections reversed. If the indicated voltage now exceeds the minimum specified voltage, the coil is okay (the adapter connections were incorrect the first time you measured coil peak voltage).

26 There is a short circuit in the engine kill switch or the kill switch is defective (see Chapter 8).

27 The ignition switch is defective (see Chapter 8).

28 The ICM connectors are loose or corroded (see Section 6).

29 There is an open circuit or bad connection in the ICM green/white (ground) wire.

30 The peak voltage adapter is defective.

31 The exciter coil is defective (see Section 4).

32 The ignition pulse generator is defective (see Section 5).

33 If everything in Steps 17 through 23 is satisfactory, the ignition control module is defective (see Section 6).

Peak voltage is satisfactory but there is no spark at the plug

34 The plug or the plug wire might be defective (see Chapter 1).

35 Secondary ignition current might be "leaking" somewhere between the coil and the plug.

36 If the plug and wire are good and there's no secondary current leak, the ignition coil is probably defective. Substitute a known good unit or have the coil tested by a Honda dealer before replacing it.

Replacement

Refer to illustration 3.40

37 Remove the left side cover (see Chapter 7).

38 Disconnect the spark plug cap from the spark plug.

39 Disconnect the coil primary connector.

40 Unbolt the coil from the frame **(see illustration)**.

41 Installation is the reverse of removal.

---

4   Exciter coil (TRX400EX models) - check and replacement

Check

Refer to illustrations 4.3 and 4.5

1   The exciter coil on these models is checked by measuring its peak voltage. This requires special test equipment not normally possessed by do-it-yourselfers. If you don't already have one of the necessary testers, it's best to have this procedure done by a Honda dealer.

4.5 TRX400EX wiring harness details

A   Connector boot (alternator 4-pin, exciter coil, starter relay,
    neutral switch)
B   Clutch diode

2   To test the coil, you'll need one of the following:

a) *Imrie diagnostic tester model 625*
b) *Digital multimeter (minimum impedance 10 meg-ohms per DC
   volt) and peak voltage adapter (Honda part no. 07HGJ-0020100 or
   equivalent)*

3   Remove the front fenders (see Chapter 7). Locate the ignition
control module (ICM) on the frame **(see illustration)**. Unplug both of its
electrical connectors, then connect the tester between the connector
terminals for the black/red and green/white wires (on the wire harness
side of the connector).

4   Crank the engine and note the peak voltage reading. **Warning:** *Do
not touch the spark plug or the tester probes while cranking or you may
get an electric shock.*

5   If peak voltage is less than the minimum listed in this Chapter's
Specifications, remove the seat and rear fenders (see Chapter 7). Disconnect the exciter coil connector (black/red wire), which is located
inside the boot above the battery **(see illustration)**. Connect the peak
voltage tester to the alternator side of the connector and to ground,
and repeat the peak voltage test.

6   If peak voltage is okay at the exciter coil connector but not at the
ICM connector, check the black/red wire for an open, a short, or a bad
connection. If it's not okay at either point, the problem is either a bad
exciter coil or a bad ICM. Further testing must be done by a Honda
dealer.

Replacement

7   The exciter coil is an integral part of the alternator stator coil (see
"Alternator – check and replacement" in Chapter 8).

---

5   Ignition pulse generator - check and replacement

TRX300EX models

Check

Refer to illustrations 5.1 and 5.3

1   Locate the four-pin pulse generator connector on the left side of
the vehicle, in front of the engine near the ignition coil **(see illustration)**. Disconnect the pulse generator connector.

2   Set the ohmmeter to R x 100. Measure the resistance between a
good ground and the blue/yellow wire terminal in the electrical connector. Compare the reading with the value listed in this Chapter's Specifications.

3   If the reading is not within the specified range, refer to Chapter 8

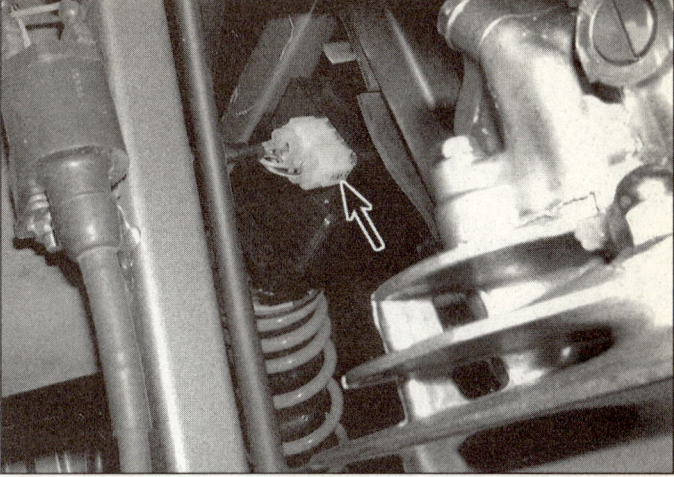

**5.1 The TRX300EX ignition pulse generator connector (arrow) is
on the left side of the machine near the ignition coil**

**5.3 Disconnect the electrical connector (center arrow), then
remove the Allen bolts (left and right arrows)**

and remove the left crankcase cover from the left side of the
crankcase. Disconnect the wire from the pulse generator, which is
mounted inside the cover **(see illustration)**. Measure the resistance
between a good ground and the pulse generator terminal. If it's now
within specifications, there's a break or bad connection in the wire. If
it's still incorrect, replace the pulse generator.

Replacement

4   Remove the left crankcase cover (see "Alternator - check and
replacement" in Chapter 8).

5   Disconnect the pulse generator connector **(see illustration 5.3)**.

6   Unscrew the pulse generator Allen bolts **(see illustration 5.3)** and
remove the pulse generator.

7   Installation is the reverse of the removal procedure. Apply non-
permanent thread locking agent to the pulse generator Allen bolts and
tighten them to the torque listed in this Chapter's Specifications.

8   Install the left crankcase cover (see "Alternator - check and
replacement" in Chapter 8).

TRX400EX models

Check

9   To test the pulse generator on these models, you'll need one of
the following:

a) *Imrie diagnostic tester model 625*

4

6.1 The TRX300EX ignition control module (ICM) is on the left side of the frame; to remove it, remove the bracket bolt (arrow)

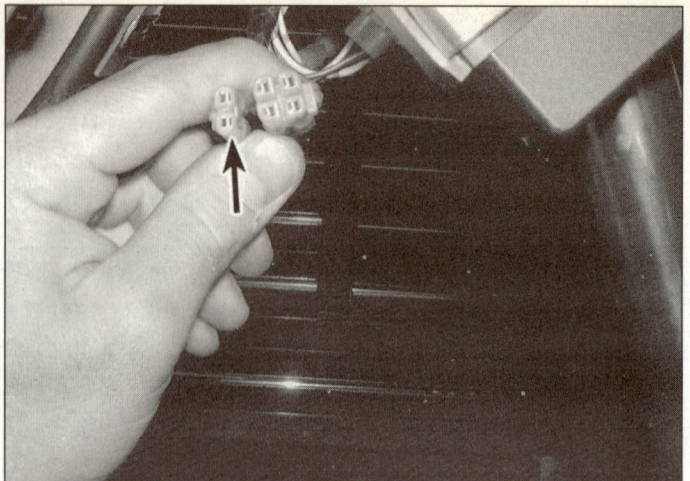

6.2 The electrical connectors for the TRX300EX ignition control module

    b) *Digital multimeter (minimum impedance 10 meg-ohms per DC volt) and peak voltage adapter (Honda part no. 07HGJ-0020100 or equivalent)*

10   Remove the front fenders (see Chapter 7). Disconnect the 4-pin connector from the ICM **(see illustration 4.3)**. Connect the peak voltage tester between the connector terminals for the blue/yellow and green/white wires (on the harness side of the connector), then crank the engine and measure peak voltage.

11   If voltage is less than the peak voltage listed in this Chapter's Specifications, remove the seat and rear fenders (see Chapter 7). Locate the 4-pin alternator connector, which is located inside the boot right above the battery **(see illustration 4.5)**. Disconnect the connector and connect the peak voltage tester to the connector terminals for the blue/yellow and green wires (on the alternator side of the connector).

12   Crank the engine and measure peak voltage again. If peak voltage is now within the Specifications, look for an open, a short or a bad connection in the blue/yellow wire. If it's still too low, the pulse generator is probably defective, but this can only be verified by substituting a known good unit.

Replacement

13   The pulse generator, which is mounted inside the left crankcase cover, is an integral component of the alternator stator coil (see "Alternator – check and replacement" in Chapter 8).

---

6   Ignition control module (ICM) – check and replacement

TRX300EX models

Check

Refer to illustrations 6.1 and 6.2

1   Remove the front fenders (see Chapter 7). Locate the ignition control module **(see illustration)** on the left side of the frame, above the left front shock absorber.

2   Disconnect the two-pin ignition switch/reverse switch connector (black/white and green/white wires) from the ICM **(see illustration)**. There are two electrical connectors for the ICM: the two-pin connector (black/white wire goes to kill switch and ignition switch; green/white wire goes to reverse switch) and a four-pin connector (black/yellow wire to ignition coil, blue/yellow wire to ignition pulse generator, green/white and yellow/white wires to alternator).

3   Turn the ignition switch to ON and the kill switch to RUN. Hook up a voltmeter to the terminals for the black/white and the green/white wires (on the harness side of the connector), and check the initial volt-

age. It should indicate battery voltage.

4   If there's no voltage, check the battery (see Chapter 1). If the battery is okay, check the ignition switch, the kill switch and the 15-amp main fuse (see Chapter 8). If those components are all okay, look for an open circuit or a bad connection in the ICM wire harness (see the wiring diagrams at the end of Chapter 8).

5   If there is battery voltage, unplug both connectors from the ICM (the other connector, from the ignition pulse generator and the alternator, is shown in **illustration 5.1**) and check the resistance, continuity and voltage at the indicated terminals (on the wire harness side of the connectors):

    a) *Ignition coil primary coil (black/yellow and green/white wires): 0.1 to 0.3 ohm*

    b) *Ignition pulse generator coil (blue/yellow and green/white wires): 300 to 500 ohms*

    c) *Ignition switch and engine stop switch - ignition switch turned to ON and kill switch set to RUN – (black/white and green/white wires): Battery voltage*

    d) *Reverse switch - gear in Reverse – (gray and green/white wires): Continuity*

    e) *AC sensor line (yellow/white and yellow wires): Continuity. Disconnect the 3-pin alternator connector and check continuity between the terminal for the yellow/white wire (ICM connector, on the wire harness side) and the terminal for the yellow wire (alternator side of the alternator connector).*

6   If the harness and all other system components are OK, the ICM is probably defective. But before buying a new one, it's a good idea to substitute a known good ICM or have it tested by a Honda dealer.

Replacement

7   Remove the front fenders (see Chapter 7).

8   Remove the bolt from the ignition control module bracket **(see illustration 6.1)**.

9   Unplug the two electrical connectors from the ICM **(see illustration 6.2)** and remove the ICM.

10   Installation is the reverse of removal.

TRX400EX models

Check

11   The ICM can only be diagnosed by a process of elimination: if all other possible causes of ignition problems have been checked and eliminated, the ICM is probably at fault. Some components of the ignition system - the ignition coil, the ignition pulse generator and the exciter coil - can only be tested with one of the following:

    a) *Imrie diagnostic tester model 625*

7.8 Point the timing light into the timing hole; at idle, the line next to the F mark on the alternator rotor should align with the notch in the timing hole

b) *Digital multimeter (minimum impedance 10 meg-ohms per DC volt) and peak voltage adapter (Honda part no. 07HGJ-0020100 or equivalent)*

Because of the special tools needed to check these components, you should take the vehicle to a dealer with the right equipment for the job.

12   If you have the proper equipment, or have access to it, troubleshoot the ICM as follows:

13   Check the ignition coil peak voltage (see Section 3).

14   Check the exciter coil peak voltage (see Section 4).

15   Check the ignition pulse generator peak voltage (see Section 5).

16   If the rest of the system checks out, the ICM is probably defective. But before buying a new one, it's a good idea to substitute a known good ICM or have it tested by a Honda dealer.

Replacement

17   Remove the front fenders (see Chapter 7). Locate the ignition control module on the left side of the frame, above the left front shock absorber.

18   Remove the ICM from its rubber mount and unplug the electrical connectors.

19   Installation is the reverse of removal.

7   Ignition timing - general information and check

General information

1   The ignition timing cannot be adjusted. But none of the ignition system parts are subject to mechanical wear, so there should be no need to check timing, unless you're troubleshooting a problem such as loss of power.

2   The ignition timing is checked with the engine running, both at idle and at a higher rpm on TRX300EX models and at idle only on TRX400EX models.

3   A (less expensive) neon timing light is probably adequate, but it might produce such dim pulses that the timing marks are hard to see. An externally-powered xenon timing light is more precise because it doesn't use the vehicle battery, which can produce an incorrect reading because of stray electrical impulses in the system.

Check

Refer to illustration 7.8

4   Warm the engine to normal operating temperature, make sure the transmission is in Neutral, and then shut the engine off.

5   Remove the timing hole plug (see "Valve clearance - check and adjustment" in Chapter 1).

6   Connect the timing light and a tune-up tachometer to the engine, following manufacturer's instructions.

7   Start the engine. Make sure it idles at the speed listed in the Chapter 1 Specifications. Adjust if necessary.

8   Point the timing light into the timing hole **(see illustration)**. At idle, the line next to the F mark on the alternator rotor should align with the notch in the timing hole. (The notch is in the bottom of the hole on TRX300EX models and in the top of the hole on TRX400EX models.)

9   On TRX300EX models, raise engine speed to the elevated rpm listed in this Chapter's Specifications. The notch at the top of the timing window should now be between the two advance timing marks on the alternator rotor.

10   If the timing is incorrect and all other ignition components have tested as good, the ICM may be defective. Have it tested by a Honda dealer.

11   When the check is complete, grease the timing window O-ring, then install the O-ring and cap and disconnect the test equipment.

4

Notes

Chapter 5
Steering, suspension and final drive

Contents

Specifications

Tie-rod balljoint spacing (distance between balljoints)	
TRX300EX	226 mm (8.9 inches)
TRX400EX	370.2 mm (14.57 inches)
Front spring free length (TRX300EX)	
Standard	181.0 mm (7.13 inches)
Limit	177.4 mm (6.98 inches)
Rear spring free length (TRX300EX)	
Standard	242.0 mm (9.53 inches)
Limit	237.2 mm (9.34 inches)
Rear spring installed length (TRX400EX)	231.5 MM (9.11 inches)
Rear axle runout limit	3.0 mm (0.12 inch)

Torque specifications

Steering

Handlebar brackets	
Upper bracket bolts	27 Nm (20 ft-lbs)
Lower bracket stud-to-steering shaft nuts	45 Nm (33 ft-lbs)
Steering shaft	
Nut at lower end of steering shaft	59 Nm (51 ft-lbs)
Steering shaft bushing bracket bolts	
TRX300EX	28 Nm (20 ft-lbs)
TRX400EX	32 Nm (24 ft-lbs)
Suspension arm balljoint-to-steering knuckle nuts	
TRX300EX	55 Nm (40 ft-lbs)
TRX400EX	29 Nm (22 ft-lbs)
Tie-rods	
Tie-rod-to-steering knuckle nuts	45 Nm (33 ft-lbs)
Tie-rod-to-steering shaft nuts	45 Nm (33 ft-lbs)

5

Front suspension

Front shock absorber mounting bolts/nuts (upper and lower)	45 Nm (33 ft-lbs)
Front suspension arm pivot bolts/nuts (upper and lower arms)	39 Nm (29 ft-lbs)

Rear suspension

Axle bearing holder pinch bolts	2 Nm (16 ft-lbs)
Rear shock absorber mounting bolts/nuts	
TRX300EX (upper and lower)	108 Nm (80 ft-lbs
TRX400EX	
Upper	108 Nm (80 ft-lbs)
Lower (shock-to-shock arm)	59 Nm (43 ft-lbs)
Shock arm and shock link (TRX400EX models)	
Shock arm-to-frame bolt/nut	59 Nm (43 ft-lbs)
Shock arm-to-shock link bolt/nut	59 Nm (43 ft-lbs)
Shock link-to-swingarm bolt/nut	45 Nm (33 ft-lbs)
Rear axle locknuts*	
Outer locknut	
Actual	88 Nm (65 ft-lbs)
Indicated	79 Nm (59 ft-lbs)
Inner locknut	
Actual	128 Nm (94 ft-lbs)
Indicated	115 Nm (85 ft-lbs)
Swingarm pivot bolt and nut	108 Nm (80 ft-lbs)

Final drive

Drive (engine) sprocket bolts	No specified torque
Driven (axle) sprocket bolts	59 Nm (44 ft-lbs)

Apply non-permanent thread locking agent to the threads.

1   General information

The steering system consists of a one-piece handlebar bolted to a steering shaft, which is connected to the steering knuckles by adjustable tie-rods. The knuckles are positioned by upper and lower balljoints in the outer ends of the front suspension arms.

The front suspension has upper and lower suspension arms on each side. The inner ends of the arms are connected to the frame by pivot bolts; the outer ends are connected to the steering knuckles by balljoints that are integral (non-removable). The front suspension uses a pair of shock absorber/coil springs units. The lower ends of the shocks are bolted to the lower suspension arms; the upper ends are bolted to brackets on the frame.

The rear suspension on all models consists of a single shock absorber with concentric coil spring and a steel swingarm. TRX400EX

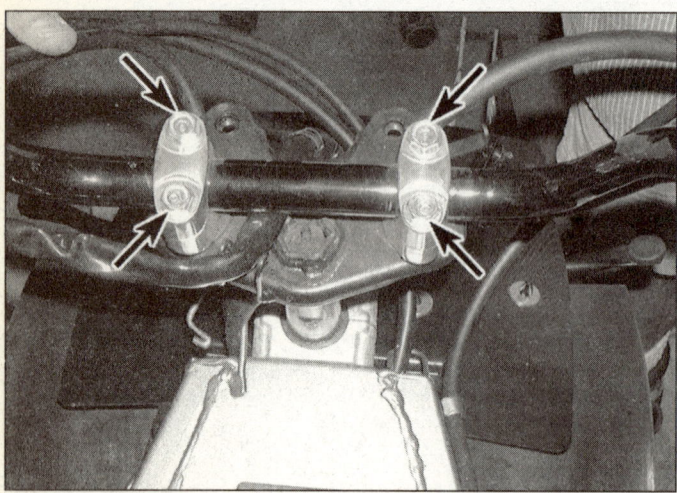

2.4 Mark the front side of each bracket, then remove the bolts (arrows), remove the brackets and lift off the handlebar (TRX300EX shown, TRX400EX similar)

models have a rising-rate linkage consisting of a shock arm and a shock link between the lower end of the shock and the swingarm.

All models use a chain-type final drive system. The procedures in this Chapter describe the removal and installation of the drive chain and sprockets. See "Drive chain and sprockets - inspection, adjustment and lubrication" in Chapter 1 for final drive maintenance.

2   Handlebar - removal and installation

1   Remove the handlebar cover (see Chapter 7).

Removing the handlebar to service other components

Refer to illustration 2.4

2   If you're removing the handlebar simply to gain access to some other component, it's not necessary to remove the clutch/parking brake lever bracket, the left switch housing, the throttle housing or the front brake master cylinder.

3   Look for a punch mark near the forward bolt on each handlebar bracket. If you can't find factory punch marks, make your own marks to ensure that the brackets are correctly oriented when they're reinstalled.

4   Remove the handlebar bracket bolts **(see illustration)**, remove the upper bracket halves, then lift the handlebar off the lower bracket halves. **Caution:** *Support the handlebar assembly with a piece of wire or rope; allowing it to hang free will damage the cables, hoses and wiring.*

5   Installation is the reverse of removal. Make sure that the factory punch marks (or the ones you made prior to disassembly) face to the front. Tighten the handlebar bracket bolts to the torque listed in this Chapter's Specifications.

Replacing the handlebar

6   If you're replacing the handlebar, remove all cable ties, then remove the following components:

a) *Remove the clutch/parking brake lever bracket* **(see illustration 17.7 in Chapter 2A)**.

b) *Remove the left switch housing (see "Handlebar switch housing - removal and installation" in Chapter 8).*
c) *Remove the throttle housing* **(see illustration 10.6 in Chapter 3)**.
d) *Remove the front brake master cylinder (see "Front brake master cylinder - removal, overhaul and installation" in Chapter 6).*

7   Follow Steps 3 through 5 above.
8   Installation is the reverse of removal. Tighten the handlebar bracket bolts to the torque listed in this Chapter's Specifications.

3   Steering shaft - removal, inspection and installation

Removal

Refer to illustrations 3.3, 3.4, 3.6, 3.7, 3.8a, 3.8b, 3.9a and 3.9b
1   Remove the front fenders (see Chapter 7).
2   On TRX300EX models, remove the headlight and headlight housing (see Chapter 8).
3   Detach the handlebar assembly from the steering shaft **(see illustration)**. (It's not necessary to remove anything from the handlebar, but be sure to support the handlebar assembly to protect the wiring, cables and hoses, and make sure that the master cylinder is upright.)
4   Remove the four bolts (TRX300EX models) or two bolts (TRX400EX models) from the upper steering shaft bushing bracket **(see illustration)**. Note the location of the brake hose guide and any cable or wiring harness guides.
5   Disconnect the inner ends of the tie-rods from the steering shaft (see Section 4).
6   Remove the cotter pin and nut from the bottom of the steering

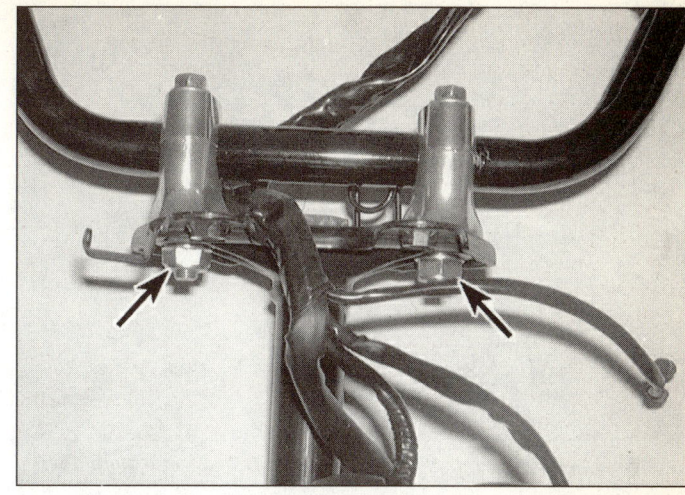

3.3 To detach the handlebar and brackets as a single assembly from the steering shaft, remove these nuts (arrows)

shaft **(see illustration)**.
7   Remove the steering shaft from the vehicle. Remove the steering shaft collar **(see illustration)**.
8   Remove the upper and lower dust seals **(see illustrations)**. Discard both seals.
9   Remove the snap-ring and drive out the steering shaft bearing **(see illustrations)**.

3.4 Note the locations of the hose and cable guides and remove the bolts (arrows) from the upper steering shaft bushing bracket (TRX300EX shown; the TRX400EX has two bolts)

3.6 To detach the lower end of the steering shaft from the frame, remove this cotter pin and nut (arrow)

3.7 Remove the steering shaft collar

5

3.8a Pry out the upper dust seal . . .

3.8b . . . and the lower dust seal; discard both seals

3.9a Remove the snap-ring . . .

3.9b . . . and drive out the steering shaft bearing from below

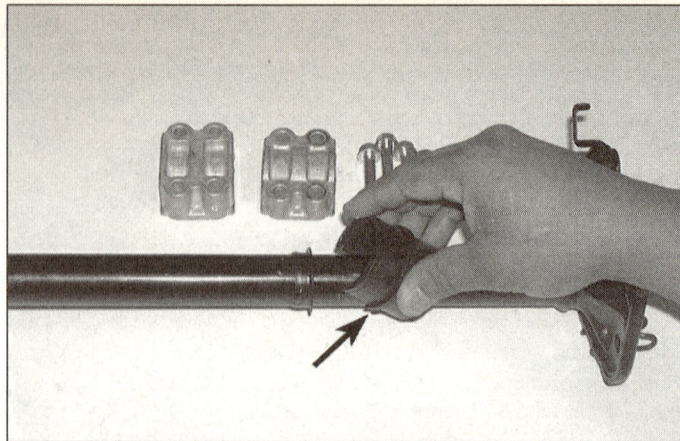

3.11a Remove the upper bushing (arrow) from the steering shaft; replace the bushing if it's worn or damaged

Inspection

Refer to illustrations 3.11a and 3.11b

10   Clean all the parts with solvent and dry them thoroughly, using compressed air, if available.

11   Inspect the steering shaft bushing and bearing for damage and excessive wear **(see illustrations)**. Replace them if they're worn or damaged.

12   Inspect the steering shaft for bending or other signs of damage. Do not attempt to repair any steering components. Replace them with new parts if defects are found.

Installation

13   Installation is the reverse of removal, with the following additions:

a)   *Lubricate the inner surface of the upper bushing and the lips of the lower bearing dust seals with multipurpose grease.*

b)   *Tighten all fasteners to the torque values listed in this Chapter's Specifications.*

c)   *Use a new cotter pin to secure the nut at the lower end of the steering shaft.*

---

4   Tie-rods - removal, inspection and installation

Removal

Refer to illustrations 4.2a, 4.2b, 4.3, 4.4a and 4.4b

1   If both of the tie-rods are to be removed, mark them "Left" and "Right" so they're not accidentally switched during reassembly.

2   Remove the cotter pin from the nut at the outer end of the tie-rod and remove the nut **(see illustrations)**.

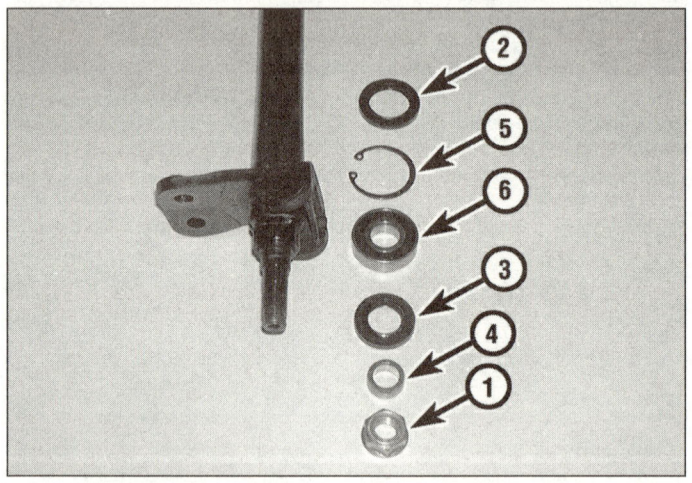

3.11b Steering shaft lower bearing and seals:

1 *Lower steering shaft nut*	4 *Steering shaft collar*
2 *Upper dust seal*	5 *Snap-ring*
3 *Lower dust seal*	6 *Bearing*

3   Separate the tie-rod stud from the knuckle with a tie rod puller **(see illustration)** or "pickle fork" balljoint separator. **Caution:** *It's very easy to damage the rubber boot on the tie-rod with a pickle fork separator. If you're going to use the tie-rod again, it's best to use another type of tool.*

4   Repeat Steps 1 and 2 to disconnect the inner end of the tie rod from the steering shaft **(see illustrations)**.

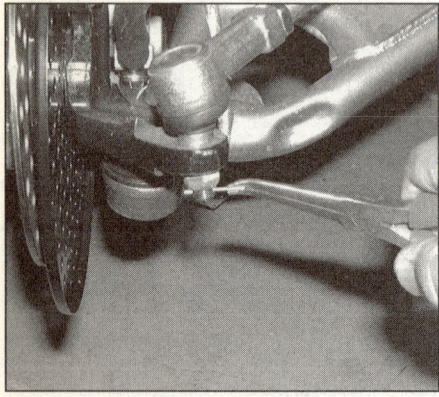

4.2a To disconnect the outer tie-rod from the steering knuckle, remove the cotter pin . . .

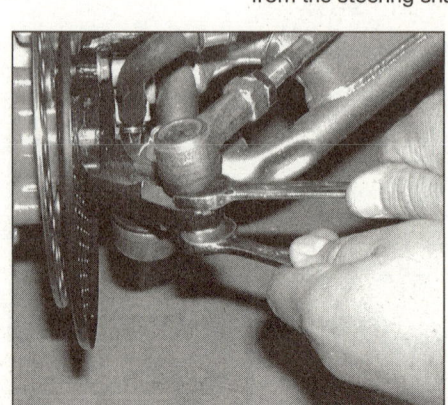

4.2b . . . and, using a back-up wrench on the flat portion of the stud above the steering knuckle, remove the locknut

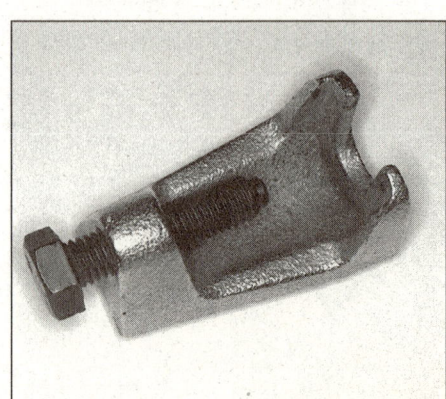

4.3 This automotive type tie-rod puller will work equally well on tie-rods and on balljoints

4.4a To disconnect the inner tie-rod end from the steering shaft, remove the cotter pin (arrow) . . .

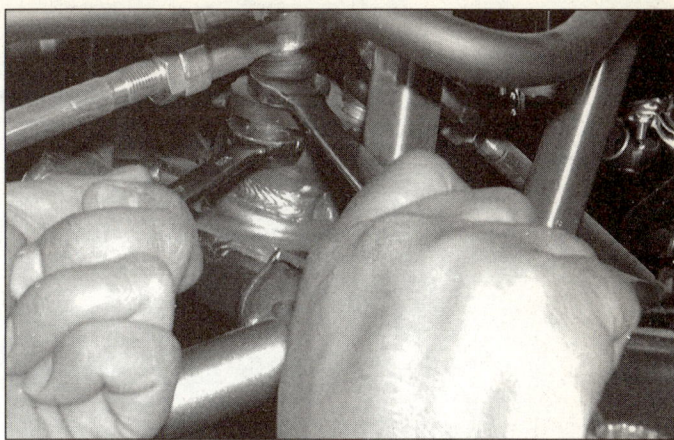

4.4b . . . and, using a back-up wrench on the flat portion of the stud above the steering shaft arm, remove the locknut

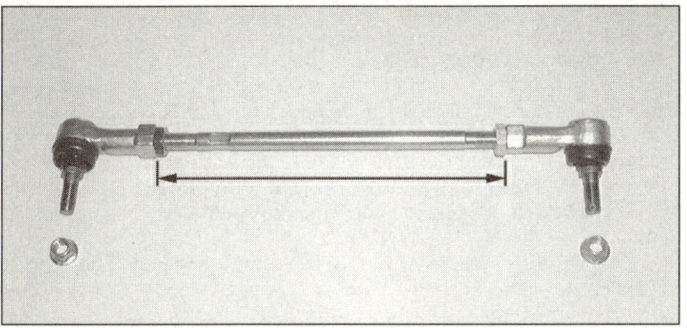

4.7 Before installing the tie-rod, make sure that the distance between the tie-rod ends (not the distance between the locknuts) is correct

Inspection

5   Inspect the tie-rod shaft for bending or other damage and replace it if any problems are found. Don't try to straighten the tie-rod shaft.
6   Check the tie-rod balljoint boots for cracks or deterioration. Twist and rotate the threaded studs. They should move easily, without roughness or looseness. If a boot or stud shows any problems, unscrew the tie-rod end from the tie-rod and install a new one.

Installation

Refer to illustration 4.7
7   Install the tie-rod end locknuts first. Make sure that the gold

locknut goes on the outer end of the tie-rod shaft (the end with the flat, for toe-in adjustment). Thread the tie-rod ends onto the tie-rod (gold-colored tie-rod end on the outer end) until the distance between the tie-rod ends **(see illustration)** is as listed in this Chapter's Specifications. Make sure that the length of the exposed threads on each end of the tie-rod is even. Tighten the locknuts securely.
8   The remainder of installation is the reverse of removal, with the following additions:

a) *Use new cotter pins and bend them to hold the nuts securely.*
b) *Check front wheel toe-in and adjust as necessary (see Chapter 1).*

5   Steering knuckles - removal, inspection, and installation

Removal

Refer to illustration 5.3, 5.5 and 5.6
1   Jack up the front end of the vehicle and support it securely on jackstands. Remove the front wheels.
2   Remove the front wheel, the hub, the brake caliper and the brake disc (see Chapter 6).
3   Remove the splash guard bolt **(see illustration)** and remove the splash guard.
4   Disconnect the outer end of the tie-rod from the steering knuckle **(see illustrations 4.2a and 4.2b)**.
5   Remove the cotter pins and nuts from the upper and lower balljoint studs **(see illustration)**.

5

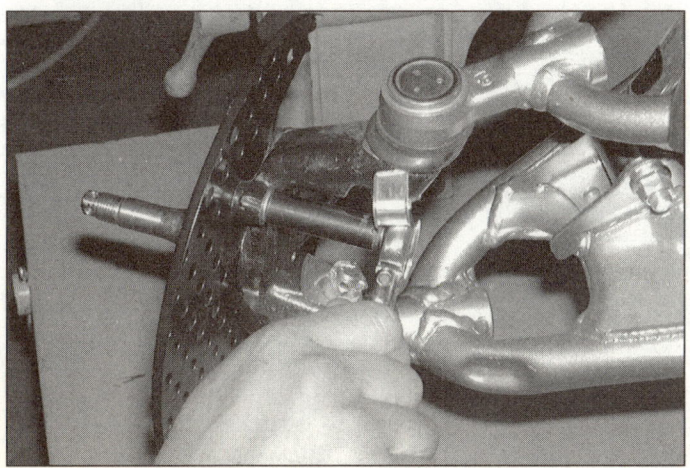

5.3 Remove the bolt and remove the splash guard from the steering knuckle

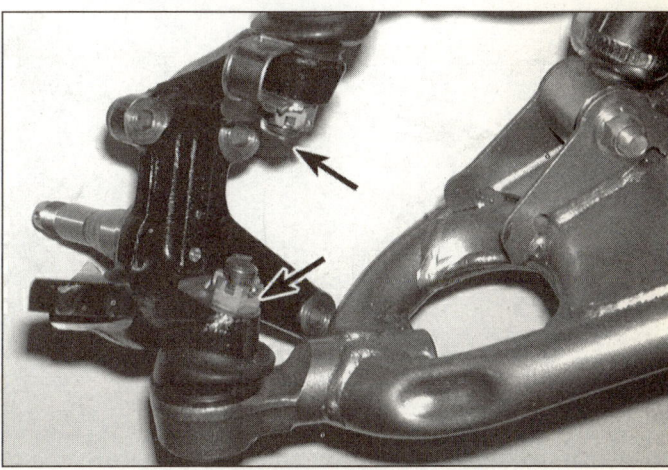

5.5 Remove the upper and lower balljoint cotter pins and castle nuts (arrows)

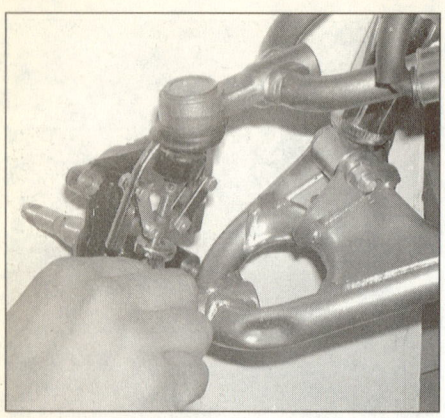

5.6 Honda has special tools for separating the balljoints from the steering knuckle, but you can obtain a small puller similar to this at many tool stores

6.3a To remove a front shock absorber, remove the upper nut and bolt (arrows) . . .

6.3b . . . and remove the lower nut and bolt (arrows)

6   Separating the balljoints from the steering knuckle requires a balljoint remover tool (Honda tool no. 07MAC-SL00200). Smaller automotive balljoint separator tools **(see illustration 4.3 and the accompanying illustration)** are suitable (try it for fit before you buy it, if possible) and can be rented from tool yards or purchased inexpensively.

7   If balljoint separation proves difficult, the knuckle and suspension arms can be removed as a single assembly, then taken to a Honda dealer for balljoint removal.

Inspection

8   Inspect the knuckle carefully for cracks, bending or other damage. Replace it if any problems are found. If the vehicle has been in a collision or has been bottomed hard, it's a good idea to have the knuckle magnafluxed by a machine shop to check for hidden cracks.

Installation

9   Installation is the reverse of removal, with the following addition: Use new cotter pins and tighten the nuts to the torque listed in this Chapter's Specifications.

---

6   Shock absorbers - removal and installation

Front shock absorbers

Refer to illustrations 6.3a and 6.3b
Note: *This procedure applies to either front shock absorber.*

1   Raise the front of the vehicle and support it securely on jackstands. Remove the front wheels (see Chapter 6). (You don't have to

remove the wheels to remove the shock absorbers, but you'll have more room to work if you remove them.)

2   Support the outer ends of the lower suspension arms with jackstands so they won't drop when the shock absorbers are removed.

3   Remove the nuts and bolts that attach the upper end of the shock to the frame bracket and the lower end to the bracket on the lower suspension arm **(see illustrations)**. Discard the old nuts. Separate the shock from the frame bracket and from the control arm bracket and lift it out.

4   Inspect the shock absorber for signs of wear or damage such as oil leaks, bending, a weak spring and worn bushings. Replace both shock absorbers as a pair if any problems are found.

5   Installation is the reverse of the removal steps. Using new upper and lower nuts, tighten the nuts and bolts to the torque listed in this Chapter's Specifications.

Rear shock absorber

Refer to illustrations 6.7, 6.8a, 6.8b, 6.9a and 6.9b
Warning: *Do not attempt to disassemble these shock absorbers. They are nitrogen-charged under high pressure. An incorrectly disassembled shock absorber could cause serious injury. Instead, take the shock to a Honda dealer service department with the right equipment to do the job.*

6   Remove the seat/rear fender assembly (see Chapter 7).

7   Jack up the rear end of the vehicle and support it securely on jackstands **(see illustration)**. Support the swingarm to prevent it from dropping when the lower end of the shock absorber is disconnected.

8   On TRX300EX models, remove the nut and bolt from the lower end of the shock absorber **(see illustration)** and disconnect the shock

6.7 Raise the rear of the machine and support it on jackstands as shown

6.8a To remove the TRX300EX rear shock absorber, remove the nut and bolt (arrows) from the lower end of the shock . . .

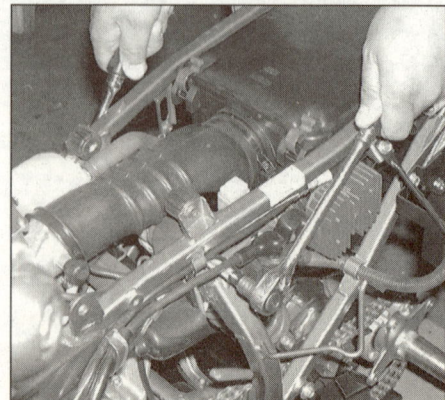

6.8b . . . and remove the nut and bolt from the upper end of the shock

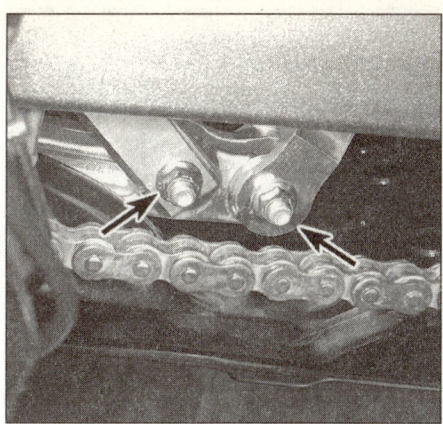

6.9a The lower end of the TRX400EX shock absorber (left arrow) is bolted to the shock link, which is bolted to the shock arm (right arrow)

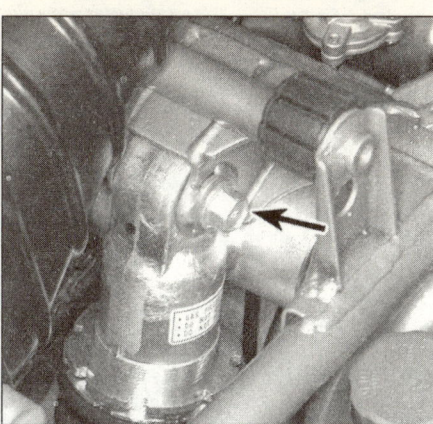

6.9b ... and remove the nut (not visible in this photo) and bolt (arrow) from the upper end of the shock

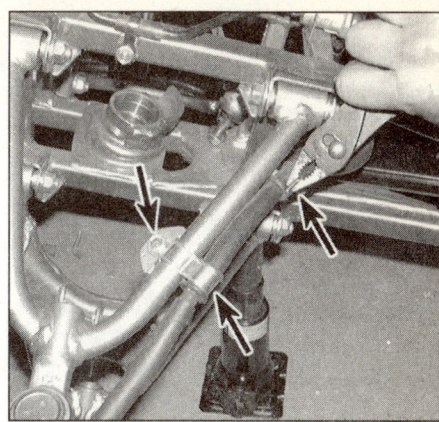

7.4 If you're removing an upper suspension arm, be sure to remove the outer hose guide bolt (arrow) and disengage the brake hose from the hose guides (arrows)

from the swingarm. Remove the nut and bolt from the upper end of the shock **(see illustration)** and disconnect the shock from the frame bracket. Remove the shock absorber assembly.

9   On TRX400EX models, remove the nut and bolt from the lower end of the shock absorber **(see illustration)** and disconnect the shock from the shock arm. Remove the nut and bolt from the upper end of

7.5a To remove an upper suspension arm, remove these nuts and bolts (arrows)

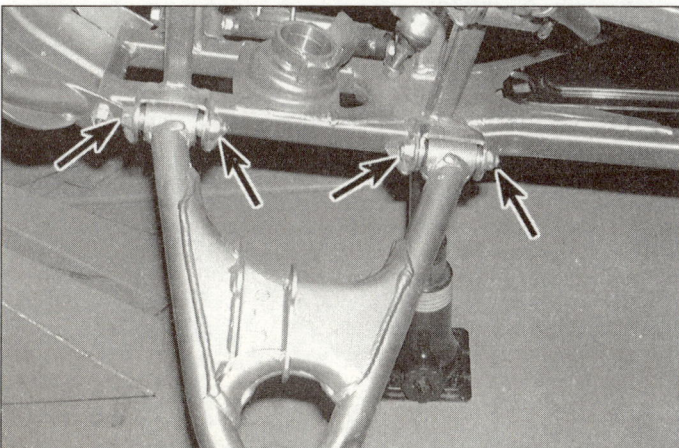

7.5b To remove a lower suspension arm, remove these nuts and bolts (arrows)

the shock **(see illustration)** and disconnect the shock from the frame bracket. Remove the shock absorber.

10   Inspect the needle bearings in the upper and lower shock absorber pivot points. If they're worn or damaged, have the old ones pressed out and new ones pressed in by a Honda dealer or other qualified machine shop. Press the bearings in to a depth of 5.5 mm (0.220 inch).

11   Installation is the reverse of the removal steps. Tighten the nuts and bolts to the torques listed in this Chapter's Specifications.

7   Suspension arms and balljoints - removal, inspection and installation

Removal

Refer to illustrations 7.4, 7.5a and 7.5b

1   Securely block both rear wheels so the vehicle won't roll. Loosen the front wheel nuts with the tires still on the ground, then jack up the front end, support it securely on jackstands and remove the front wheels.

2   Remove the front fender and the front bumper (see Chapter 7).

3   Remove the steering knuckle (see Section 5).

4   If you're removing an upper arm, disengage the brake hose from the arm **(see illustration)**.

5   Remove the pivot bolts and nuts and remove the suspension arm **(see illustrations)**.

Inspection

6   Inspect the suspension arm(s) for bending, cracks or corrosion. Replace damaged parts. Don't attempt to straighten them.

7   Inspect all rubber bushings for cracks or deterioration. Check the inner collars on the lower suspension arms for damage or corrosion. Inspect the pivot bolts for wear as well. Replace the bushings and pivot bolts if they're worn or deteriorated.

8   Check the balljoint boot for cracks or deterioration. Twist and rotate the threaded stud. It should move easily, without roughness or looseness. The balljoints can't be replaced separately from the suspension arms. If the boot or stud show any problems, replace the suspension arm together with the balljoint.

Installation

9   Installation is the reverse of removal, but don't torque the suspension fasteners while the vehicle is off the ground. Tighten the nuts and bolts *slightly* while the vehicle is jacked up, and then tighten them to the torque listed in this Chapter's Specifications after the vehicle is resting on its wheels.

5

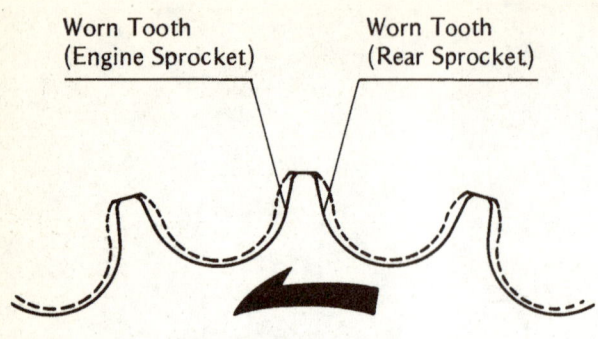

9.1 Check the sprocket teeth in the indicated areas to see if they're worn excessively

8   Drive chain - removal, cleaning, inspection and installation

Removal

1   Remove the drive sprocket cover (see Section 6).
2   Remove the swingarm (see Section 13).
3   Lift the chain off the driven sprocket and remove it from the vehicle.
4   Inspect the chain guide and rollers for wear or damage (see Chapter 1) and replace them as necessary.

Cleaning and inspection

5   The drive chain has small rubber O-rings between the chain plates. Soak the chain in kerosene and use a brush to work the solvent into the spaces between the links and plates. **Caution:** *Do NOT use steam, high-pressure washes or solvents, all of which can damage the O-rings, to clean the chain. Use only kerosene.*
6   Wipe the chain dry, then inspect it carefully for worn or damaged links. Replace the chain if wear or damage is found at any point.
7   If the chain is worn or damaged, inspect the sprockets (see Section 9). If they're worn or damaged, replace them. **Caution:** *Do NOT install a new chain on worn sprockets; it will wear out quickly.*
8   Lubricate the chain (Honda recommends Pro Honda Chain Lube or an equivalent aftermarket chain lubricant designed specifically for use on O-ring chains).

Installation

9   Installation is the reverse of removal.
10   Adjust the chain when you're done (see Chapter 1).

9.3 Remove the bolts (arrows) and the sprocket cover (on TRX300EX models, the upper bolt attaches the clamp for the reverse shifter rod)

9   Sprockets - check and replacement

Check

Refer to illustration 9.1

1   Inspect the teeth on the drive (engine) sprocket and the driven (axle) sprocket for wear **(see illustration)**. The engine sprocket is visible through the cover slots. If either sprocket is worn, replace it.
2   Inspect the drive chain (see Section 8) and replace it if it's worn. Installing a worn chain on new sprockets will cause them to wear quickly.

Replacement

Drive (engine) sprocket

Refer to illustrations 9.3, 9.4a and 9.4b

3   Remove the drive sprocket cover **(see illustration)**.
4   Remove the sprocket bolts and the fixing plate and pull the sprocket off the transmission shaft **(see illustrations)**.
5   Inspect the seal behind the engine sprocket. If it has been leaking, pry it out (take care not to scratch the seal bore) and tap in a new seal with a socket the same diameter as the seal.
6   Installation is the reverse of removal. Make sure that the OUTSIDE mark on the sprocket faces out. Don't forget to install the fixing plate. Tighten the bolts securely.

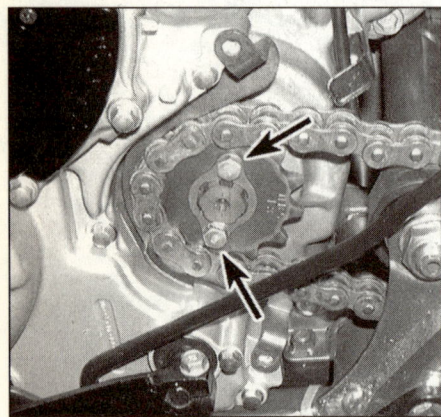

9.4a Remove the drive sprocket bolts (arrows), remove the fixing plate . . .

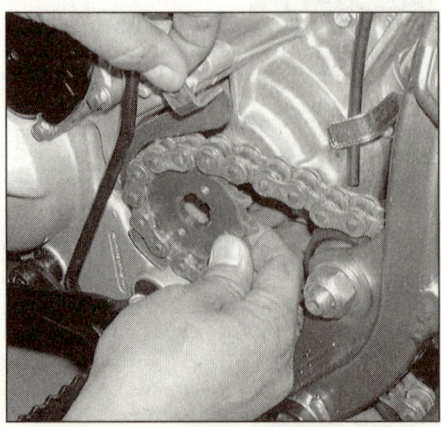

9.4b . . . and slide the sprocket off the splined transmission shaft

9.7 The driven sprocket is secured by four bolts (fourth bolt hidden) and nuts

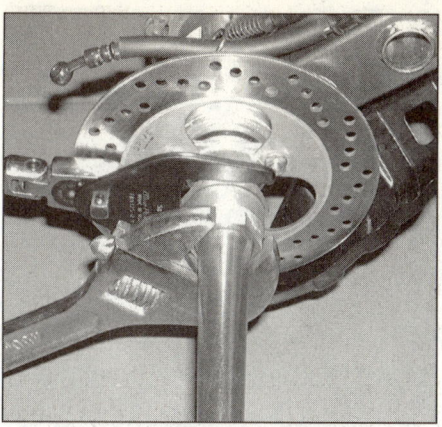

10.4a Turn the inner axle locknut clockwise to loosen it (it has left-hand threads), then unscrew the outer locknut counterclockwise . . .

10.4b . . . remove the stopper ring . . .

10.4c . . . and remove both locknuts

Driven (axle) sprocket

Refer to illustration 9.7

7   Have an assistant apply the rear brake while you loosen the driven sprocket bolts **(see illustration)** and nuts.

8   Remove the left rear wheel and hub (see Chapter 6).

9   Remove the drive chain (see Section 8).

10   Unbolt the driven sprocket from the sprocket flange and slip the sprocket off the axle.

11   Installation is the reverse of removal. After the drive chain, hub and wheel have been installed and the vehicle is back on the ground, tighten the driven sprocket bolts and nuts to the torque listed in this Chapter's Specifications.

12   Adjust the chain when you're done (see Chapter 1).

10 Rear axle - removal, inspection and installation

Removal

Refer to illustrations 10.4a, 10.4b, 10.4c, 10.6a, 10.6b, 10.8 and 10.9

1   Have an assistant apply the rear brake while you loosen the driven sprocket bolts **(see illustration 9.7)**.

2   Block the front wheels so the vehicle won't roll. Jack up the rear end of the vehicle and support it securely, positioning the jackstands so they won't obstruct removal of the axle. The supports must be secure enough so the vehicle won't be knocked off of them while the

axle is removed. **Warning:** *The axle nuts are secured with Loctite and tightened to a very high torque. When tightening or loosening them with the vehicle jacked up, make sure the vehicle is securely positioned on jackstands so it can't fall.*

3   Remove the rear wheels, the rear brake caliper and the rear wheel hubs (see Chapter 6).

4   Loosen the inner axle locknut **(see illustration)** by turning it *clockwise* (the inner axle locknut has left-hand threads) to unlock the outer nut. Unscrew the outer nut by turning it *counterclockwise*, then remove the stopper ring and the outer and inner axle locknuts **(see illustrations)**. **Note:** *You'll need special wrenches for the inner and outer axle locknuts: the inner nut takes a 56 mm wrench, the outer locknut a 45 mm wrench. Honda sells special tools for this job (56 mm wrench, tool no. 07916-HA20000 or 07916-HA2010A; 45 mm wrench, tool no. 07916-1870101). You can also obtain aftermarket wrenches similar to the Honda tools. If you can't find anything else, you can always make your own tools from 1/4-inch or thicker steel plate. You can also use a pair of pipe wrenches, but they should be used only as a last resort, because they will mar the axle locknuts.*

5   On TRX400EX models, unbolt the skid plate from the swingarm (see Chapter 7).

6   Slide the brake disc hub off the axle **(see illustration)**. Remove the old O-ring from the hub **(see illustration)** and discard it. It's not necessary to detach the disc from the hub, but this is a good time to inspect the disc (see "Rear brake disc – inspection, removal and installation" in Chapter 6).

5

10.6a Slide the rear disc and disc hub off the axle

10.6b Remove the old disc hub O-ring (arrow) from its groove in the hub and discard it

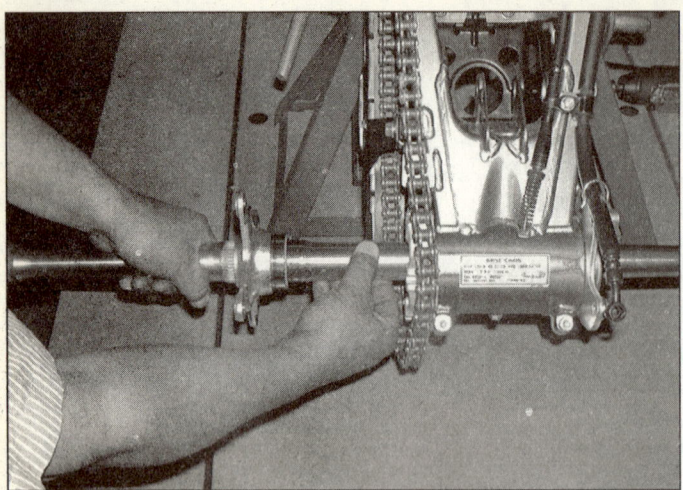

10.8 Pull the axle out of the left end of the axle housing

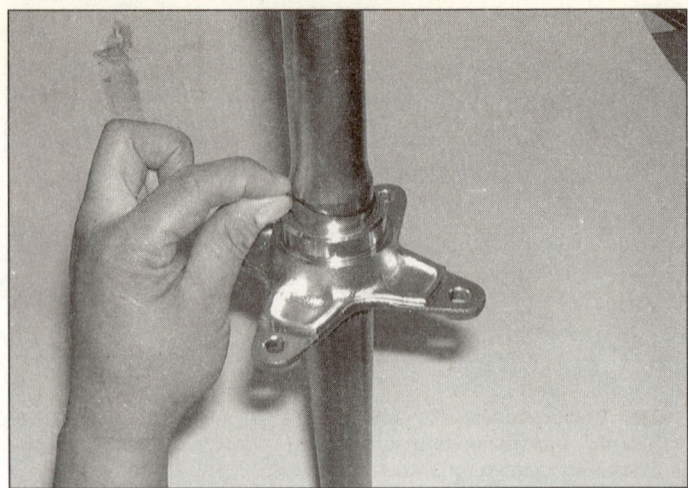

10.9 Remove the old O-ring from the axle and discard it; always use a new O-ring when installing the axle

7   Unbolt the driven sprocket from the driven sprocket flange and detach the sprocket and chain from the flange. Disengage the sprocket from the chain and slip it off the axle. Drape the chain over the chain adjuster for now.

8   Tap on the right end of the axle with a soft faced hammer to free it, then pull it out of the left end of the axle housing **(see illustration)**. Hang the chain with a piece of wire so it doesn't pick up dirt from, or deposit chain lube on, the floor.

9   Remove the old O-ring from the driven sprocket flange and discard it **(see illustration)**. Remove the driven sprocket flange if it's damaged or if you're replacing the axle.

Inspection

Refer to illustrations 10.10 and 10.11

10   Inspect the axle for obvious damage, such as step wear of the splines or bending, and replace it as necessary **(see illustration)**.

11   Place the axle in V-blocks and set up a dial indicator to contact each of the outer ends in turn **(see illustration)**. Rotate the axle and compare runout to the value listed in this Chapter's Specifications. If runout is excessive, replace the axle.

12   Inspect the brake disc (see Chapter 6) and replace it if necessary.

13   Inspect the drive and driven sprockets and the drive chain (Sections 8 and 9).

14   Inspect and, if necessary, replace the axle bearings and seals (see Section 11).

Installation

Refer to illustration 10.20

15   Install the rear axle bearings and seals, if removed (see Section 11).

16   Coat a new O-ring with oil and install the O-ring between the

driven sprocket flange and the axle **(see illustration 10.9)**.

17   Install the driven sprocket flange, if removed (see Section 9). Install the driven sprocket on the flange and tighten the bolts securely.

18   Insert the axle through the axle bearings until it is fully seated. Make sure the splines for the rear brake disc flange hub don't damage the lips of the axle bearing seals.

19   Coat a new O-ring with oil and install the O-ring in the groove in the rear disc hub for the rear brake disc. Grease the splines for the rear disc hub and install the hub on the axle. If you removed the disc from the hub, be sure to install the disc with the stamped MIN TH (minimum thickness) facing out, and tighten the disc retaining nuts to the torque listed in the Chapter 6 Specifications.

20   Install the inner and outer axle locknuts. Apply a non-permanent thread-locking agent to the threads of the locknuts **(see illustration)**. Install the stopper ring onto the axle but don't install it in the ring groove yet.

21   Turn the outer axle locknut clockwise until the stopper ring groove is covered with the outer locknut. Turn the outer axle locknut counter-clockwise until the ring groove is visible, and then install the stopper ring into the groove. Grease the seating surface of the outer locknut and stopper ring, and then turn the outer locknut to seat it against the stopper ring. Tighten the outer axle locknut to the torque listed in this Chapter's Specifications. **Note:** *Because the indicated torque doesn't equal the actual torque when the torque wrench is offset from the turning axis, Honda uses a correction factor for this torque value. That's why there are two torque specifications listed. The first is the actual specified torque. The second is the indicated torque you should use (when using the Honda special tools) to achieve the actual specified torque.*

22   Tighten the inner axle locknut to the torque listed in this Chapter's Specifications.

23   Install the rear brake caliper (see Chapter 6).

24   On TRX400EX models, install the swingarm skid plate (see Chapter 7).

25   Install the rear wheel hubs and the rear wheels.

26   Lower the machine to the ground.

27   Check drive chain freeplay and adjust as necessary (see Chapter 1).

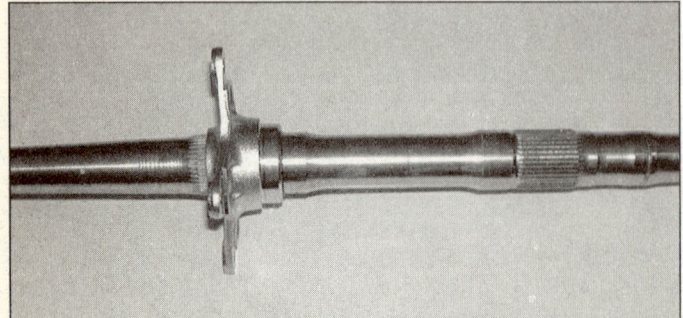

10.10 Inspect the axle splines and the axle surfaces in contact with the axle bearing seals

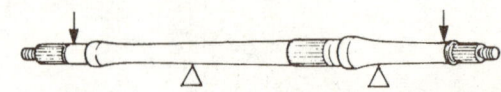

10.11 Place the axle in a pair of V-blocks and measure runout with a dial indicator at the ends (arrows)

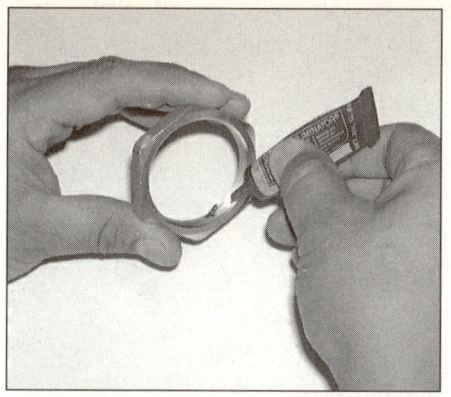

10.20 Be sure to coat the threads of the axle locknuts with a non-permanent thread-locking agent before installing them on the axle

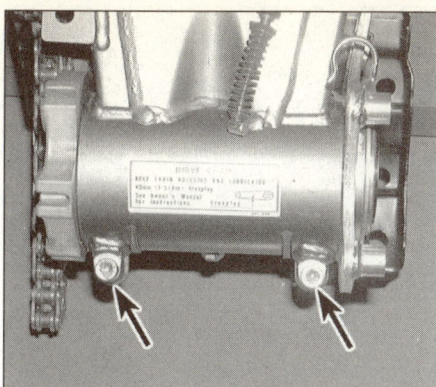

11.3a On TRX300EX models, remove these two axle bearing holder pinch bolts (arrows) . . .

11.3b . . . and remove the two pinch bolt rubber seals

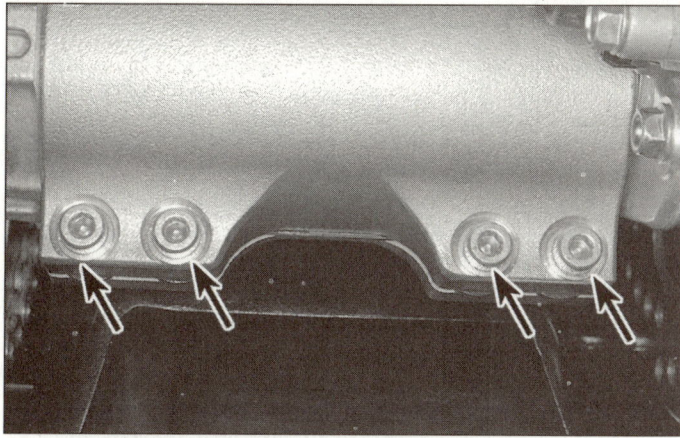

11.3c On TRX400EX models, remove these four axle bearing holder pinch bolts (arrows) and remove the single-piece pinch bolt rubber seal

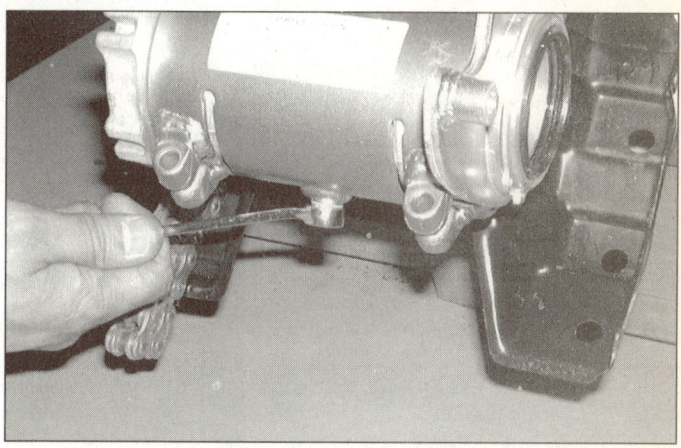

11.4 On TRX300EX models, remove the axle holder stopper bolt and washer

11 Rear axle bearings - removal, inspection and installation

Removal and inspection

Refer to illustrations 11.3a, 11.3b, 11.3c, 11.4, 11.5, 11.6a, 11.6b, 11.7a, 11.7b, 11.8, 11.9, 11.11, 11.12a and 11.12b

1   Jack up the rear end of the vehicle and support it securely on jackstands.

2   Remove the rear axle (see Section 10).

3   Loosen the axle bearing holder pinch bolts and the pinch bolt rubber seal(s) **(see illustrations)**.

4   On TRX300EX models, remove the axle holder stopper bolt and washer **(see illustration)**.

5   Remove the snap-ring **(see illustration)**.

6   Remove the rear brake caliper bracket and stopper bushing **(see illustrations)**.

5

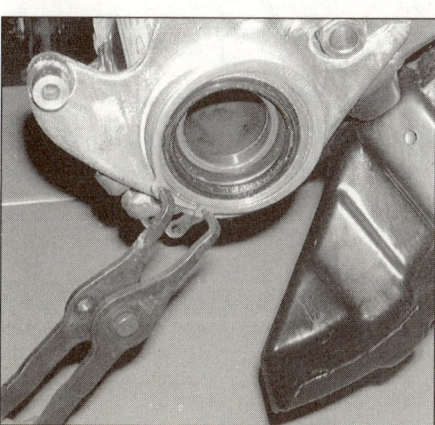

11.5 Remove the snap-ring from the right end of the bearing holder

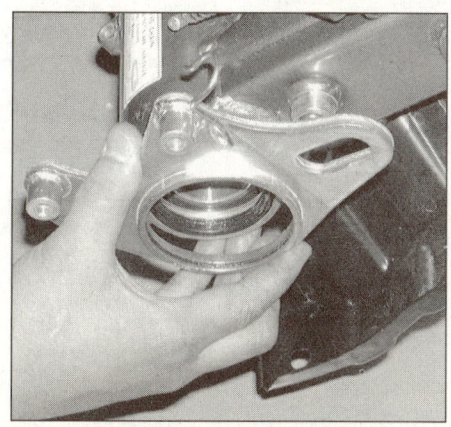

11.6a Remove the rear brake caliper bracket . . .

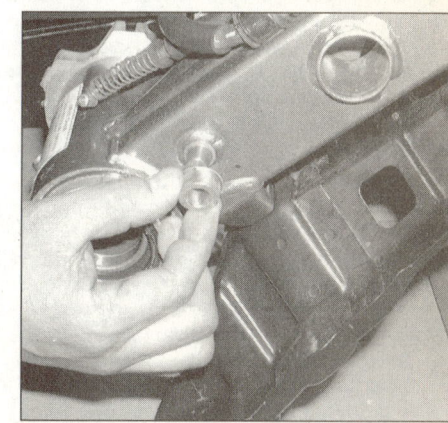

11.6b . . . and the stopper bushing

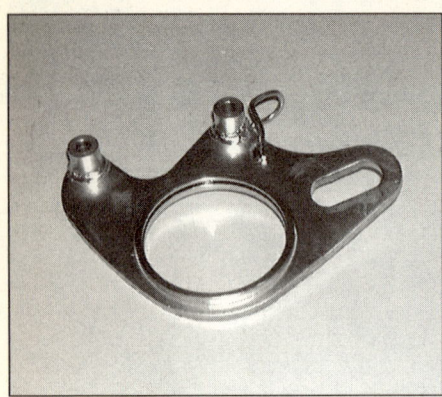

11.7a Remove the O-rings from the caliper bracket . . .

11.7b . . . and from the bearing holder

11.8 Tap the bearing holder from the right side of the swingarm and pull it out from the left

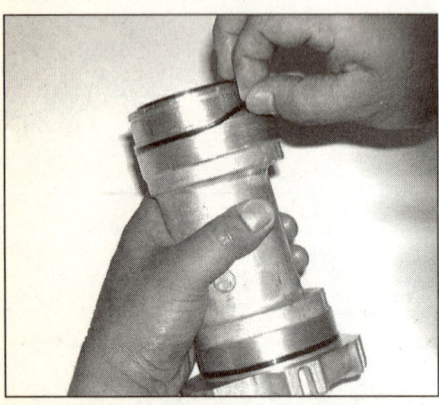

11.9 Remove the two O-rings from the bearing holder

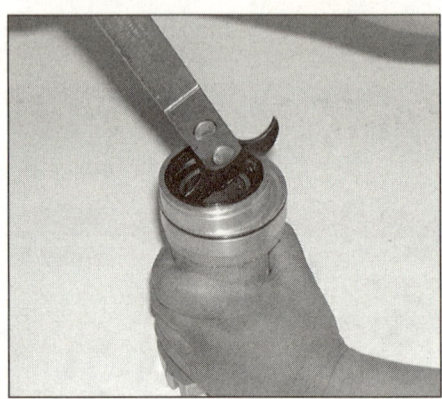

11.11 Pry the dust seals out of the ends of the bearing holder

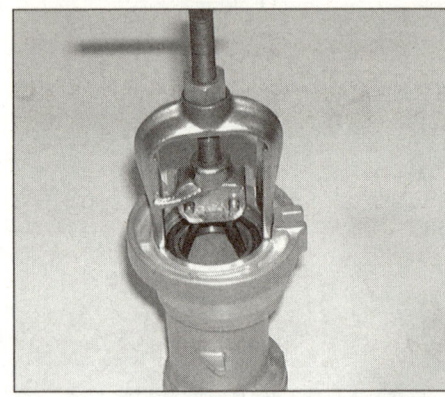

11.12a Remove the bearings from the holder with a suitable puller (shown), or tap them out with a brass punch

7   Remove the O-rings from the caliper bracket and from the bearing holder **(see illustrations)**.

8   Tap the bearing holder from the right side of the swingarm **(see illustration)** and pull it out from the left.

9   Remove the O-rings from the bearing holder **(see illustration)**.

10   Spin the bearing inside each end of the hub and check for roughness, looseness or noise.

11   Pry the dust seals out of the ends of the bearing holder **(see illustration)**.

12   Remove the bearings from the holder with a suitable puller **(see illustration)**. If you don't have a suitable puller, insert a brass punch into the hub and tilt it so it pushes the spacer aside and catches the edge of the bearing on the far side. To drive the bearing from the hub, tap gently against the outer race of the bearing, working the punch around the circumference of the bearing. Insert the drift from the other side and drive the other bearing out in the same way. Remove the spacer **(see illustration)**.

Installation

Refer to illustration 11.19

13   Clean the holder thoroughly inside and outside. Clean the surface of the recess in the swingarm where the holder is housed.

14   Place one of the new bearings in the holder, with the part number on the bearing facing out, and tap it into position with a bearing driver or socket the same diameter as the bearing outer race. Install the spacer from the other end of the holder, and then install the other bearing, again with its part number facing out.

15   Coat the new axle bearing holder O-rings with oil and install them in their grooves in the axle bearing holder **(see illustration 11.9)**.

16   Grease the outer surface of the axle bearing holder and insert it into the bearing holder recess in the swingarm. Make sure you don't

damage the O-rings.

17   Coat the new O-rings for the caliper bracket and the right end of the axle bearing holder and install them into the groove in the caliper bracket and onto the end of the bearing holder **(see illustrations 11.7a and 11.7b)**.

18   Grease the inner surface of the rear caliper bracket and install the bracket on the right end of the axle bearing holder and install the stop-

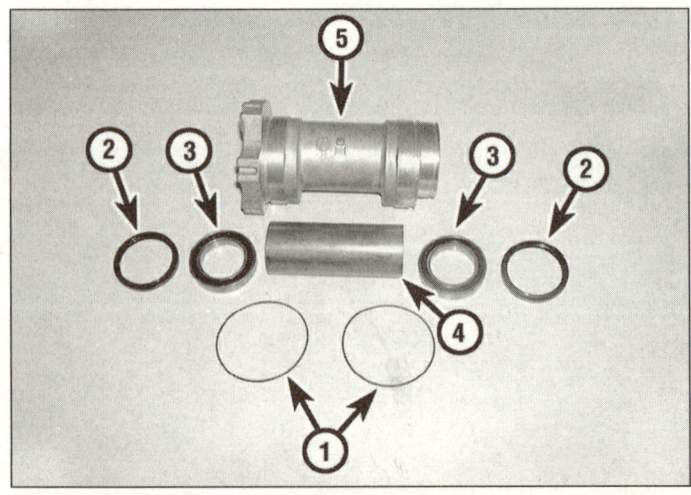

11.12b Rear axle bearing holder details

1	*O-rings*	*4 Spacer*
2	*Seals*	*5 Axle bearing holder*
3	*Bearings*	

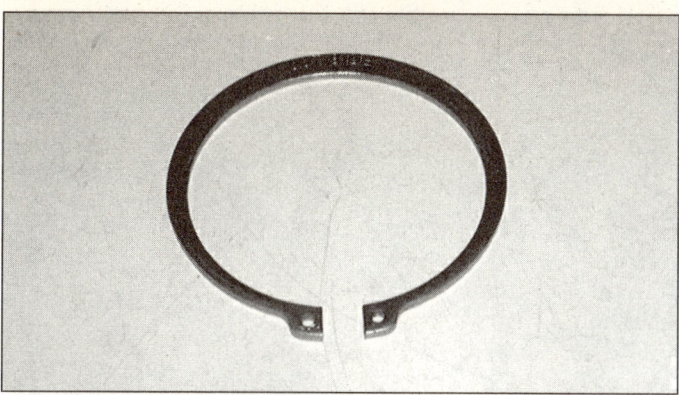

11.19 When installing the snap-ring in the caliper bracket, make sure that the OUTSIDE marking faces out

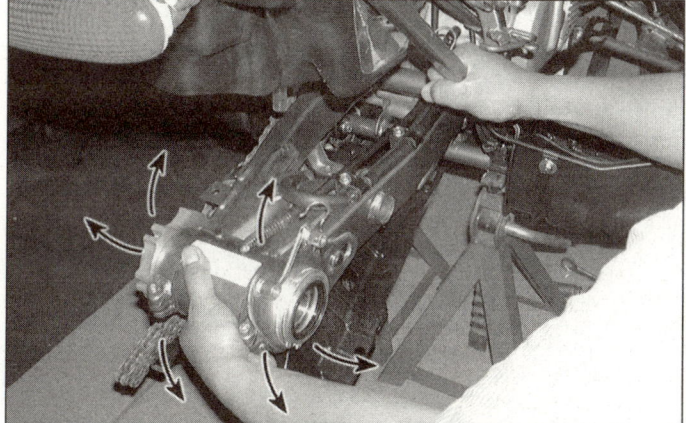

12.2 Move the swingarm up-and-down and from side-to-side to check the bearings

per bushing on the swingarm **(see illustrations 11.6a and 11.6b)**.

19   Install the large snap-ring into the groove in the axle bearing holder **(see illustration 11.5)** with its OUTSIDE marking facing out **(see illustration)**.

20   Pack the dust seal cavities with about 1 gram (0.04 ounce) of grease per cavity. Install a new seal in each end of the holder with its sealed side (the side with the part number on it) facing out. Tap in the seals until they're flush with axle bearing holder. Coat the seal lips with multipurpose grease.

21   Installation is otherwise the reverse of removal. Be sure to tighten

the hub nuts and bolts to the torque listed in this Chapter's Specifications.

22   Adjust the drive chain when you're done (see Chapter 1).

12 Swingarm bearings - check

Refer to illustration 12.2

1   Remove the rear wheels (see Chapter 6), then remove the rear shock absorber (see Section 6).

2   Grasp the rear of the swingarm with one hand and place your other hand at the junction of the swingarm and the frame **(see illustration)**. Try to move the rear of the swingarm from side-to-side. If the bearings are worn, they will allow some freeplay, which produces movement between the swingarm and the frame at the front (the swingarm will move forward and backward at the front, not from side-to-side). If there's any play, remove the swingarm (see Section 13) and replace the swingarm bearings (see Section 14).

3   Move the swingarm up and down through its full travel. It should move freely, without any binding or rough spots. If it does not move freely, remove the swingarm (see Section 13) and inspect the bearings (see Section 14). You might be able to clean and lubricate the bearings, if they're simply dried out. If they're damaged, however, replace them.

13 Swingarm - removal and installation

Removal

Refer to illustrations 13.9, 13.10, 13.11a and 13.11b

1   Raise the rear end of the vehicle and support it securely on jackstands.

2   Remove the rear wheels, hubs, brake caliper and brake disc (see Chapter 6).

3   Remove the skid plate from the swingarm (see Chapter 7).

4   Remove the final drive chain (see Section 8).

5   Remove the driven sprocket (see Section 9).

6   Remove the rear axle (see Section 10).

7   If you are going to replace the swingarm, remove the axle bearing holder (see Section 11). If you're only replacing the swingarm bearings, it's not necessary to remove the axle bearing holder.

8   Support the swingarm so it won't drop, then detach the lower end of the shock absorber from the swingarm or shock linkage (see Section 6).

9   If you are going to replace the swingarm on a TRX400EX model, remove the shock arm and the shock link (see Section 15) **(see illustration)**.

10   Detach the brake hose/parking brake cable clamps **(see illustration)**.

5

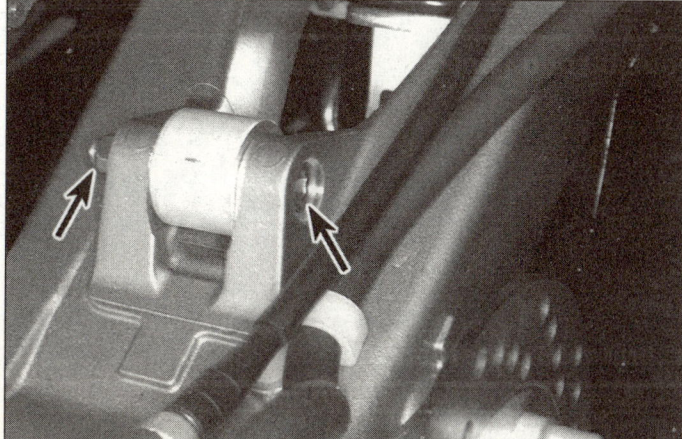

13.9 Remove the nut and bolt that connect the shock link to the swingarm

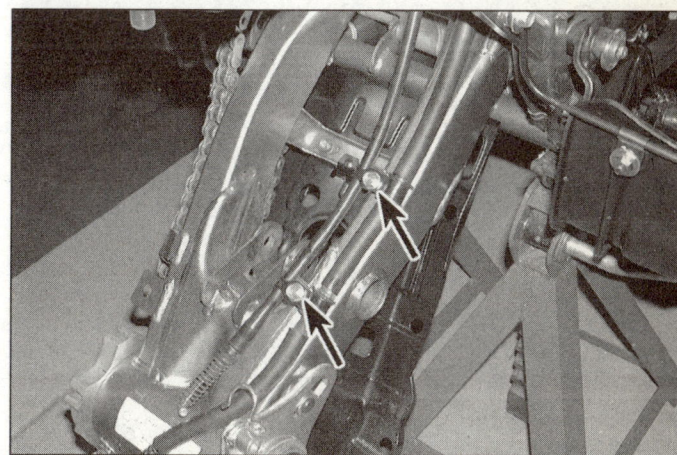

13.10 Before unbolting the swingarm from the frame, unbolt the clamps for the rear brake hose and parking brake cable (TRX300EX shown, TRX400EX similar)

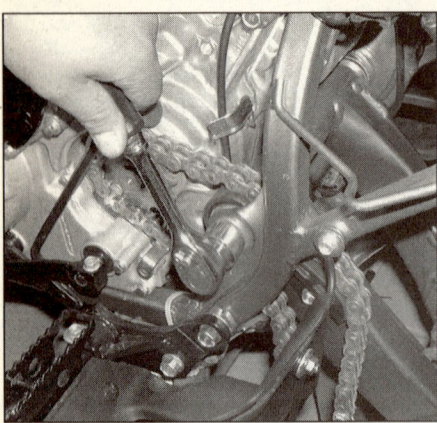

13.11a Remove the swingarm pivot bolt nut (this is a TRX300EX) . . .

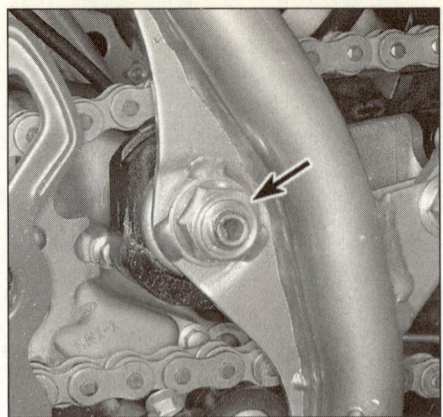

13.11b . . . and this is a TRX400EX

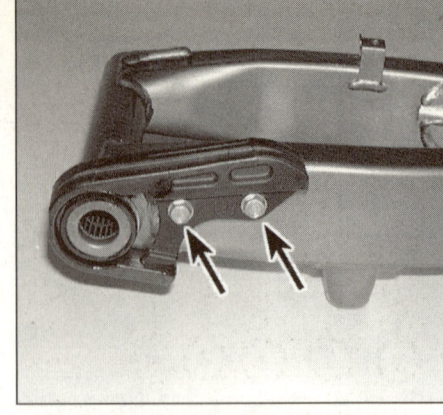

14.3 To detach the drive chain slider from the swingarm, remove these two bolts (arrows)

11   Remove the swingarm pivot bolt nut **(see illustration)** and then pull the pivot bolt out of the swingarm **(see illustration)**.
12   Pull the swingarm back and away from the vehicle.
13   Inspect the pivot bearings in the swingarm for dryness and deterioration (see Section 14). If they're damaged or worn, have them replaced by a Honda dealer service department or by a machine shop.

Installation

14   Lift the swingarm into position in the frame. Install the pivot bolt (from the right side on all models) and nut to hold the swingarm in the frame, but don't tighten them yet.
15   Raise and lower the swingarm several times, moving it through its full travel to seat the bearings and pivot bolt.
16   Tighten the pivot bolt and nut to the torque listed in this Chapter's Specifications.
17   The remainder of installation is the reverse of the removal steps, with the following additions:

   a) *On TRX400EX models, be sure to grease the needle bearings in the shock arm and shock link before reassembling them.*
   b) *On TRX400EX models, when reassembling the shock absorber, shock arm and shock link, make sure that all bolts are installed from the right side of the machine.*
   c) *On TRX400EX models, be sure to tighten the shock arm and shock link fasteners to the torque listed in this Chapter's Specifications.*

14 Swingarm bearings - replacement

Refer to illustrations 14.3, 14.5, 14.6, 14.7, 14.8 and 14.9
1   The swingarm pivots on two needle roller bearings. Bushing are mounted inside the bearings.
2   Remove the swingarm (see Section 13).
3   Inspect the drive chain slider **(see illustration)**. If it's damaged or excessively worn, replace it. (It's a good idea to remove the slider anyway while servicing the swingarm, but not absolutely necessary.)
4   Remove the dust seal cap, the dust seal and the washer from each side of the swingarm pivot bore.
5   Tap out the pivot thrust bushings with a punch **(see illustration)**. Use a soft brass punch if you plan to reuse them and make sure you don't nick the area around the ends of the swingarm pivot bore).
6   Push out the collar. Remove the swingarm needle bearing from each side with a small blind-hole puller and slide hammer **(see illustration)**. If either bearing is rough or loose, or has excessive play, replace the bearings as a set. If you don't have the right tools, have the bearings replaced by a Honda dealer or by a machine shop.
7   Clean the collar and other parts thoroughly. Inspect the collar for burrs, scoring and other damage; inspect the other parts too **(see illustration)**. If the collar or any other part is damaged or worn, replace it.
8   Make sure that the pivot bore is clean, then tap a new bearing into position with a bearing driver **(see illustration)** or with a socket just

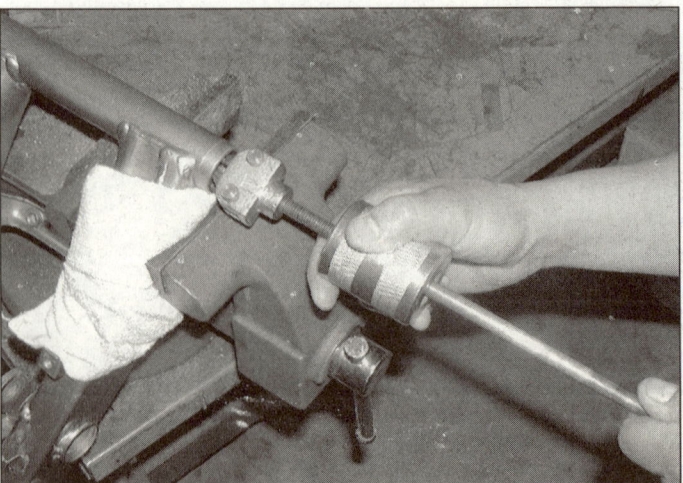

14.5 Tap out the pivot thrust bushings with a punch

14.6 Remove the swingarm needle bearing from each side with a small blind-hole puller and slide hammer

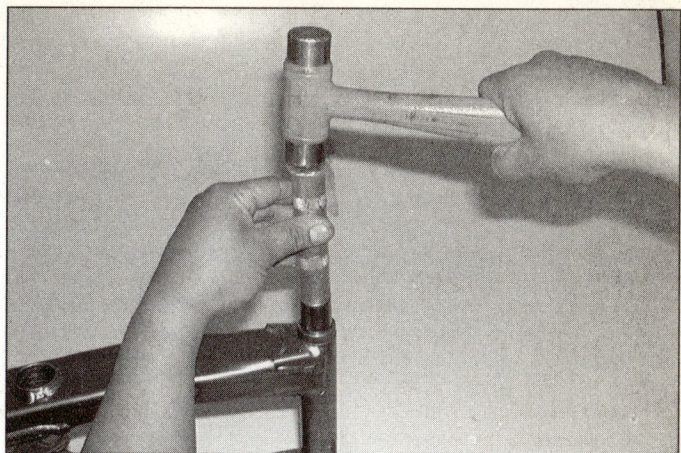

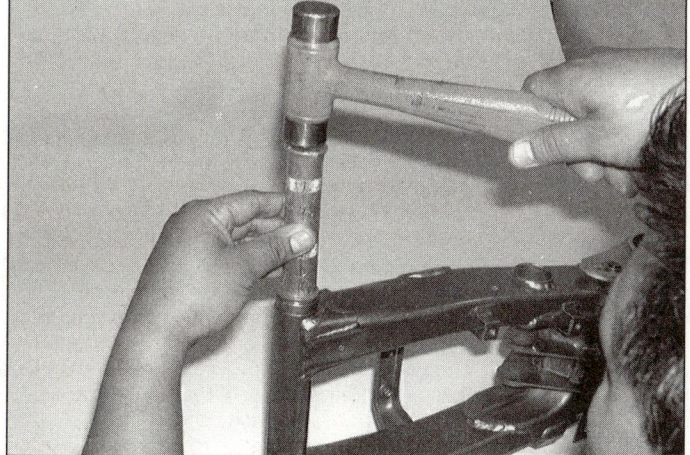

14.7 The disassembled right end of the swingarm pivot bearing assembly:

1 *Dust seal cap*	4 *Pivot thrust bushing*
2 *Dust seal*	5 *Needle bearing*
3 *Washer*	6 *Collar*

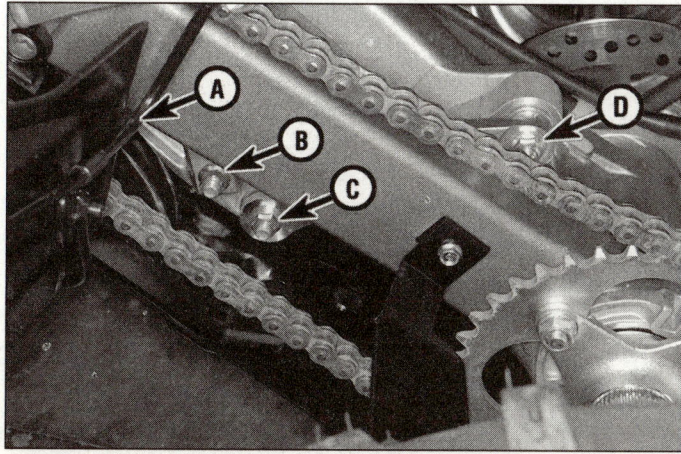

14.9 Tap new pivot thrust bushings into position with a driver too, or with a socket just slightly smaller than the outer diameter of the bushings

14.8 Tap the bearings into position carefully

slightly smaller than the diameter of the outer race. Tap the other new bearing into place from the other side. Lubricate the collar with grease and install it. **Caution:** *Be very careful not to collapse the bearing when you drive it in. A shouldered driver, with a shaft that fits inside the bearing and a shoulder that rests against it, is recommended.*

9   Tap new pivot thrust bushings into position with a driver **(see illustration)** or with a socket just slightly smaller than the outer diameter of the bushing.

10   Pack the pivot bearing cavity with waterproof lithium-based wheel bearing grease.

11   On each end, install the washer, tap a new seal into position with a seal driver or a socket just slightly smaller than the outside diameter of the seal, and install the dust seal cap.

12   Install the swingarm. Make sure that the pivot bolt is clean and free of corrosion or anything that might mar the inner surface of the collar.

15   Shock linkage (TRX400EX) – removal, inspection and installation

1   TRX400EX models are equipped with a progressive rising rate shock linkage, consisting of a shock link and a shock arm. The shock arm is attached to a pivot on the frame at its front end and to the shock link at its rear end. The bottom end of the shock absorber is attached to the shock arm. The shock link connects the shock arm to the swingarm.

Removal

Refer to illustration 15.4

2   Remove the seat and rear fenders (see Chapter 7).

3   Remove the rear wheels and support the swingarm securely.

4   Unbolt the shock link and shock arm from the frame, the shock absorber and the swingarm **(see illustration)**.

5   Remove the shock arm and shock link from the vehicle and unbolt them from each other.

Inspection

6   Remove the dust seal from each side of the shock link and shock arm pivot points. Push the bushings out of the needle bearings.

7   Inspect the needle bearings, bushings and dust seals for wear or damage. The needle bearings can be pressed out with a hydraulic press and drift. Press new ones in to a depth of 5.5 mm (0.220 inch). Push new dust seals in until they seat against the bearings, then slide the bushings in.

Installation

8   Installation is the reverse of removal. Install all of the bolts from the right side of the vehicle. Tighten the nuts and bolts to the torques listed in this Chapter's Specifications.

15.4 Shock linkage details (TRX400EX)

A *Shock arm pivot in frame (hidden)*	
B *Shock absorber lower end*	
C *Shock arm-to-shock link pivot*	
D *Shock link-to-swingarm pivot*	

5

Notes

Chapter 6
Brakes, wheels and tires

Contents

Specifications

Brakes

Brake pedal height	See Chapter 1
Brake pedal maximum travel	See Chapter 1
Brake pad thickness limit	See Chapter 1
Brake disc	
Maximum runout	0.30 mm (0.012 inch)
Minimum allowable thickness	
TRX300EX	
Front	
Standard	3.5 mm (0.14 inch)
Limit	3.0 mm (0.12 inch)
Rear	
Standard	4.0 mm (0.16 inch)
Limit	3.0 mm (0.12 inch)
TRX400EX	
Front	
Standard	2.8 to 3.2 mm (0.11 to 0.13 inch)
Limit	2.5 mm (0.10 inch)
Rear	
Standard	3.8 to 4.2 mm (0.15 to 0.17 inch)
Limit	3.5 mm (0.14 inch)

6

Master cylinder
  Master cylinder inside diameter
    Standard ....   12.700 to 12.743 mm (0.5000 to 0.5017 inch)
    Limit ....   12.755 mm (0.5022 inch)
  Piston outside diameter
    Standard ....   12.657 to 12.684 mm (0.4983 to 0.4994 inch)
    Limit ....   12.645 mm (0.4978 inch)
Caliper
  Piston bore diameter
    Standard ....   33.960 to 34.010 mm (1.3370 to 1.3390 inch)
    Limit ....   34.020 mm (1.3394 inch)
  Piston outside diameter
    Standard
      TRX300EX ....   33.878 to 33.928 mm (1.3338 to 1.3357 inch)
      TRX400EX ....   33.895 to 33.928 mm (1.3344 to 1.3357 inch)
    Limit ....   33.870 mm (1.3335 inch)

Wheels and tires

Tire pressures ....   See Chapter 1
Tire tread depth ....   See Chapter 1

Torque specifications

Wheels and hubs

Wheel lug nuts ....   64 Nm (47 ft-lbs)
Front wheel hub nuts
  TRX300EX ....   59 to 79 Nm (43 to 58 ft-lbs)
  TRX400EX ....   59 Nm (43 ft-lbs)
Rear wheel hub nuts
  TRX300EX ....   140 to 160 Nm (101 to 116 ft-lbs)
  TRX400EX ....   137 Nm (101 ft-lbs)

Brake calipers (front and rear)

Caliper bolts..   30 Nm (22 ft-lbs)
Brake hose-to-caliper banjo bolts ....   34 Nm (25 ft-lbs)
Brake pad pins ....   18 Nm (156 in-lbs)
Caliper bracket pin ....   18 Nm (156 in-lbs)
Caliper slide pin ....   23 Nm (17 ft-lbs)

Brake discs

Front brake disc retaining bolts...   43 Nm (31 ft-lbs)
Rear brake disc retaining bolts..   27 Nm (20 ft-lbs)

Master cylinders

Brake hose banjo bolt..   34 Nm (25 ft-lbs)
Handlebar clamp bolts (front master cylinder) ....   No specified torque
Master cylinder mounting bolts (rear master cylinder) ....   No specified torque

Refer to marks cast into the drum (they supersede information printed here)

1   General information

The vehicles covered by this manual are equipped with three hydraulically-operated disc brakes: a disc at each front wheel and a single disc on the rear axle. A lever-actuated master cylinder on the right end of the handlebar operates the front brakes; the rear brake is operated by a pedal-actuated master cylinder on the right side of the vehicle, right behind the right footrest. The rear brake caliper is equipped with a parking brake system, which is actuated by a lever on the left end of the handlebar.

All models are equipped with wheels that require very little maintenance and allow tubeless tires to be used. **Caution:** *Brake components rarely require disassembly. Do not disassemble components unless absolutely necessary.*

2   Wheels - inspection, removal and installation

Inspection

1   Clean the wheels thoroughly to remove mud and dirt that may interfere with the inspection procedure or mask defects. Make a general check of the wheels and tires as described in Chapter 1.
2   The wheels should be visually inspected for cracks, flat spots on the rim and other damage. Since tubeless tires are involved, look very closely for dents in the area where the tire bead contacts the rim. Dents in this area may prevent complete sealing of the tire against the rim, which leads to deflation of the tire over a period of time.
3   If damage is evident, the wheel will have to be replaced with a new one. Never attempt to repair a damaged wheel.

2.5 To remove a wheel, remove these four nuts (arrows)

4.2 Remove the disc splash guard

Removal

Refer to illustration 2.5

4   Securely block the wheels at the opposite end of the vehicle from the wheel being removed, so it can't roll.

5   Loosen the lug nuts **(see illustration)** on the wheel being removed. Jack up one end of the vehicle and support it securely on jackstands.

6   Remove the lug nuts and pull the wheel off.

Installation

7   Position the wheel on the studs. Make sure the directional arrow on the tire points in the forward rotating direction of the wheel.

8   Install the wheel nuts with their tapered sides toward the wheel. This is necessary to locate the wheel accurately on the hub.

9   Snug the wheel nuts evenly in a criss-cross pattern.

10   Remove the jackstands, lower the vehicle and tighten the wheel nuts, again in a criss-cross pattern, to the torque listed in this Chapter's Specifications.

3   Tires - general information

1   Tubeless tires are used as standard equipment on this vehicle. Unlike motorcycle tires, they run at very low air pressures and are completely unsuited for use on pavement. Inflating ATV tires to excessive pressures will rupture them, making replacement of the tire necessary.

2   The force required to break the seal between the rim and the

bead of the tire is substantial, much more than required for motorcycle tires, and is beyond the capabilities of an individual working with normal tire irons or even a normal bead breaker. A special bead breaker is required for ATV tires; it produces a great deal of force and concentrates it in a relatively small area.

3   Also, repair of the punctured tire and replacement on the wheel rim requires special tools, skills and experience that the average do-it-yourselfer lacks.

4   For these reasons, if a puncture or flat occurs with an ATV tire, the wheel should be removed from the vehicle and taken to a dealer service department or a repair shop for repair or replacement of the tire. The accompanying illustrations can be used as a guide to tire replacement in an emergency, provided the necessary bead breaker is available.

4   Front wheel hub and bearing - removal and installation

Removal

Refer to illustrations 4.2, 4.3, 4.4 and 4.6

Note: *This procedure applies to either front hub.*

1   Remove the front wheel (see Section 2).

2   Remove the disc splash guard **(see illustration)**.

3   Bend back the cotter pin and pull it out of the hub nut **(see illustration)**.

4   Have an assistant apply the front brake while you loosen the hub nut **(see illustration)**. Remove the hub nut and the washer.

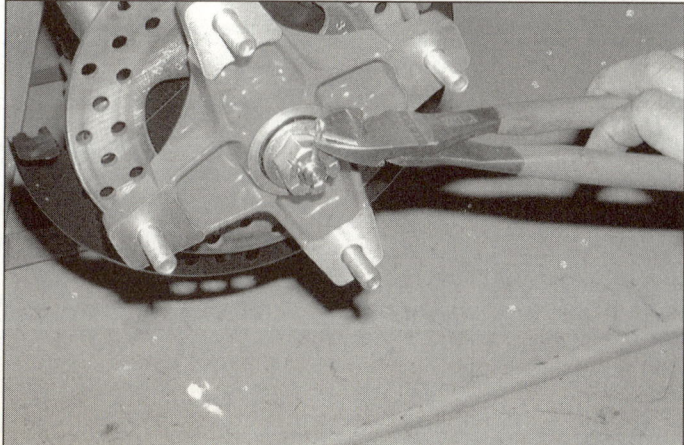

4.3 Bend back the cotter pin and pull it out of the hub nut

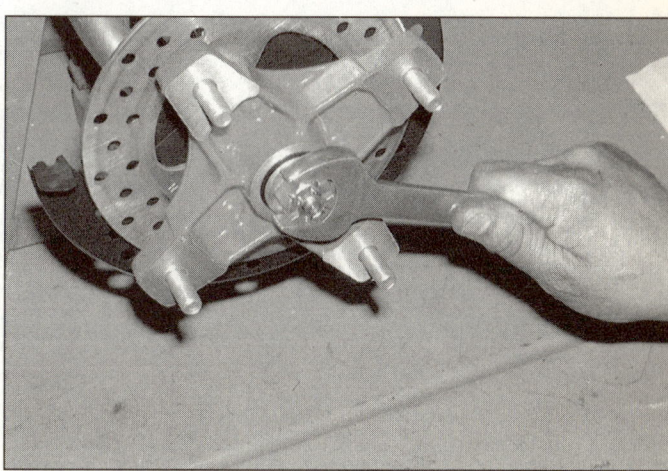

4.4 While an assistant applies the front brake, loosen the hub nut

6

5   Remove the front brake caliper (see Section 6). It's not necessary to disconnect the brake hose from the caliper; set the caliper aside and hang it from the suspension with rope or wire.

6   Pull the hub off the spindle. If the hub is stuck, remove it with a puller **(see illustration)**.

Bearing inspection and replacement

Refer to illustrations 4.8, 4.9a, 4.9b, 4.9c, 4.10a, 4.10b, 4.11, 4.12 and 4.13

7   Remove the front brake disc from the hub and inspect the disc (see Section 7). Wipe off the spindle and hub. **Caution:** *Do NOT immerse the hub in any kind of cleaning solvent. The sealed hub bearings, which cannot be disassembled and repacked, could be damaged if any solvent enters them.*

8   Insert your fingers into each hub bearing and turn the bearing **(see illustration)**. If the bearing feels rough or dry, replace it.

9   Remove the outer collar and pry out the old seals **(see illustrations)**. Discard the old seals.

10   The easiest way to remove the hub bearings is to use a suitable puller, similar to the one shown in the accompanying illustration, to remove the outer bearing **(see illustration)**. Then drive out the inner bearing with a bearing removal tool or with a brass drift **(see illustration)**. If you don't have a suitable puller, but you do have a suitable brass drift, you can drive both bearings out with the drift as follows. Lay the hub on a workbench, with the outer side of the hub facing down, insert the drift into the hub from the inner side of the hub, push the collar to the side and tap gently against the inner face of the outer bearing. Then flip over the hub, inner side facing down, insert the drift from the outer side of the hub, and drive out the inner bearing the

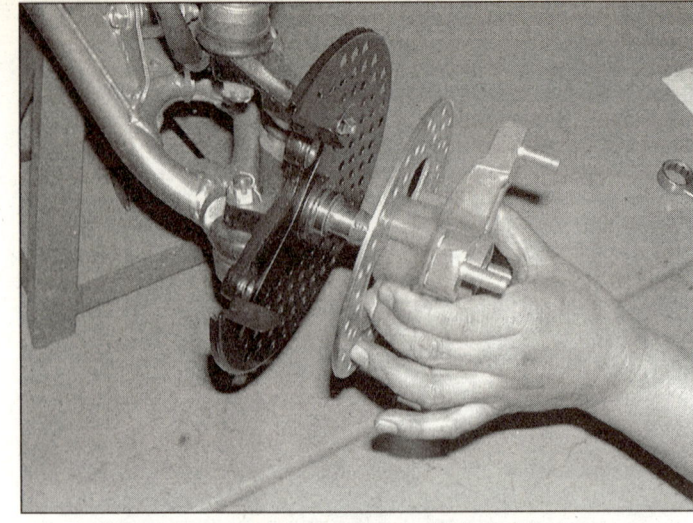

4.6 Pull the hub off the spindle (if the hub is stuck, remove it with a puller)

same way. If you don't have a suitable drift, have the bearings removed by a Honda dealer service department or by a machine shop.

11   Before reassembling the hub, inspect all the parts **(see illustration)** for wear and damage. Replace any damaged or excessively worn parts.

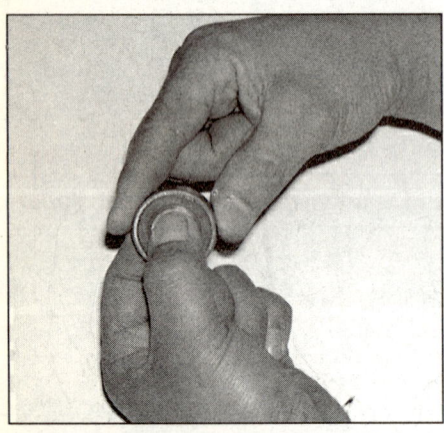

4.8 Spin the bearing with fingers; if it feels rough or dry, replace it (bearing removed from hub for clarity)

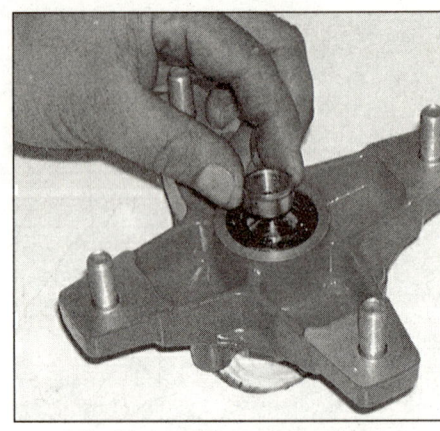

4.9a Remove the outer collar . . .

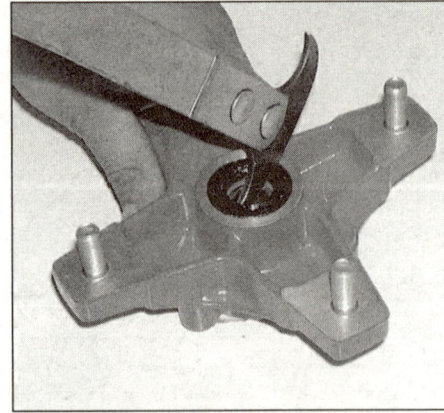

4.9b . . . pry out the outer dust seal . . .

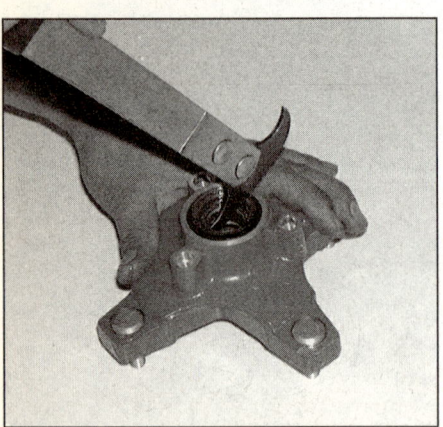

4.9c . . . and pry out the inner seal

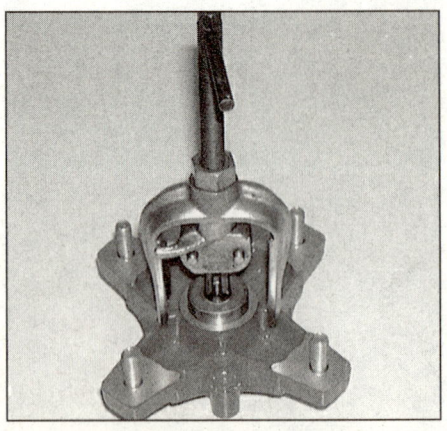

4.10a A puller like this one is the easiest way to pull out the outer bearing . .

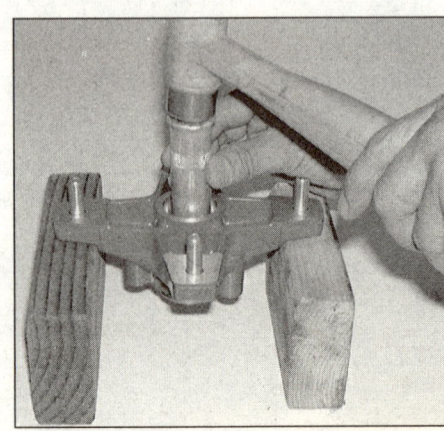

4.10b . . . drive out the inner bearing with a bearing removal tool or a brass drift

TIRE CHANGING SEQUENCE - TUBELESS TIRES

Deflate tire. After releasing beads, push tire bead into well of rim at point opposite valve. Insert lever next to valve and work bead over edge of rim.

Use two levers to work bead over edge of rim. Note use of rim protectors.

When first bead is clear, remove tire as shown.

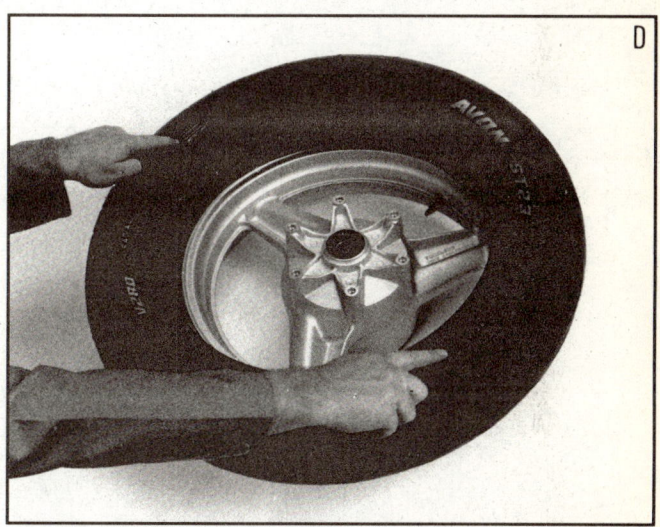

Before installing, ensure that tire is suitable for wheel. Take note of any sidewall markings such as direction of rotation arrows.

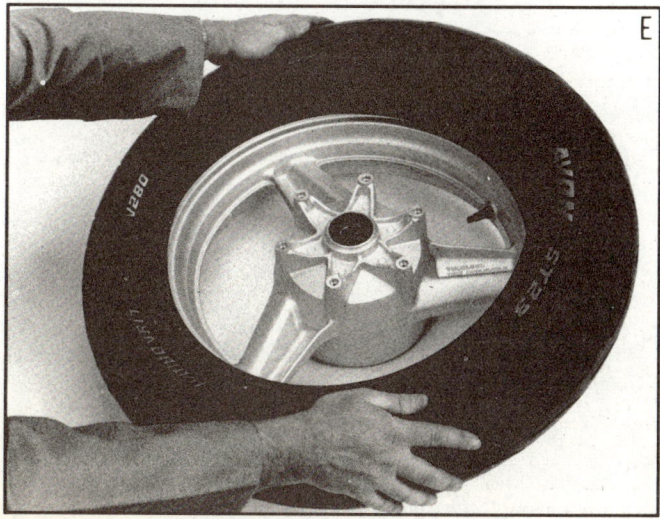

Work first bead over the rim flange.

Use a tire lever to work the second bead over rim flange.

12   To install the new bearings, drive them into place with a bearing installer tool or an old socket **(see illustration)**. The socket must have an outside diameter that's the same, or slightly smaller than, the outer diameter of the bearings. **Caution:** *Do NOT strike the center race or the ball bearings*.

13   Tap new inner and outer seals into place with a block of wood **(see illustration)**. Do NOT use the old seals.

Installation

14   Installation is the reverse of removal. Lubricate the spindle with wheel bearing grease. Tighten the hub nut to the torque listed in this Chapter's Specifications. If necessary, tighten it an additional amount to align the cotter pin slots. Don't loosen the nut to align the slots. Install a new cotter pin and bend it to secure the nut.

5   Front brake pads - replacement

Warning: *The dust created by the brake system may contain asbestos, which is harmful to your health. Never blow it out with compressed air and don't inhale any of it. An approved filtering mask should be worn when working on the brakes. Do not, under any circumstances, use petroleum-based solvents to clean brake parts. Use brake cleaner only!*
Note: *Always replace both pairs of brake pads at the same time.*
Refer to illustrations 5.3a, 5.3b, 5.4, 5.5a, 5.5b, 5.5c, 5.6, 5.7a and 5.7b

1   Remove the front wheel (see Section 2).
2   Disengage the brake hose from the hose clamp.
3   Remove the brake pad pin plugs **(see illustration)**. Loosen the

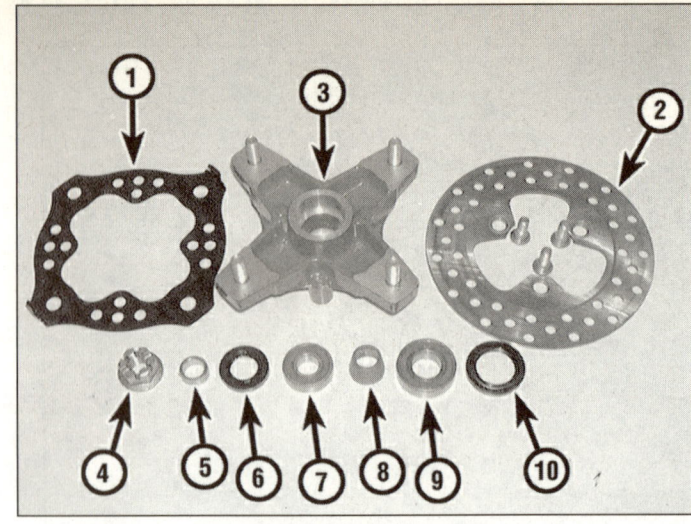

4.11 Front wheel hub details

1	*Splash guard*	*6*	*Outer dust seal*
2	*Front brake disc*	*7*	*Outer bearing*
3	*Front wheel hub*	*8*	*Collar*
4	*Hub nut*	*9*	*Inner bearing*
5	*Collar*	*10*	*Inner dust seal*

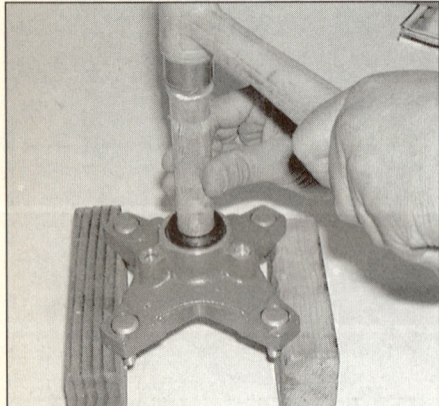

4.12 To install the new bearings, drive them into place with a bearing installer tool or an old socket

4.13 Tap new inner and outer seals into place with a block of wood

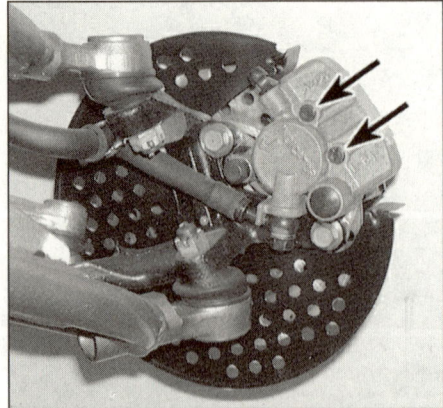

5.3a Remove the brake pad pin plugs (arrows) . . .

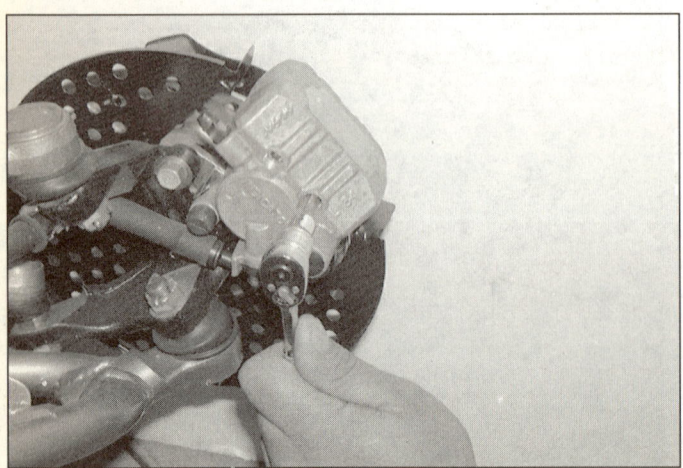

5.3b . . . and loosen the pins (they're easier to loosen while the caliper is bolted to the steering knuckle)

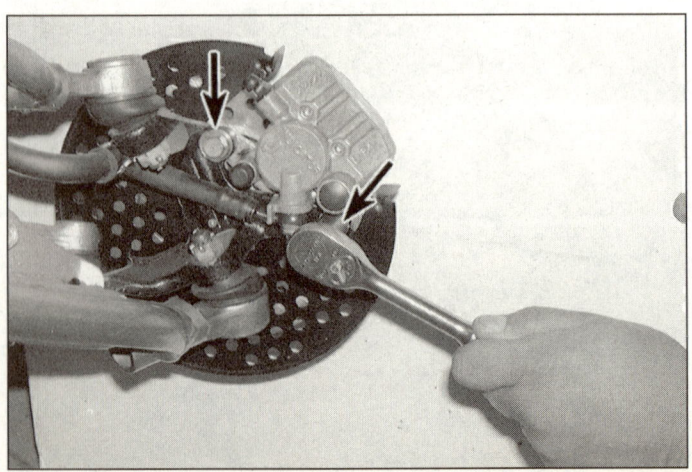

5.4 Unbolt the caliper from the steering knuckle

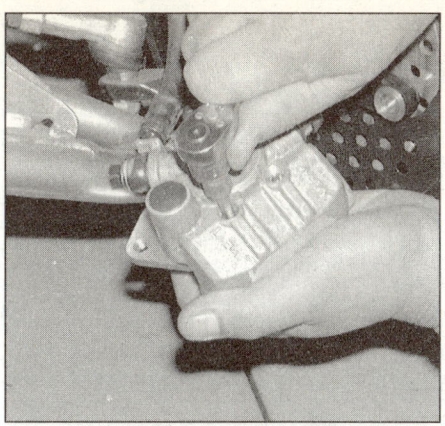

5.5a Remove the brake pad pins from the caliper . . .

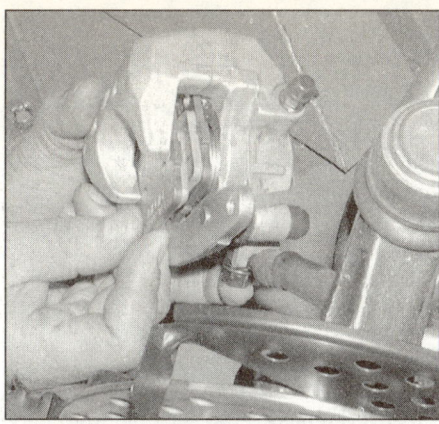

5.5b . . . then remove the outer brake pad . . .

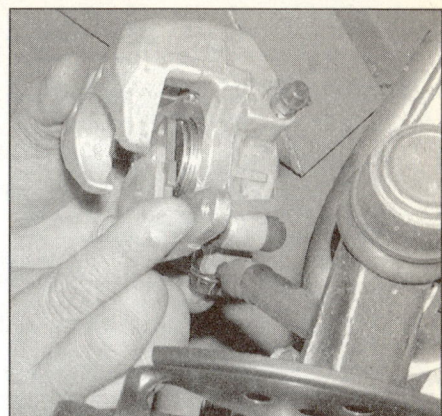

5.5c . . . and the inner brake pad

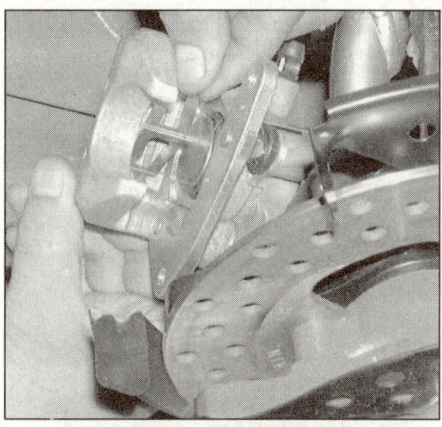

5.6 Remove the pad spring from the caliper; if it's damaged or distorted, replace it

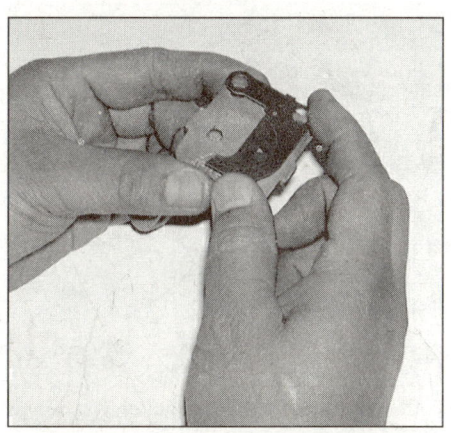

5.7a Remove the shim from the old inner brake pad . . .

5.7b . . . and install it on the new inner pad

brake pad pins **(see illustration)**. (The pad retaining bolts are easier to loosen while the caliper is still bolted to the steering knuckle.)

4   Remove the caliper mounting bolts **(see illustration)**.

5   Remove the caliper, remove the brake pad pins and then remove the brake pads **(see illustrations)**.

6   Remove the pad spring from the caliper **(see illustration)** and inspect it. If it's damaged or distorted, replace it.

7   Remove the shim from the old inner brake pad and install it on the

new inner pad **(see illustrations)**.

8   Installation is the reverse of removal. Using a C-clamp, depress the piston back into the caliper bore to provide enough room for the new pads to clear the disc. Be sure to tighten the brake pad pins and caliper bolts to the torque listed in this Chapter's Specifications.

9   Replace the brake pads on the other front caliper.

6   Front brake caliper - removal, overhaul and installation

Warning: *The dust created by the brake system may contain asbestos, which is harmful to your health. Never blow it out with compressed air and don't inhale any of it. An approved filtering mask should be worn when working on the brakes. Do not, under any circumstances, use petroleum-based solvents to clean brake parts. Use brake cleaner only!* **Note:** *This procedure applies to both front calipers.*

Removal

Refer to illustration 6.1

1   Place a drain pan under the front caliper, and then disconnect the brake hose from the caliper **(see illustration)**. Disregard this step if you're only removing the caliper to service some other component, such as the wheel hub, the brake disc or the steering knuckle.

2   Remove the caliper from the steering knuckle and, if you're going to overhaul the caliper, remove the brake pads, shim and spring (see Section 5). If you're removing the caliper only to service some other component, leave it assembled and hang it out of the way with a piece of wire (NOT with the brake hose!).

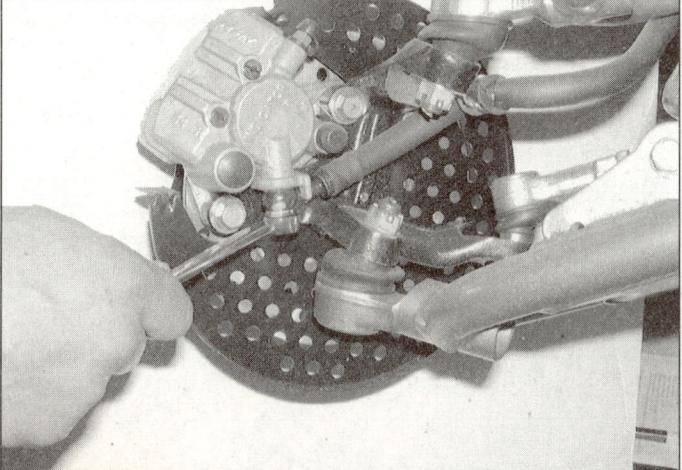

6.1 Remove the brake hose banjo bolt and both sealing washers; discard the washers

6

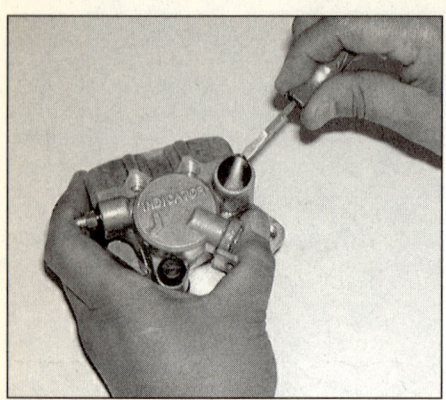

6.3a Pry the dust plug from the slide pin hole in the caliper . . .

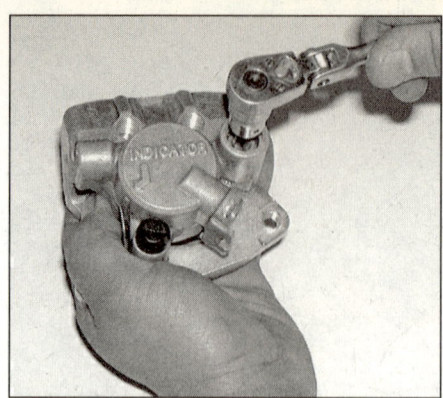

6.3b . . . unscrew the slide pin from the caliper bracket, remove the bracket . . .

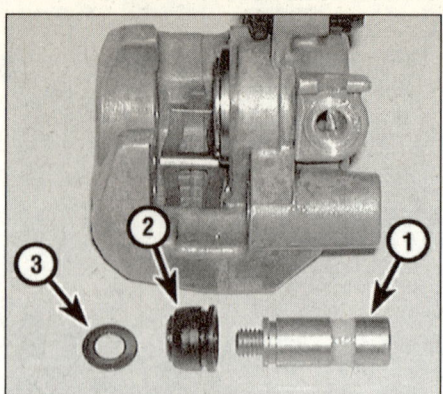

6.3c . . . and remove the slide pin (1), the dust boot (2) and washer (3)

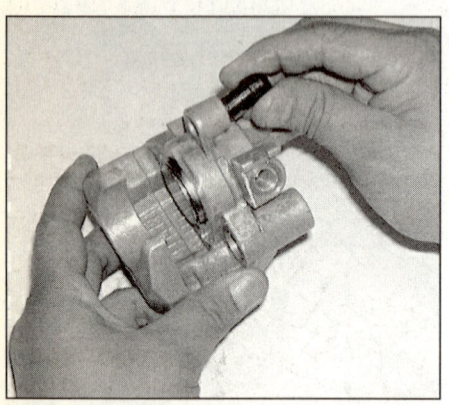

6.4a Remove the dust boot for the pin bolt . . .

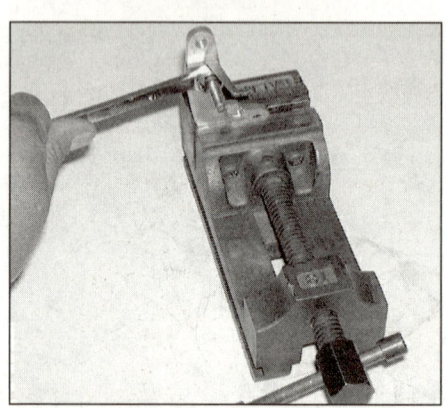

6.4b . . . put the caliper bracket in a bench vise, unscrew the pin bolt and remove the washer and indicator plate

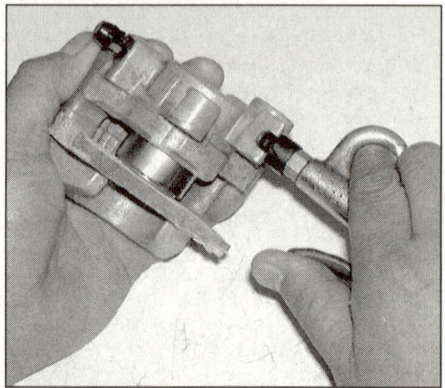

6.5 Slowly apply compressed air to push the piston out of the caliper – DO NOT get your fingers in the way!

Overhaul

Refer to illustrations 6.3a, 6.3b, 6.3c, 6.4a, 6.4b, 6.5, 6.7, 6.8, 6.9 and 6.10

3   Remove the dust plug from the slide pin hole in the caliper and unscrew the slide pin from the caliper bracket. Remove the caliper bracket and remove the slide pin dust boot and washer **(see illustrations)**.

4   Remove the pin bolt dust boot **(see illustration)**. Put the caliper bracket in a bench vise, unscrew the pin bolt **(see illustration)** and remove the washer and indicator plate.

5   Place a few rags between the piston and the caliper frame to act as a cushion. Lay the caliper on the work bench so that the piston is facing down, toward the work bench surface, then use compressed air,

directed into the fluid inlet to remove the piston **(see illustration)**. Use only small quick blasts of air to ease the piston out of the bore. If a piston is blown out with too much force, it might be damaged. **Warning:** *Never place your fingers in front of the piston in an attempt to catch or protect it when applying compressed air. Doing so could result in serious injury.* Once the piston is protruding from the caliper, remove it.

6   If compressed air isn't available, reconnect the caliper to the brake hose and pump the brake lever until the piston is free. (You'll have to put brake fluid in the master cylinder reservoir and get most of the air out of the hose to use this method.)

7   Remove the old dust seal **(see illustration)**.

8   Using a wood or plastic tool, remove the piston seal **(see illustration)**.

9   Remove the bleeder valve and dust cap **(see illustration)**.

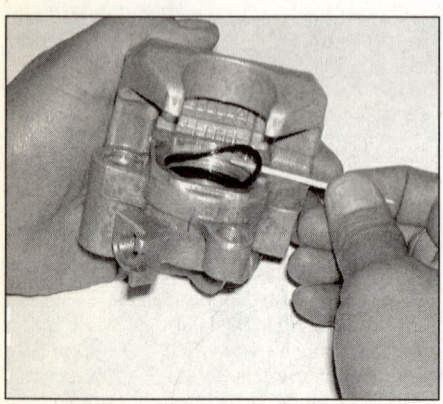

6.7 Using a wood or plastic tool, remove the dust seal

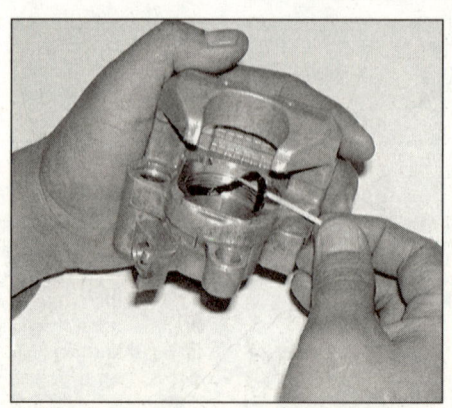

6.8 Using a wood or plastic tool, remove the piston seal

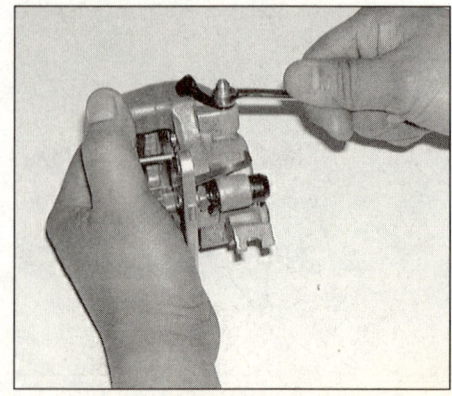

6.9 Remove the bleeder valve and dust cap

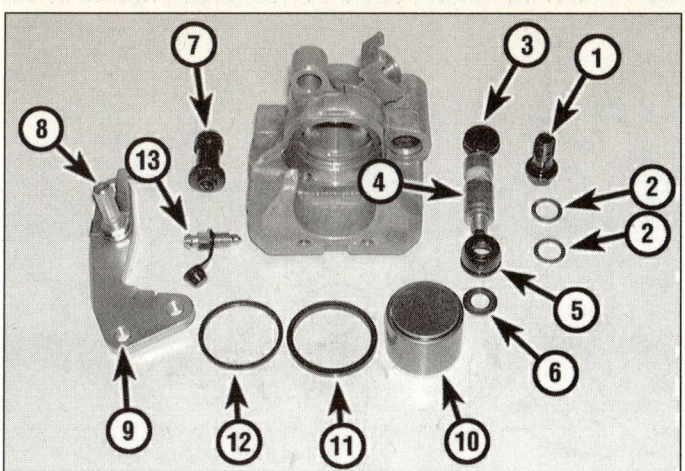

6.10 Front brake caliper details

1	Brake hose banjo bolt	8	Pin bolt
2	Banjo bolt sealing	9	Caliper bracket
	washers (always replace)	10	Piston
3	Slide pin plug	11	Piston dust seal
4	Slide pin	12	Piston seal
5	Slide pin dust boot	13	Bleeder valve and dust
6	Slide pin washer		cap
7	Pin bolt dust boot		

10   Clean the caliper parts with denatured alcohol, fresh brake fluid or brake system cleaner. Dry them off with filtered, unlubricated compressed air and inspect them carefully **(see illustration)**. Replace any obviously worn or damaged parts. It's a good idea to replace all rubber parts (dust boots, seals, etc.) anytime you disassemble the caliper. Inspect the surfaces of the piston and the piston bore for rust, corrosion, nicks, burrs and loss of plating. If you find defects on the surface of either piston or piston bore, replace the piston and caliper assembly (the piston is matched to the caliper). If the caliper is in bad shape, inspect the master cylinder too (see Section 12).

11   Lubricate the new piston seal with clean brake fluid and install it in its groove in the caliper bore. Make sure it's not twisted and is fully and correctly seated.

12   Install the new dust seal. Make sure that the inner lip of the seal is seated in its groove in the piston and the outer circumference of the seal is seated in its groove in the caliper bore.

13   Lubricate the piston with clean brake fluid and install it into its bore in the caliper. Using your thumbs, push the piston all the way in; make sure it doesn't become cocked in the bore.

14   Install the spring, the inner brake pad and shim, the outer pad, and the pad pins and plugs (see Section 5).

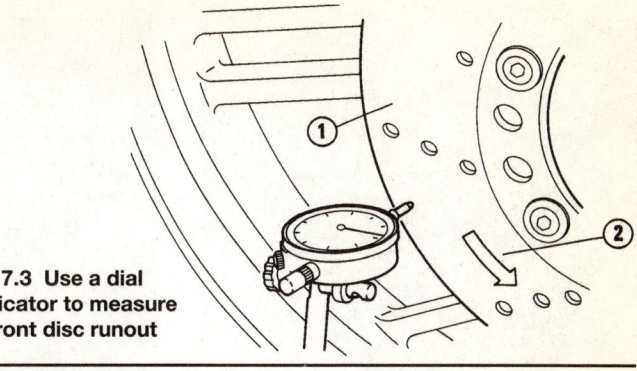

7.3 Use a dial indicator to measure front disc runout

Installation

15   Install the caliper and brake pads (see Section 5).

16   Connect the brake hose-to-caliper banjo bolt **(see illustration 6.1)**. Be sure to use new sealing washers. Tighten the banjo bolt to the torque listed in this Chapter's Specifications.

17   Remove and overhaul the other front brake caliper.

18   Bleed the front brake system (see Section 15).

---

7   Front brake disc - inspection, removal and installation

Note: *This procedure applies to both front discs.*

1   Remove the front wheels (see Section 2).

Inspection

Refer to illustrations 7.3, 7.4a and 7.4b

2   Visually inspect the surface of the disc for score marks and other damage. Light scratches are normal after use and won't affect brake operation, but deep grooves and heavy score marks will reduce braking efficiency and accelerate pad wear. If the disc is badly grooved it must be machined or replaced.

3   To check disc runout, mount a dial indicator with the plunger on the indicator touching the surface of the disc about 1/2-inch from the outer edge **(see illustration)**. Slowly turn the wheel hub and watch the indicator needle, comparing your reading with the disc runout limit listed in this Chapter's Specifications. If the runout is greater than allowed, replace the disc.

4   The disc must not be thinner than, or machined below, the minimum allowable thickness listed in this Chapter's Specifications. The minimum thickness is also stamped into the disc itself **(see illustration)**. If the minimum thickness stamped into the disc differs from the minimum thickness listed in this Chapter's Specifications, it supersedes the information in this book. Check the thickness of the disc with a micrometer **(see illustration)**. If the disc is thinner than the minimum allowable thickness, replace it.

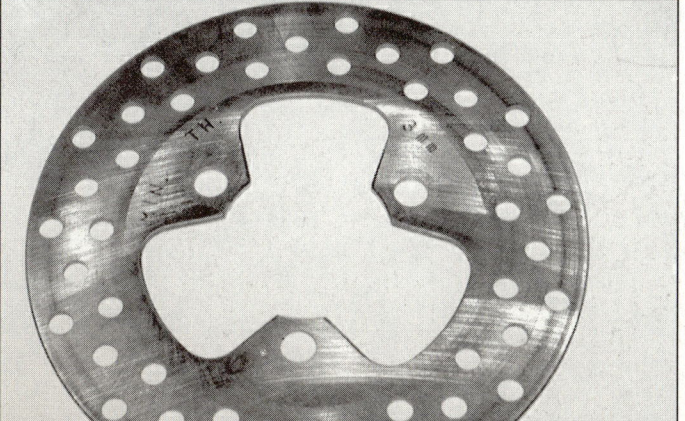

7.4a The minimum thickness of each front disc is stamped into the disc (disc removed for clarity)

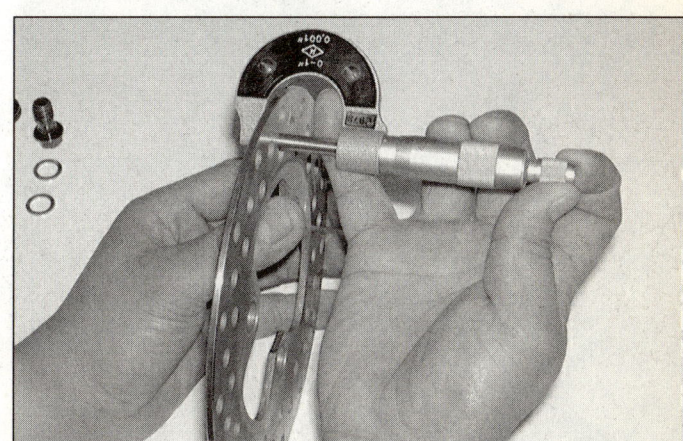

7.4b Check the thickness of the disc with a micrometer (disc removed for clarity)

6

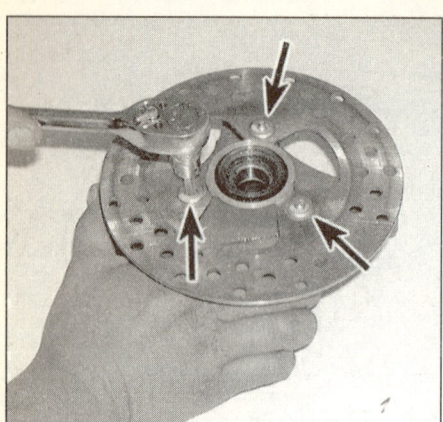

7.6 Remove the disc-to-hub bolts (arrows)

8.1 Bend back the tangs on the lockwasher with a small punch and loosen the brake pad pins (arrows)

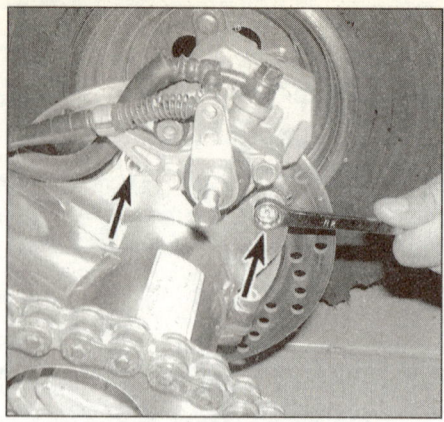

8.2 Remove the caliper mounting bolts (arrows)

Removal and installation

Refer to illustration 7.6

5   Remove the front wheel hub (see Section 4).

6   To detach the brake disc from the hub, remove the three disc retaining bolts **(see illustration)**.

7   Installation is the reverse of removal. Tighten the disc retaining bolts to the torque listed in this Chapter's Specifications.

8   Rear brake pads - replacement

Refer to illustrations 8.1, 8.2, 8.3a, 8.3b, 8.3c, 8.4, 8.5a and 8.5b

Warning: *The dust created by the brake system may contain asbestos, which is harmful to your health. Never blow it out with compressed air and don't inhale any of it. An approved filtering mask should be worn when working on the brakes. Do not, under any circumstances, use petroleum-based solvents to clean brake parts. Use brake cleaner only!*

1   Bend back the tangs on the lock washer **(see illustration)**, and then loosen the brake pad pins. (The pins are easier to loosen while the caliper is still bolted onto its bracket.)

2   Remove the caliper mounting bolts **(see illustration)**.

3   Remove the brake pad pins **(see illustration)**, lift off the caliper and pull out the brake pads **(see illustrations)**.

4   Remove the pad spring from the caliper **(see illustration)** and inspect it. If it's damaged or distorted, replace it.

5   Remove the shim from the old inner brake pad and install it on the new inner pad **(see illustrations)**.

6   Installation is the reverse of removal. Using a C-clamp, depress the piston back into the caliper bore to provide enough room for the new pads to clear the disc. Be sure to tighten the brake pad pins and caliper bolts to the torque listed in this Chapter's Specifications.

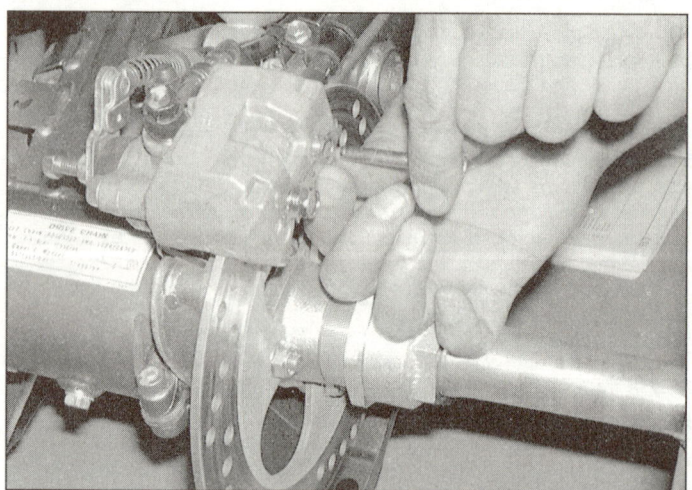

8.3a Remove the brake pad pins, remove the caliper . . .

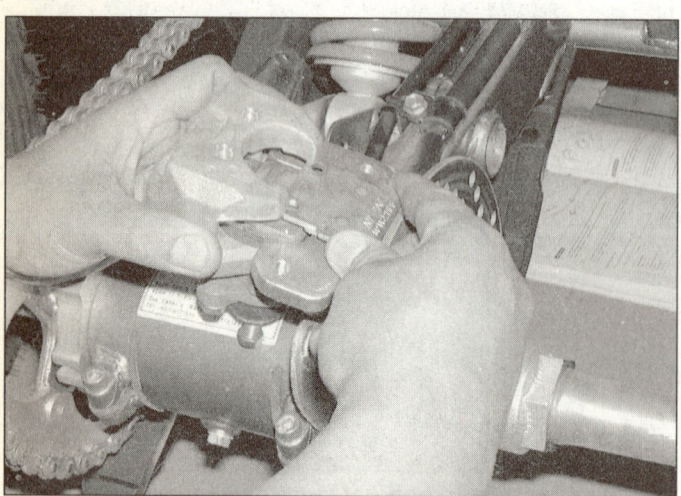

8.3b . . . remove the outer brake pad . . .

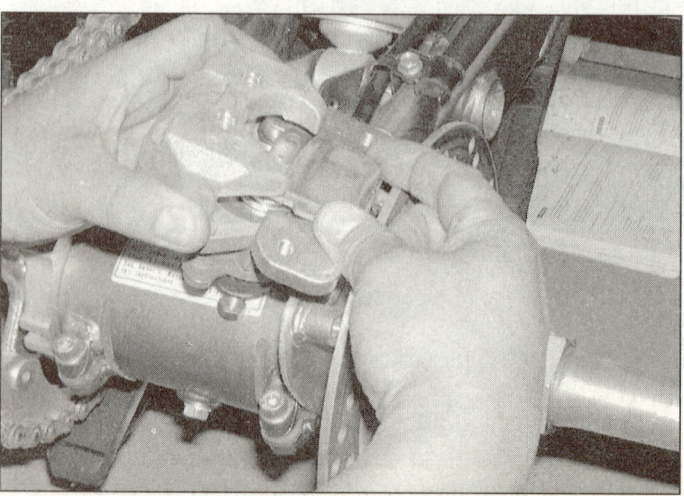

8.3c . . . and remove the inner pad

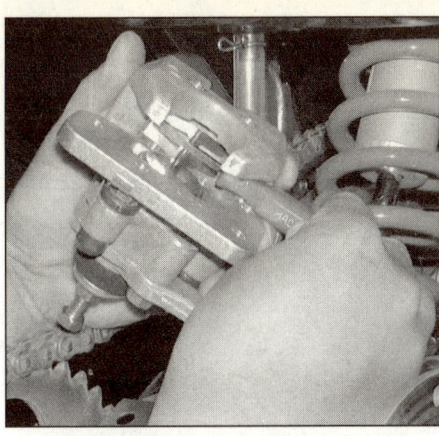

8.4 Remove the pad spring from the caliper; if it's damaged or distorted, replace it

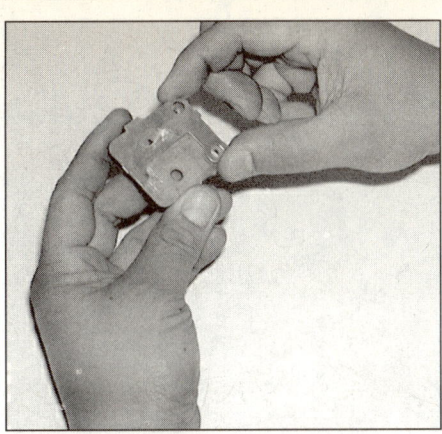

8.5a Remove the shim from the old inner brake pad . . .

8.5b . . . and install it on the new inner pad

9 Rear brake caliper - removal, overhaul and installation

Warning: *The dust created by the brake system may contain asbestos, which is harmful to your health. Never blow it out with compressed air and don't inhale any of it. An approved filtering mask should be worn when working on the brakes. Do not, under any circumstances, use petroleum-based solvents to clean brake parts. Use brake cleaner only!*

Removal

Refer to illustrations 9.1, 9.2a and 9.2b

1 Place a drain pan under the rear caliper brake hose fitting, then disconnect the brake hose from the caliper **(see illustration)**. (Disregard this step if you're only removing the caliper to service the disc, the rear axle, the axle bearings, the swingarm or some other component.)

2 Loosen the locknut, remove the parking brake adjusting bolt and remove the parking brake arm **(see illustrations)**. (Disregard this step if you're only removing the caliper to service the disc, the rear axle, the axle bearings, the swingarm or some other component.)

3 Remove the caliper and, if you're going to overhaul the caliper, remove the brake pads, shim and spring (see Section 8). If you're only removing the caliper to service some other component, leave it assembled and hang it out of the way with a piece of wire (NOT with the brake hose!). **Note:** *If you plan to overhaul the caliper and you don't have a source of compressed air as described in Step 9, you can use*

the vehicle's hydraulic system to remove the caliper piston as long as it's in good enough condition to produce some pressure. Just operate the brake pedal with the pads removed; the hydraulic pressure should push the piston out.

9.1 With a drain pan under the rear caliper brake hose fitting, remove the banjo bolt and sealing washers

9.2a Loosen the locknut (arrow), unscrew and remove the parking brake adjusting bolt . . .

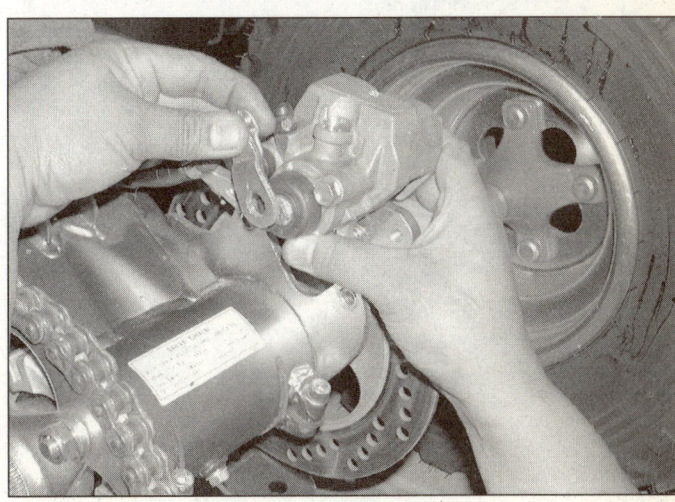

9.2b . . . and remove the parking brake arm

6

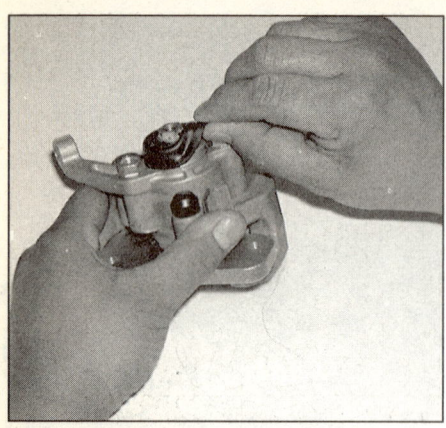

9.4 Remove the parking brake shaft dust boot

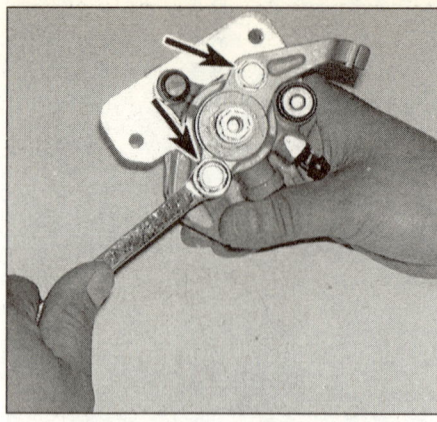

9.5a Remove the parking brake base bolts (arrows) . . .

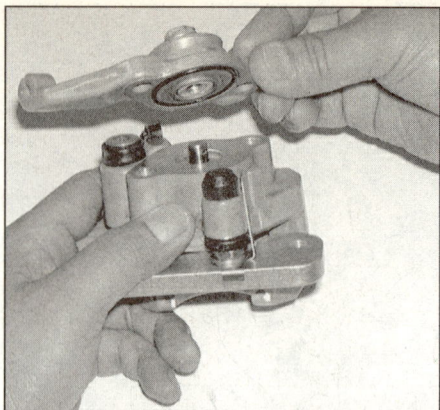

9.5b . . . and detach the base from the caliper

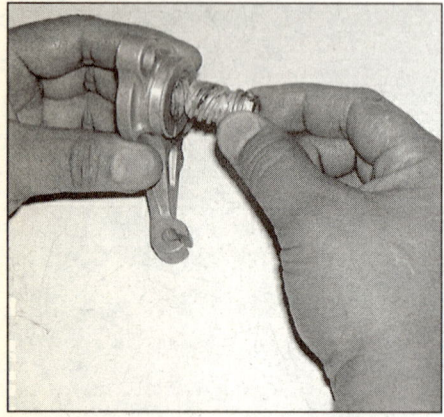

9.6a Remove the parking brake shaft . . .

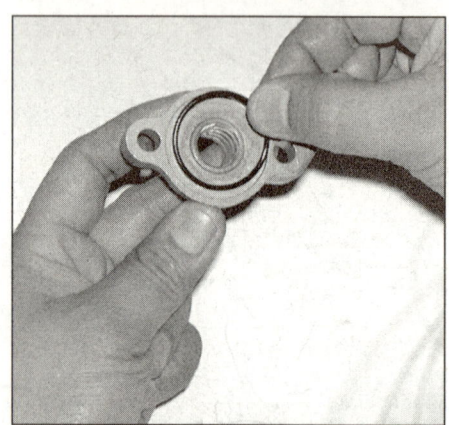

9.6b . . . and the old O-ring from the parking brake base

9.7a Remove the slide pin dust seal . . .

Overhaul

Refer to illustrations 9.4, 9.5a, 9.5b, 9.6a, 9.6b, 9.7a, 9.7b, 9.7c, 9.8a, 9.8b, 9.9, 9.10, 9.11a, 9.11b, 9.12a, 9.12b and 9.13

4   Remove the parking brake shaft dust boot **(see illustration)**.

5   Remove the parking brake base bolts and detach the base from the caliper **(see illustrations)**.

6   Remove the parking brake shaft and the old O-ring from the parking brake base **(see illustrations)**.

7   Remove the slide pin dust seal, the slide pin bolt boot and the slide pin bolt **(see illustrations)**.

8   To detach the caliper bracket from the caliper, pull the bracket pin bolt out of the caliper **(see illustration)**. Remove the bracket pin bolt

dust boot from the caliper **(see illustration)**.

9   Place a few rags or a block of wood between the piston and the caliper frame to act as a cushion. Lay the caliper on the work bench so that the piston is facing down, toward the work bench surface, then use compressed air, directed into the fluid inlet to remove the piston **(see illustration)**. Use only small quick blasts of air to ease the piston out of the bore. If a piston is blown out with too much force, it might be damaged. **Warning:** *Never place your fingers in front of the piston in an attempt to catch or protect it when applying compressed air. Doing so could result in serious injury.*

10   Once the piston is protruding from the caliper, remove it **(see illustration)**.

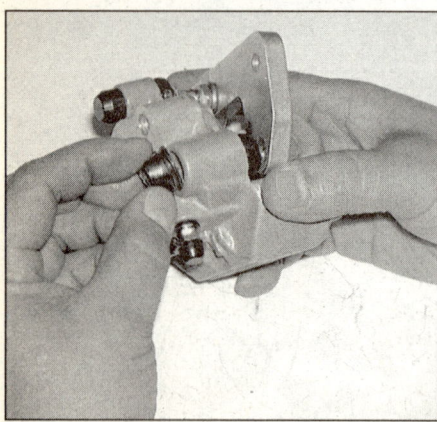

9.7b . . . remove the slide pin bolt boot . . .

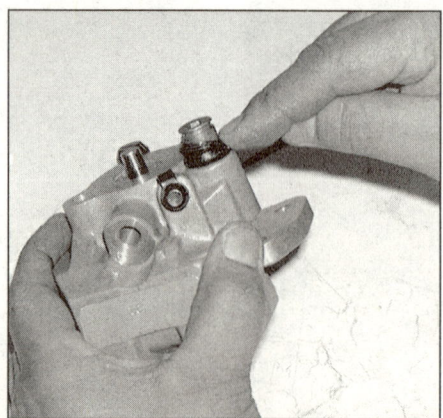

9.7c . . . and remove the slide pin bolt

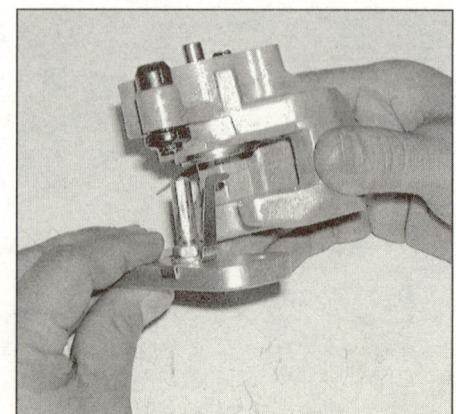

9.8a To detach the bracket from the caliper, pull out the bracket pin bolt

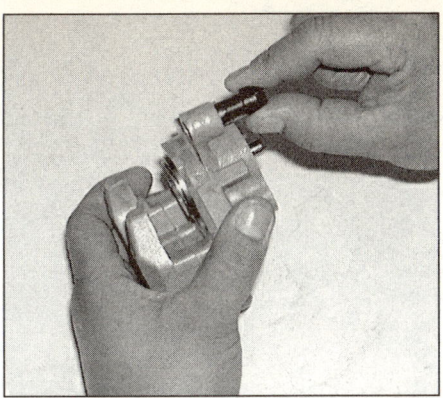

9.8b Remove the bracket pin bolt dust boot from the caliper

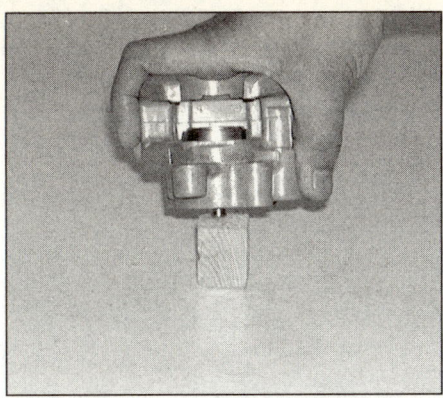

9.9 Slowly apply compressed air to push the piston out of the caliper – DO not get your fingers in the way!

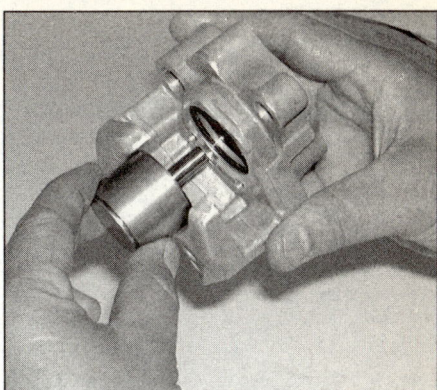

9.10 Once the piston is protruding from the caliper, remove it

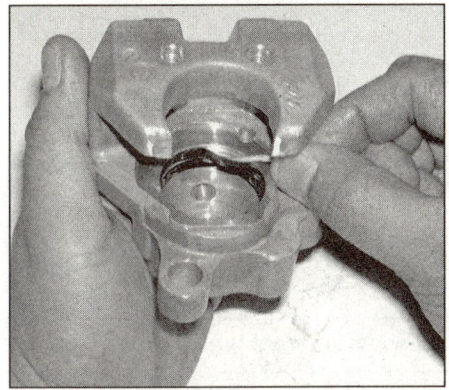

9.11a Using a wood or plastic tool, remove the dust seal . . .

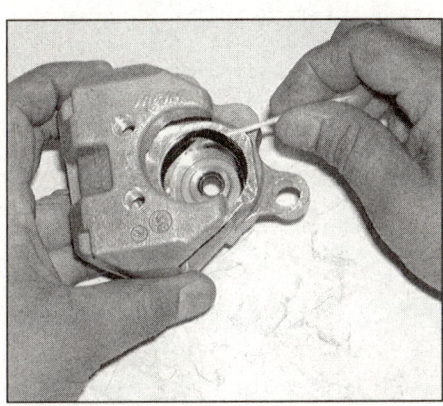

9.11b . . . and remove the piston seal

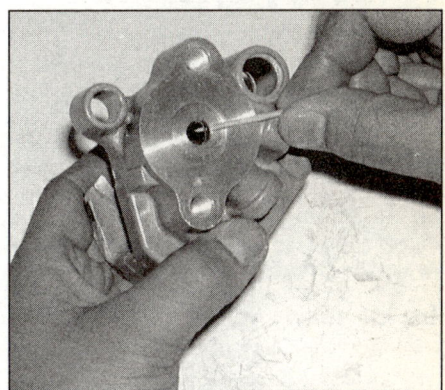

9.12a Remove the parking brake shaft O-ring . . .

11 Using a wood or plastic tool, remove the piston dust seal and the piston seal **(see illustrations)**.

12 Remove the parking brake shaft O-ring and the parking brake shaft seal **(see illustrations)**.

13 Clean the caliper parts with denatured alcohol, fresh brake fluid or brake system cleaner and dry them off with filtered, unlubricated compressed air. Inspect the caliper components **(see illustration)**. Replace all damaged or worn parts. It's a good idea to replace all rubber parts regardless of their apparent condition. Inspect the surfaces of the piston and the piston bores for rust, corrosion, nicks, burrs and loss of plating. If you find defects on the surface of either piston or piston bore, replace the piston and caliper assembly (the piston is matched to the caliper). If the caliper is in bad shape, inspect the master cylinder too.

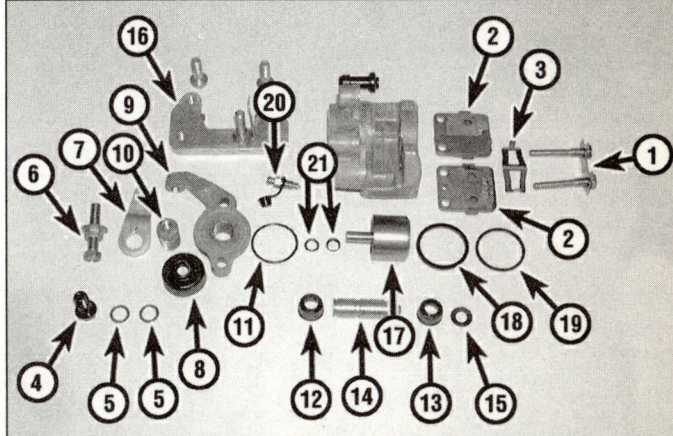

9.13 Rear brake caliper details

1 Lockwasher and brake pad pins	10 Parking brake shaft
2 Brake pads	11 Parking brake base O-ring
3 Brake pad spring	12 Slide pin bolt dust seal
4 Brake hose banjo bolt	13 Slide pin bolt dust boot
5 Banjo bolt sealing washers (always replace)	14 Slide pin bolt
	15 Slide pin bolt washer
6 Parking brake adjusting bolt and locknut	16 Caliper bracket/bracket pin bolt
7 Parking brake arm	17 Piston
8 Parking brake shaft dust boot	18 Piston seal
	19 Piston dust seal
9 Parking brake base	20 Bleeder valve and dust cap

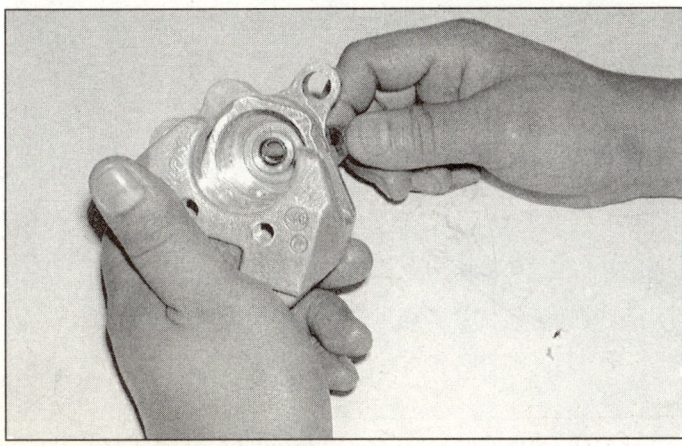

9.12b . . . and remove the parking brake shaft seal

6

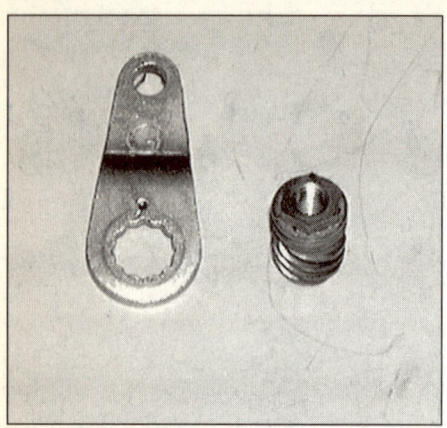

9.19 Align the punch marks on the parking brake arm and shaft when you assemble them

10.2 Remove the cotter pin from the hub nut

10.3 You can lock the rear axle by wedging a prybar between the socket and a wheel stud

14   Lubricate the new piston seal and piston dust seal with clean brake fluid and install them in their grooves in the caliper bore. Make sure both seals are fully and correctly seated.

15   Lubricate the piston with clean brake fluid and install it into its bore in the caliper. Using your thumbs, push the piston all the way in; make sure it doesn't become cocked in the bore.

16   Reassembly of the parking brake base is essentially the reverse of disassembly. Tighten the parking brake base bolts to the torque listed in this Chapter's Specifications.

Installation

Refer to illustration 9.19

17   Install the brake pads and install the caliper (see Section 8).

18   Connect the brake hose-to-caliper banjo bolt **(see illustration 9.1)**. Be sure to use new sealing washers. Tighten the banjo bolt to the torque listed in this Chapter's Specifications.

19   When installing the parking brake lever, make sure that the punch mark on the parking brake arm is aligned with the punch mark on the parking brake shaft **(see illustration)**.

20   Bleed the front brake system (see Section 15).

10   Rear wheel hubs - removal and installation

Refer to illustrations 10.2, 10.3 and 10.4

1   Remove the rear wheel (see Section 2).

2   Bend back the cotter pin and pull it out of the hub nut **(see**

illustration).

3   Have an assistant apply the rear brake while you loosen the hub nut. If you have no help available, apply the parking brake to lock the rear axle. If the caliper has been removed, use a prybar wedged between the socket and one of the wheel studs **(see illustration)**. Unscrew the hub nut and remove the washer.

4   Pull the hub off the axleshaft **(see illustration)**. Clean the hub and the axle splines.

5   Installation is the reverse of removal.

11   Rear brake disc - inspection, removal and installation

Inspection

Refer to illustrations 11.2, 11.3a and 11.3b

1   Visually inspect the surface of the disc for score marks and other damage. Light scratches are normal after use and won't affect brake operation, but deep grooves and heavy score marks will reduce braking efficiency and accelerate pad wear. If the disc is badly grooved it must be machined or replaced.

2   To check disc runout, mount a dial indicator with the plunger on the indicator touching the surface of the disc about 1/2-inch from the outer edge **(see illustration)**. Slowly turn the wheel hub and watch the indicator needle, comparing your reading with the disc runout limit listed in this Chapter's Specifications. If the runout is greater than allowed, replace the disc.

10.4 Pull off the hub and clean the splines

11.2 Position a dial indicator with its probe about 1/2-inch from the edge of the rear disc

11.3a The minimum thickness is stamped into the rear disc

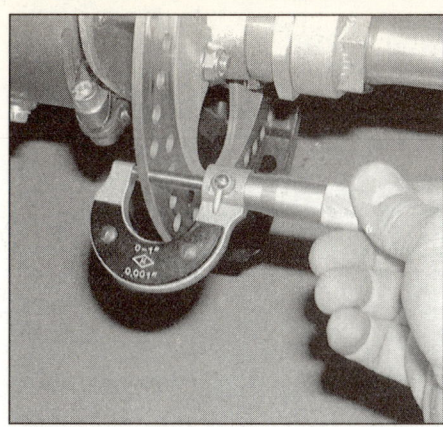

11.3b Measure the thickness of the rear disc with a micrometer

11.6 Remove the disc-to-hub nuts (arrows)

3   The disc must not be machined or allowed to wear down to a thickness less than the minimum allowable thickness, listed in this Chapter's Specifications. The minimum thickness is also stamped into the rear disc **(see illustration)**. If the minimum thickness listed in the Specifications and the minimum thickness stamped into the disc are different, the information on the disc supersedes the information in this book. Measure the thickness of the disc with a micrometer **(see illustration)**. If the disc is thinner than the allowable minimum, replace it.

Removal and installation

Refer to illustration 11.6

4   Remove the right rear wheel hub (see Section 10).
5   Remove the rear caliper (see Section 9).
6   Remove the disc retaining nuts **(see illustration)** and remove the disc.
7   Installation is the reverse of removal. Be sure to tighten the disc retaining bolts to the torque listed in this Chapter's Specifications.

---

12   Front brake master cylinder - removal, overhaul and installation

---

Removal

Refer to illustrations 12.4a, 12.4b and 12.5

1   If the front brake master cylinder is leaking fluid, or if the lever does not produce a firm feel when applied, and bleeding the brakes does not help, master cylinder overhaul is recommended.
2   Before disassembling the master cylinder, read through the entire procedure and make sure that you have the correct rebuild kit. Also, you will need some new, clean brake fluid of the recommended type, some clean rags and internal snap-ring pliers. **Caution:** *Disassembly, overhaul and reassembly of the brake master cylinder must be done in a spotlessly clean work area to avoid contamination and possible failure of the brake hydraulic system components. To prevent damage to the paint from spilled brake fluid, always cover the top cover or upper fuel tank when working on the master cylinder.*

3   Remove the reservoir cover, plate and diaphragm **(see illustrations 3.4a and 3.4b in Chapter 1)**. Siphon as much brake fluid from the reservoir as you can to avoid spilling it on the vehicle.
4   Pull back the rubber dust boot, remove the brake hose banjo bolt **(see illustration)**, discard the old sealing washers and separate the brake hose from the master cylinder. To prevent excessive loss of brake fluid, fluid spills and system contamination, plug the banjo fitting with a short section of rubber hose **(see illustration)**. If you don't have any hose of a suitable diameter, wrap the banjo fitting in a clean rag and suspend the hose in an upright position, or bend it down carefully and place the open end in a clean container.
5   Remove the master cylinder mounting bolts **(see illustration)** and separate the master cylinder from the handlebar. **Caution:** *Do not tip the master cylinder upside down or any brake fluid still in the reservoir will run out.*

Overhaul

Refer to illustrations 12.6a, 12.6b, 12.7a, 12.7b and 12.7c

6   Remove the dust cover from the brake lever pivot bolt, then remove the nut from the lever pivot bolt **(see illustrations)**. Remove the pivot bolt and lever.

12.4a Pull back the dust boot, remove the banjo bolt and discard the sealing washers

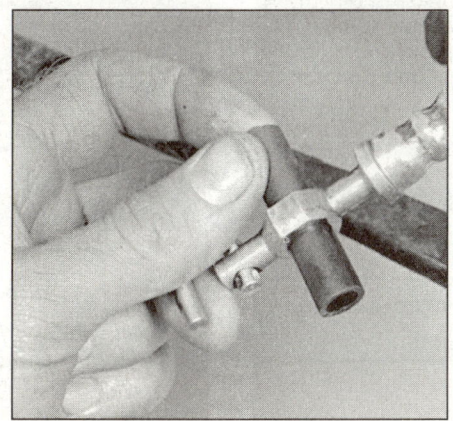

12.4b A short piece of rubber hose makes a good plug for the brake hose banjo fitting

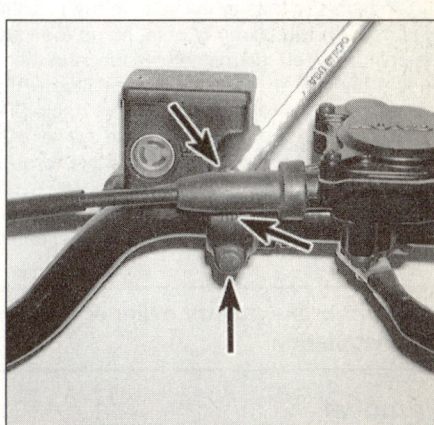

12.5 Remove the clamp bolts (arrows) and detach the master cylinder from the handlebar; the UP mark (arrow) must face upward on installation

6

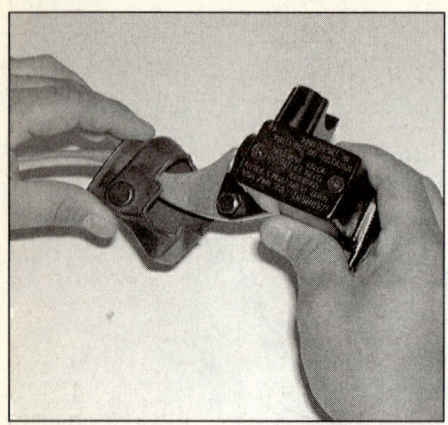

12.6a Remove the brake lever pivot bolt dust cover . . .

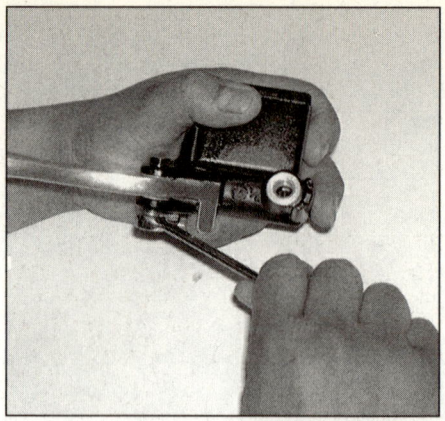

12.6b . . . remove the lever pivot bolt nut, the pivot bolt and the lever

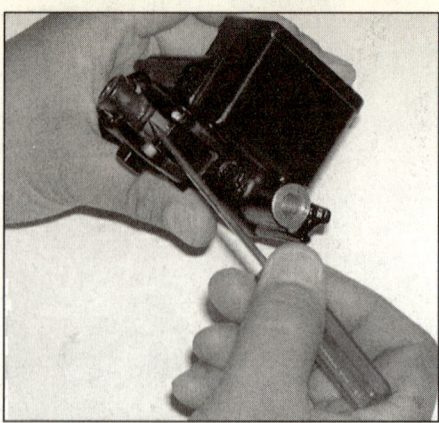

12.7a Remove the rubber dust boot from the end of the piston . . .

7   Carefully remove the rubber dust boot from the end of the piston **(see illustration)**. Using snap-ring pliers, remove the snap-ring. Pull out the piston and spring **(see illustrations)**.

8   Clean all of the parts with brake system cleaner (available at motorcycle dealerships and auto parts stores), isopropyl alcohol or clean brake fluid. **Caution:** *Do not, under any circumstances, use a petroleum-based solvent to clean brake parts. If compressed air is available, use it to dry the parts thoroughly (make sure it's filtered and unlubricated). Check the master cylinder bore for corrosion, scratches, nicks and score marks. If damage is evident, the master cylinder must be replaced with a new one. If the master cylinder is in poor condition, then the calipers should be checked as well.*

9   Lay out the parts neatly to prevent confusion during reassembly.

10   The dust seal, piston assembly and spring are included in the rebuild kit. Use all of the new parts, regardless of the apparent condition of the old ones.

11   Before reassembling the master cylinder, soak the piston and the rubber cup seals in clean brake fluid for ten or fifteen minutes. Lubricate the master cylinder bore with clean brake fluid, then carefully insert the piston and related parts in the reverse order of disassembly. Make sure the lips on the cup seals do not turn inside out when they are slipped into the bore.

12   Depress the piston, then install the snap-ring (make sure the snap-ring is properly seated in the groove). Install the rubber dust boot (make sure the lip is seated properly in the piston groove).

13   Lubricate the brake lever pivot bolt and the friction surface on the lever that pushes against the piston assembly.

Installation

14   Attach the master cylinder to the handlebar. Make sure that the UP arrow on the clamp is pointing up **(see illustration 12.5)** and that the split between the master cylinder and the clamp is aligned with the punch mark on the handlebar **(see illustration 10.3d in Chapter 3)**. Tighten the bolts to the torque listed in this Chapter's Specifications.

15   Connect the brake hose to the master cylinder, using new sealing washers. Tighten the banjo bolt to the torque listed in this Chapter's Specifications.

16   Fill the master cylinder with the recommended brake fluid (see Chapter 1), then bleed the front brake system (see Section 15).

13   Rear brake master cylinder - removal, overhaul and installation

Removal

Refer to illustrations 13.4a, 13.4b, 13.5, 13.6a and 13.6b

1   If the rear brake master cylinder is leaking fluid, or if the pedal does not produce a firm feel when it's applied, and bleeding the brakes does not help, master cylinder overhaul is recommended.

12.7b . . . remove the snap-ring, piston and spring

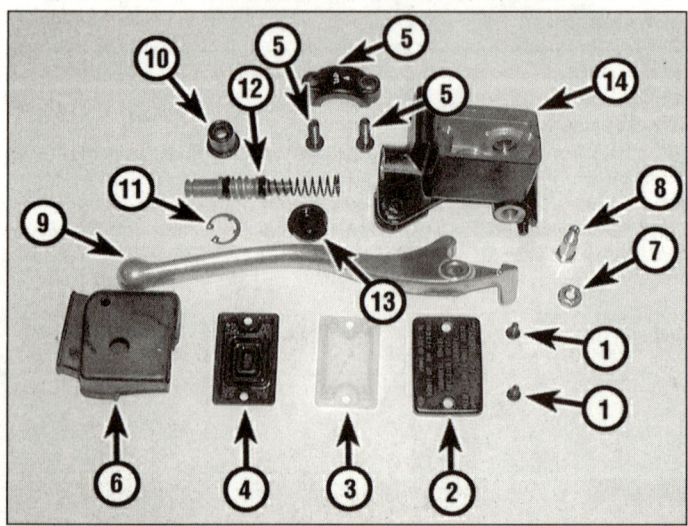

12.7c Front brake master cylinder details

1	*Reservoir cover screws*	*8*	*Brake lever pivot bolt*
2	*Reservoir cover*	*9*	*Brake lever*
3	*Diaphragm plate*	*10*	*Dust boot*
4	*Diaphragm*	*11*	*Snap-ring*
5	*Master cylinder clamp and bolts*	*12*	*Piston and spring*
6	*Brake lever dust cover*	*13*	*Reservoir strainer (fits inside bottom of reservoir)*
7	*Brake lever pivot bolt nut*		

13.4a Rear master cylinder mounting details (TRX300EX)

| 1 | Clevis pin | 3 | Fluid hose banjo fitting |
| 2 | Reservoir hose clamp | 4 | Mounting bolts |

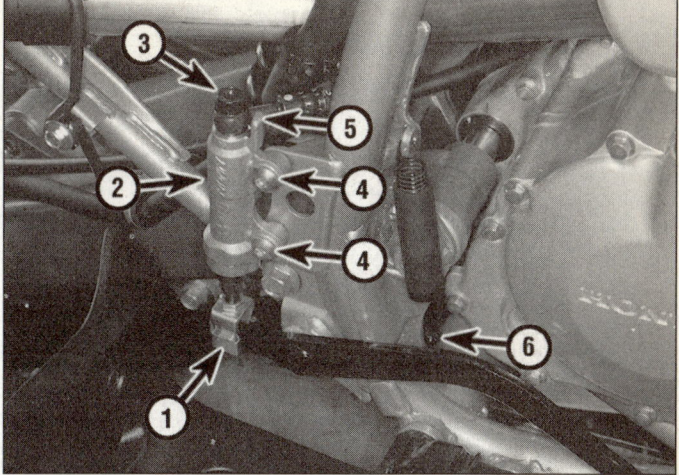

13.4b Rear master cylinder mounting details (TRX400EX)

1	Pushrod clevis	4	Mounting bolts
2	Reservoir hose clamp (hidden)	5	Fluid hose stopper
3	Fluid hose banjo fitting	6	Brake pedal spring

2   Before disassembling the master cylinder, read through the entire procedure and make sure that you have the correct rebuild kit. Also, you will need some new, clean brake fluid of the recommended type, some clean rags and internal snap-ring pliers. **Caution:** *Disassembly, overhaul and reassembly of the brake master cylinder must be done in a spotlessly clean work area to avoid contamination and possible failure of the brake hydraulic system components. To prevent damage to the paint from spilled brake fluid, always cover the top cover or upper fuel tank when working on the master cylinder.*

3   Unscrew the reservoir cover and siphon as much brake fluid from the reservoir as you can to avoid spilling it on the machine **(see illustrations 3.2b and 3.2c in Chapter 1)**.

4   Remove the brake hose banjo bolt **(see illustrations)** and discard the old sealing washers. Loosen the hose clamp that secures the hose coming from the master cylinder reservoir and disconnect both hoses from the master cylinder. To prevent excessive loss of brake fluid, fluid spills and system contamination, plug the banjo fitting with a short section of rubber hose **(see illustration 12.4b)**. If you don't have any hose of a suitable diameter, wrap the banjo fitting in a clean rag and suspend the hose in an upright position, or bend it down carefully and place the open end in a clean container.

5   Remove the clevis cotter pin and disconnect the pushrod clevis from the rear brake pedal **(see illustration)**.

6   Remove the master cylinder mounting bolts **(see illustration 13.4a or 13.4b)** and detach the master cylinder from the frame. Remove the reservoir mounting bolt **(see illustrations)** and collar and remove the reservoir from the frame.

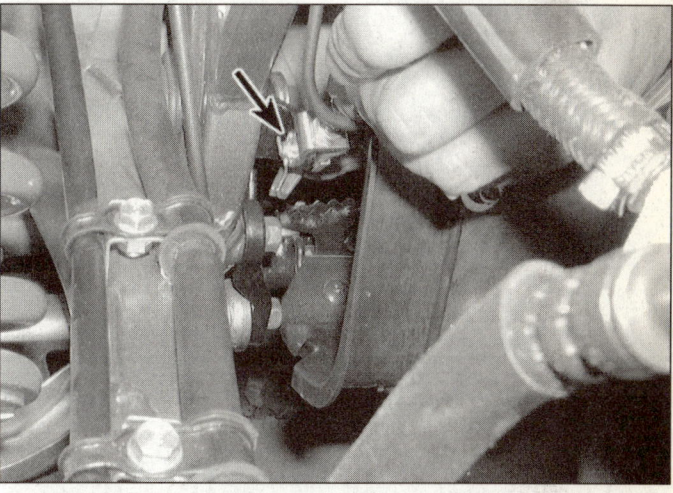

13.5 Remove the cotter pin (arrow) and clevis pin to disconnect the pushrod clevis from the rear brake pedal

13.6a TRX300EX master cylinder reservoir mounting bolt (right arrow) and reservoir hose clamp (left arrow)

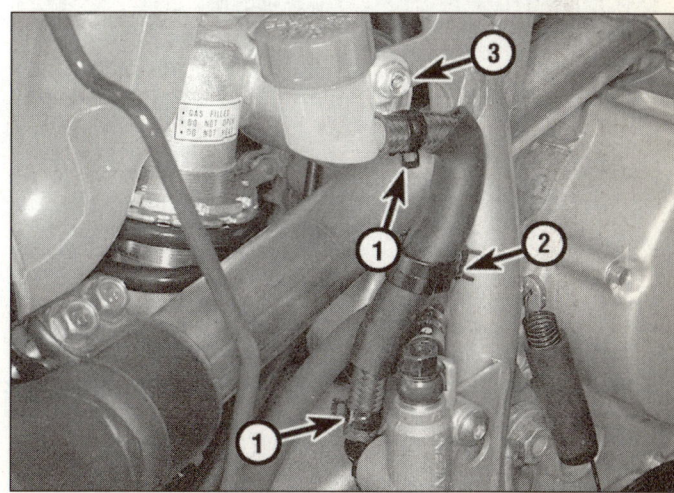

13.6b TRX400EX master cylinder reservoir hose clamps (1), hose retainer (2) and reservoir mounting bolt (3)

6

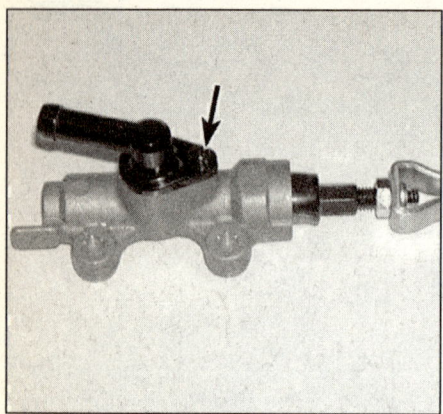

13.7 Remove the screw (arrow), pull off the hose adapter and discard its O-ring

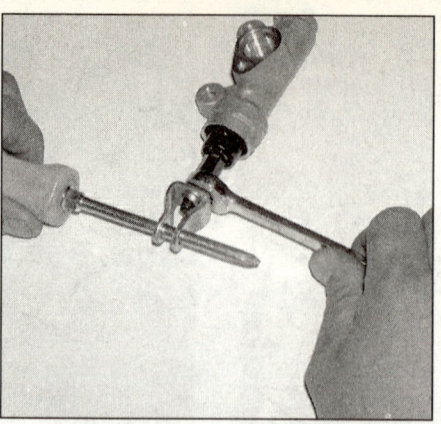

13.8a Mark the position of the clevis threads, hold the clevis and loosen its locknut, then unscrew the clevis and locknut

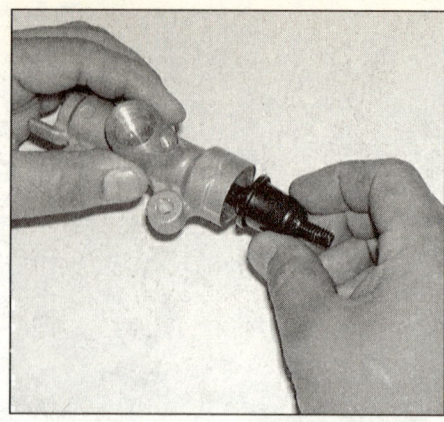

13.8b Remove the rubber dust boot from the pushrod . . .

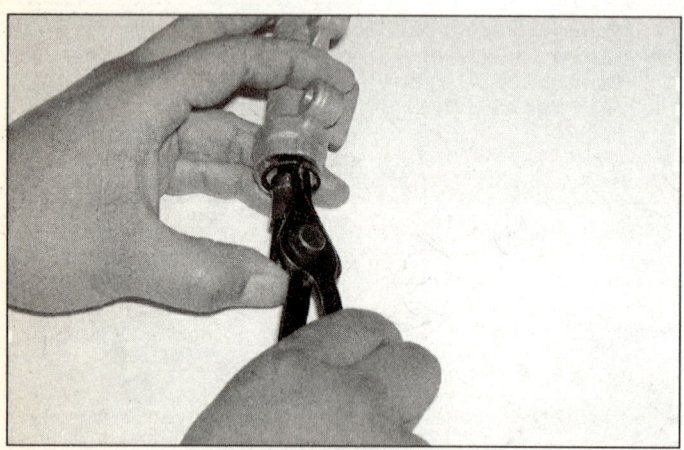

13.8c . . . using snap-ring pliers, remove the snap-ring . . .

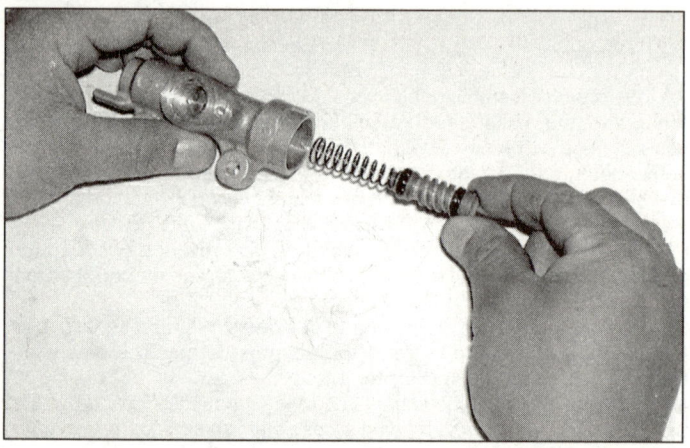

13.8d . . . slide out the piston and spring and separate the spring from the piston

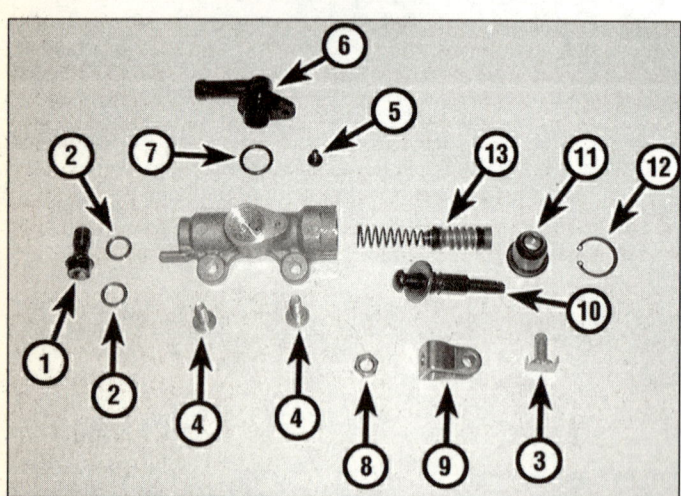

13.10 Rear master cylinder details

1	Brake hose banjo bolt	
2	Sealing washers (always replace)	
3	Pushrod clevis pin	
4	Master cylinder mounting bolts	
5	Reservoir hose adapter screw	
6	Reservoir hose adapter	
7	Reservoir hose adapter O-ring	
8	Pushrod/clevis locknut	
9	Clevis	
10	Pushrod	
11	Dust boot	
12	Snap-ring	
13	Piston and spring assembly	

Overhaul

Refer to illustrations 13.7, 13.8a, 13.8b, 13.8c, 13.8d and 13.10

7   Remove the reservoir hose adapter screw **(see illustration)**, pull off the adapter and remove and discard the old adapter O-ring.

8   Back off the pushrod clevis locknut **(see illustration)**, mark the position of the clevis by marking the threads with typing whiteout or some other suitable marking agent, then unscrew the clevis from the pushrod. Unscrew and remove the locknut from the pushrod. Carefully remove the rubber dust boot from the pushrod **(see illustration)**. Remove the snap-ring with snap-ring pliers, then slide out the piston and spring **(see illustrations)**.

9   Clean all of the parts with brake system cleaner (available at motorcycle dealerships and auto parts stores), isopropyl alcohol or clean brake fluid. **Caution:** *Do not, under any circumstances, use a petroleum-based solvent to clean brake parts. If compressed air is available, use it to dry the parts thoroughly (make sure it's filtered and unlubricated). Check the master cylinder bore for corrosion, scratches, nicks and score marks. If damage is evident, the master cylinder must be replaced with a new one. If the master cylinder is in poor condition, then the calipers should be checked as well.*

10   Lay out the parts neatly to prevent confusion during reassembly **(see illustration)**.

11   The dust seal, piston assembly and spring are included in the rebuild kit. Use all of the new parts, regardless of the apparent condition of the old ones.

12   Before reassembling the master cylinder, soak the piston and the rubber cup seals in clean brake fluid for ten or fifteen minutes. Lubricate the master cylinder bore with clean brake fluid, then carefully

14.2a Inspect the brake hoses and make sure they're secure in their clamps (TRX400EX left front brake hose shown)

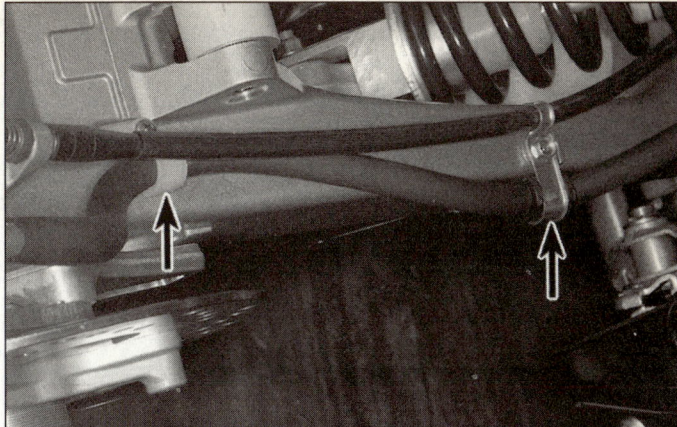

14.2b Make sure the rear brake hose retainers, including the one shared with the parking brake cable, are tight

insert the piston and related parts in the reverse order of disassembly. Make sure the lips on the cup seals do not turn inside out when they are slipped into the bore.

13 Depress the piston, then install the snap-ring (make sure the snap-ring is properly seated in the groove).

14 Install the rubber dust boot onto the pushrod, push it all the way on and make sure the lip is seated correctly over the ridge on the end of the master cylinder. Install the locknut on the pushrod. Screw it on beyond the mark you made for the clevis prior to disassembly. Screw the clevis onto the pushrod. Make sure that it's aligned with the mark you made before disassembly. Tighten the locknut securely.

15 Install a new O-ring on the adapter for the reservoir hose, install the adapter on the master cylinder and tighten the screw securely.

Installation

16 Connect the rear caliper brake hose to the master cylinder, using new sealing washers. Tighten the banjo bolt to the torque listed in this Chapter's Specifications.

17 Install the master cylinder on the frame and install - but don't tighten - the master cylinder retaining bolts.

18 Reattach the pushrod clevis to the rear brake pedal. Secure the clevis pin with a new cotter pin.

19 Tighten the rear brake master cylinder retaining bolts to the torque listed in this Chapter's Specifications.

20 Install the reservoir, if removed, and tighten the reservoir retaining bolt securely. Reattach the reservoir hose to the master cylinder. Use a new hose clamp.

21 Fill the master cylinder with the recommended brake fluid (see Chapter 1), then bleed the front brake system (see Section 15).

14 Brake hoses - inspection and replacement

Inspection

Refer to illustrations 14.2a and 14.2b

1 Once a week or, if the motorcycle is used less frequently, before every ride, check the condition of the brake hoses.

2 Twist and flex the rubber hoses while looking for cracks, bulges and seeping fluid **(see illustrations)**. Check extra carefully around the areas where the rubber hoses connect with the metal banjo fittings, as these are common areas for hose failure.

3 Inspect the metal banjo fittings on the ends of the brake hoses. If the fittings are rusted, scratched or cracked, replace them.

Replacement

Refer to illustrations 14.5a and 14.5b

4 Cover the surrounding area with plenty of rags, then disconnect the ends of the hose. Remove the banjo bolt from the caliper and disconnect the hose. Discard the old sealing washers.

5 To disconnect a front brake hose, unscrew the banjo bolt **(see illustration)** and discard the sealing washers. There is a sealing washer on each side of the fitting; where two fittings are together, as shown in the illustration, there's a sealing washer between them as well. Where the hose meets a metal brake line, hold the hex on the hose with a backup wrench and unscrew the flare nut on the metal line with a flare nut wrench **(see illustration)**.

6 Rear brake hoses are attached to the swingarm in several places **(see illustration 14.2b)**. Like the front hoses, they're also attached to the rear master cylinder and to the rear caliper by banjo bolts.

14.5a The front brake hoses join the metal line at this junction on the frame; use a flare nut wrench to disconnect the metal line

14.5b At the steering shaft bushing bracket, use a backup wrench and a flare nut wrench to disconnect the metal line

6

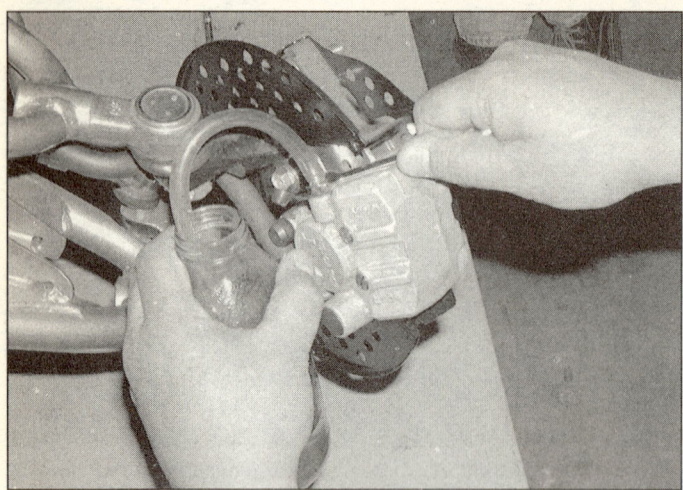

**15.5 Bleed the brakes at the caliper bleeder
valve (front caliper shown)**

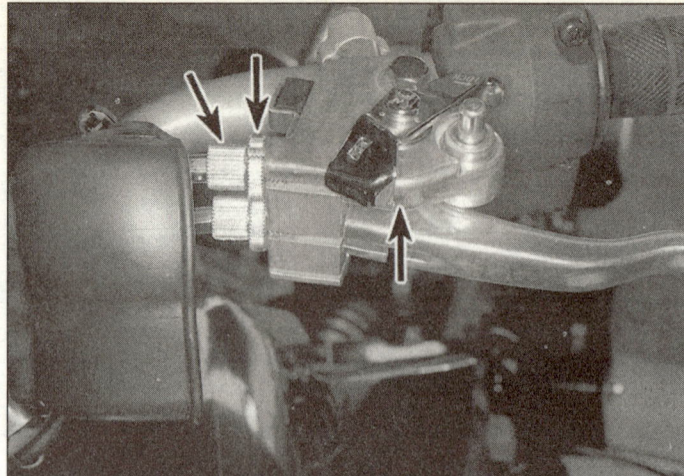

**16.2 Line up the slots in the adjuster and locknut (left arrows) and
remove the cable end from the lever (right arrow)**

7   Position the new hose, making sure it isn't twisted or otherwise strained, between the two components. Use new sealing washers on both sides of each banjo fitting, and tighten banjo bolts to the torque listed in this Chapter's Specifications. On non-banjo fittings, tighten the fitting securely.

8   Flush the old brake fluid from the system, refill the system with the recommended fluid (see Chapter 1) and bleed the air from the system (see Section 15). Check the operation of the brakes carefully before riding the motorcycle.

15 Brake system - bleeding

Refer to illustration 15.5

1   Bleeding the brake system removes all the air bubbles from the brake fluid reservoirs, the lines and the brake calipers. Bleeding is necessary whenever a brake system hydraulic connection is loosened, when a component or hose is replaced, or when the master cylinder or caliper is overhauled. Leaks in the system may also allow air to enter, but leaking brake fluid will reveal their presence and warn you of the need for repair.

2   To bleed the brakes, you will need some new, clean brake fluid of the recommended type (see Chapter 1), a length of clear vinyl or plastic tubing, a small container partially filled with clean brake fluid, some rags and a wrench to fit the brake caliper bleeder valves.

3   Cover the fuel tank and any other painted surfaces near the reservoir to prevent damage in the event that brake fluid is spilled.

4   Remove the reservoir cover screws and remove the cover and diaphragm (or, on rear TRX400EX rear brakes, unscrew the reservoir cover). Slowly pump the brake lever (or brake pedal) a few times, until no air bubbles can be seen floating up from the holes at the bottom of the reservoir. Doing this bleeds the air from the master cylinder end of the line. Top up the reservoir with new fluid, then install the reservoir diaphragm and cover, but don't tighten the screws (or cap); you may have to remove them several times during the procedure.

5   Remove the rubber dust cover from the bleeder valve on the caliper and slip a box wrench over the bleeder. Attach one end of the clear vinyl or plastic tubing to the bleed valve and submerge the other end in the brake fluid in the container **(see illustration)**.

6   Carefully pump the brake lever or brake pedal three or four times and hold it while opening the caliper bleeder valve. When the valve is opened, brake fluid will flow out of the caliper into the clear tubing and the lever will move toward the handlebar (or the pedal will move down). Retighten the bleeder valve, then release the brake lever or pedal.

7   Repeat this procedure until no air bubbles are visible in the brake fluid leaving the caliper and the lever or pedal is firm when applied. **Note:** *Remember to add fluid to the reservoir as the level drops. Use*

only new, clean brake fluid of the recommended type. Never re-use the fluid lost during bleeding.

8   Keep an eye on the fluid level in the reservoir, especially if there's a lot of air in the system. Every time you crack open the bleeder valve, the fluid level in the reservoir drops a little. Do not allow the fluid level to drop below the lower mark during the bleeding process. If the level looks low, remove the reservoir cover and add some fluid.

9   When you're done, inspect the fluid level in the reservoir one more time, add some fluid if necessary, then install the diaphragm and reservoir cover and tighten the screws securely. Wipe up any spilled brake fluid and check the entire system for leaks. **Note:** *If bleeding is difficult, it may be necessary to let the brake fluid in the system stabilize for a few hours (it may be aerated). Repeat the bleeding procedure when the tiny bubbles in the system have settled out.*

16 Parking brake cable - removal and installation

Refer to illustrations 16.2, 16.3, 16.4a and 16.4b

1   Remove the fuel tank (see Chapter 3).

2   At the clutch/parking brake lever bracket on the left end of the handlebar, peel back the rubber dust cover from the clutch and parking brake lever cable adjusters. Loosen the locknut at the parking brake lever on the handlebar **(see illustration)** and screw in the adjuster all the way to remove all tension from the cable. Line up the

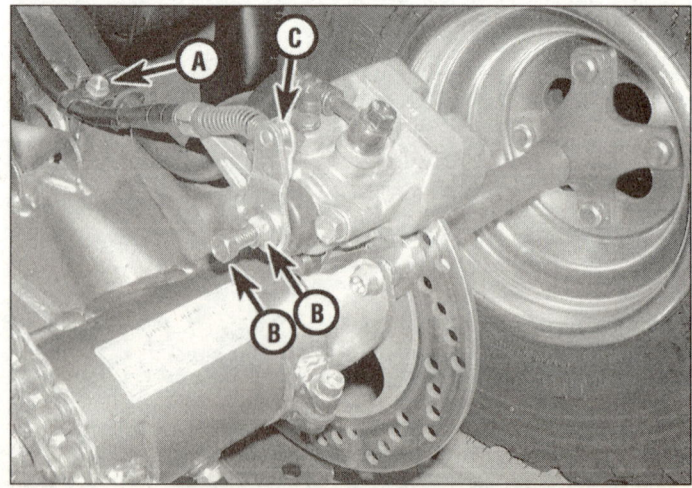

**16.3 Unbolt the cable clamp (A), loosen the locknut and adjuster
bolt (B) and disengage the cable end (C)**

16.4a The TRX300EX parking brake cable (arrow) is routed along the inside of the upper right frame member

16.4b The TRX400EX parking brake cable (arrow) is routed along the underside of the upper right frame member

slots in the adjuster and locknut with each other and with the slot in the clutch/parking brake lever. Pull the cable out of the adjuster, locknut and clutch/parking brake lever bracket. Rotate the cable so that it's aligned with the vertical slot in the parking brake lever and disengage the cable end plug from its recess in the parking brake lever.

3   Loosen the locknut and adjuster bolt at the parking brake arm on the rear brake caliper **(see illustration)** and disengage the cable from the parking brake arm.

4   Before removing the parking brake cable, carefully note the routing of the cable **(see illustration 14.2b and accompanying illustrations)**.

5   Installation is the reverse of removal.

6   When you're done, adjust the parking brake cable (see Chapter 1).

6

Notes

Chapter 7
Bodywork and frame

Contents

1 General information

This Chapter covers the procedures necessary to remove and install the body panels and other body parts. Since many service and repair operations on these vehicles require removal of the panels and/or other body parts, the procedures are grouped here and referred to from other Chapters.

In the case of damage to the panels or other body parts, it is usually necessary to remove the broken component and replace it with a new (or used) one. The material that the plastic body parts are composed of doesn't lend itself to conventional repair techniques. There are, however, some shops that specialize in "plastic welding", so it would be advantageous to check around first before throwing the damaged part away.

Note: *When attempting to remove any body panel, first study the panel closely, noting any fasteners and associated fittings, to be sure of returning everything to its correct place on installation. In some cases, the aid of an assistant will be required when removing panels, to help*

avoid damaging the surface. Once the visible fasteners have been removed, try to lift off the panel as described but DO NOT FORCE the panel - if it will not release, check that all fasteners have been removed and try again. Where a panel engages another by means of tabs and slots, be careful not to break the tabs or to damage the bodywork. Remember that a few moments of patience at this stage will save you a lot of money in replacing broken panels!

2 Seat and rear fenders - removal and installation

Refer to illustrations 2.1, 2.4, 2.5, 2.6, 2.8a and 2.8b

1 To unlock the seat and rear fenders, flip up the latch lever **(see illustration)** and lift the back end of the seat.

2 To remove the seat and rear fenders from the vehicle, pull the unit to the rear.

3 Replace either the seat or rear fenders as follows.

4 Remove the four nuts **(see illustration)**, remove the two seat lock lever brackets and remove the seat lock lever mechanism.

2.1 To release the seat, pull this lever (TRX300EX shown, TRX400EX similar)

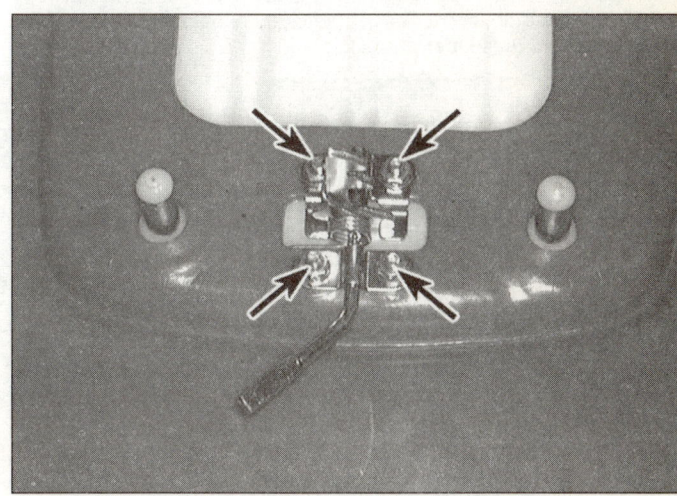

2.4 Remove the lock lever mechanism nuts (arrows) and two brackets (TRX300EX shown, TRX400EX similar)

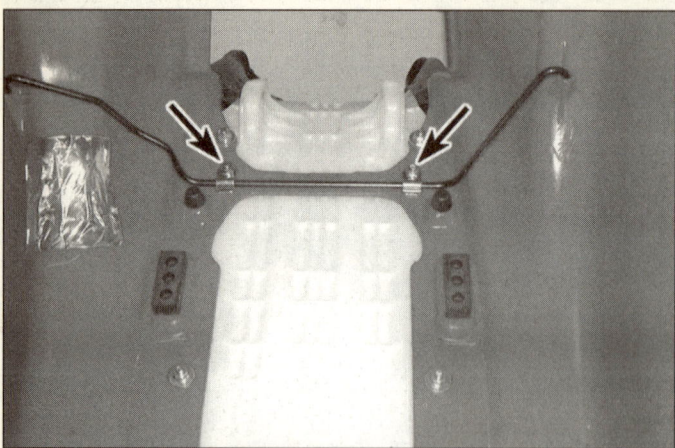

2.5 Remove the support rod nuts (arrows) and brackets (TRX300EX shown, TRX400EX similar)

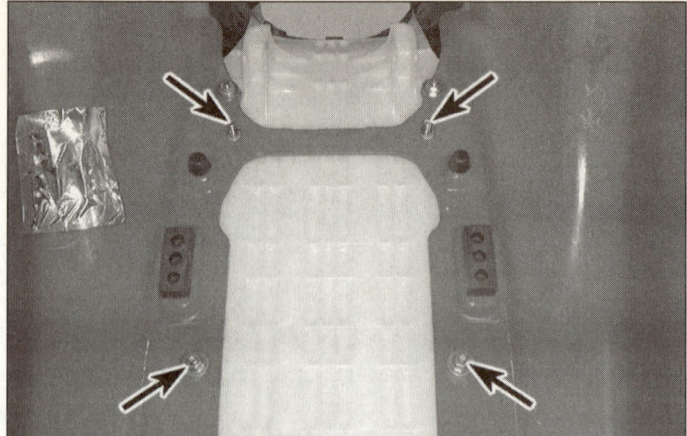

2.6 Four nuts and washers secure the seat to the fenders (TRX300EX shown, TRX400EX similar)

2.8a Insert the prongs at the front of the seat under the front insulators (arrows) (TRX300EX shown, TRX400EX similar) . . .

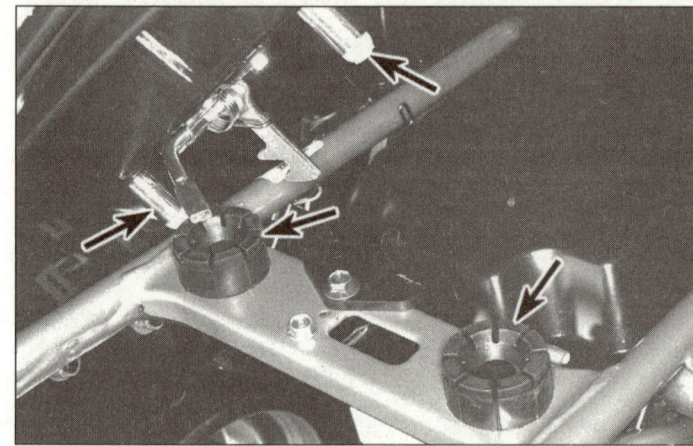

2.8b . . . and insert the pins under the rear of the seat (upper arrows) into the rear insulators (lower arrows) (TRX400EX shown, TRX300EX similar)

5   Remove the two nuts (see illustration), remove the support rod brackets and remove the support rod.

6   Remove the four nuts (see illustration), remove the washers and detach the seat from the rear fenders.

7   Reassembly of the seat and rear fenders is the reverse of disassembly.

8   Installation is the reverse of removal. Make sure that the prongs under the front of the seat are inserted under the front rubber insulators and the pins under the rear of the seat are inserted into the rear rubber insulators (see illustrations).

3   Handlebar cover – removal and installation

TRX300EX models

Refer to illustrations 3.2a, 3.2b and 3.2c

1   Remove the headlight (see Chapter 8).

2   Pry out the handlebar cover retaining screw caps, remove the cover screws, pull the fuel tank breather hose out of the cover, pull the

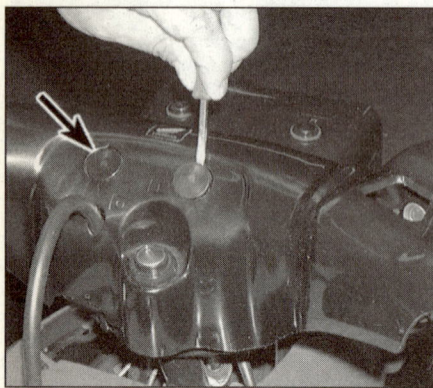

3.2a Pry the two retaining screw caps out of the handlebar cover . . .

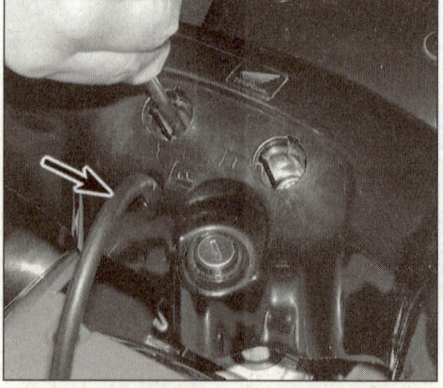

3.2b . . . remove both cover screws, pull the fuel tank breather hose (arrow) out of the cover . . .

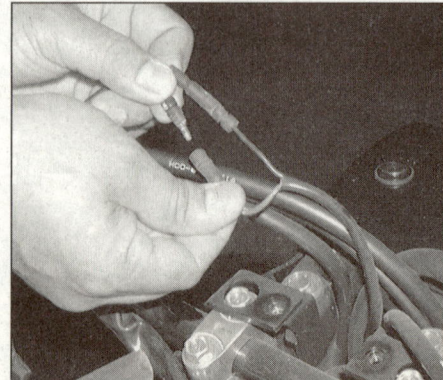

3.2c . . . lift off the cover and disconnect the ignition switch connectors

4.2 Remove the fuel tank center cover bolts (arrows) and the center cover (TRX300EX) . . .

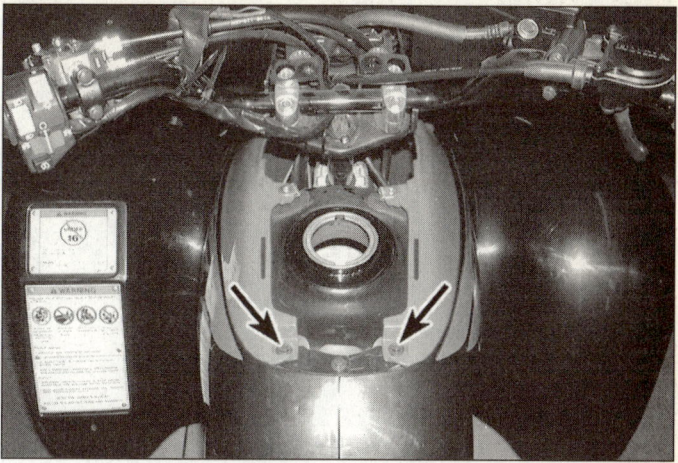

4.3 . . . remove the fuel tank cover bolts (arrows) and the fuel tank cover (TRX300EX)

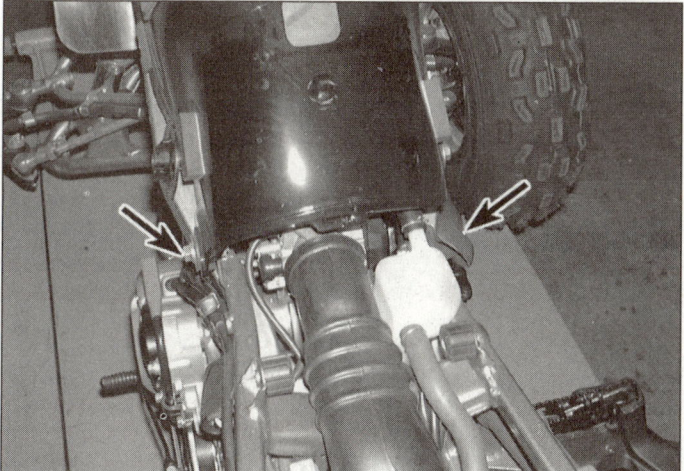

4.4 Remove the flange bolts (arrows) from the rear of the fender (TRX300EX)

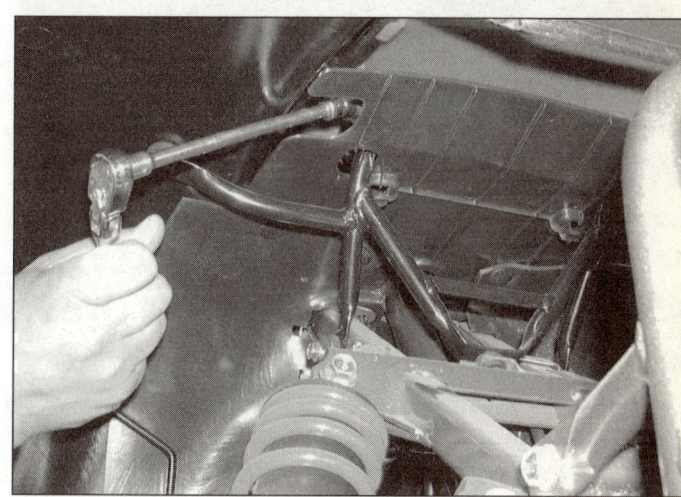

4.5a Remove both front fender mounting bolts (right bolt shown) . . .

cover off the handlebar and unplug the ignition switch electrical connectors **(see illustrations)**.

3   If you're replacing either the ignition switch or the handlebar cover, remove the ignition switch from the handlebar cover (see Chapter 8).

4   Installation is the reverse of removal.

4.5b . . . and remove both front fender support bolts (arrows) and washers (TRX300EX)

TRX400EX models

5   Pull the fuel tank breather hose out of the handlebar cover.
6   Remove the handlebar cover retaining bolt.
7   Remove the handlebar cover.
8   Installation is the reverse of removal.

4   Front fenders - removal and installation

TRX300EX models

Refer to illustrations 4.2, 4.3, 4.4, 4.5a and 4.5b

1   Remove the seat and rear fenders (see Section 2).
2   Remove the fuel tank breather hose from the handlebar cover **(see illustration 3.2b)**. Remove the two fuel tank center cover bolts **(see illustration)** and remove the fuel tank center cover.
3   Remove the two fuel tank cover bolts **(see illustration)** and remove the fuel tank cover.
4   Remove the two flange bolts from the rear of the fender unit **(see illustration)**.
5   Remove the two front fender mounting bolts and the two front fender support bolts and washers **(see illustrations)**.
6   Spread the rear part of the front fender unit just enough to clear the fuel tank and remove the fenders by pulling them to the front.
7   Installation is the reverse of removal.

7

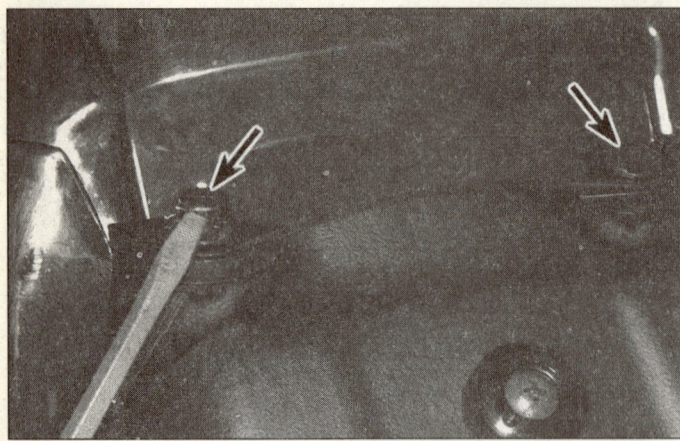

4.10a Pry up the center pins of these two trim clips (arrows), pull the clips out . . .

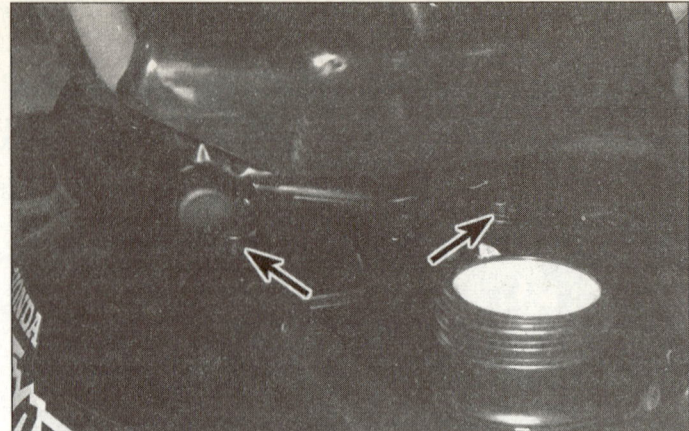

4.10b . . . and do the same with the two front clips (TRX400EX)

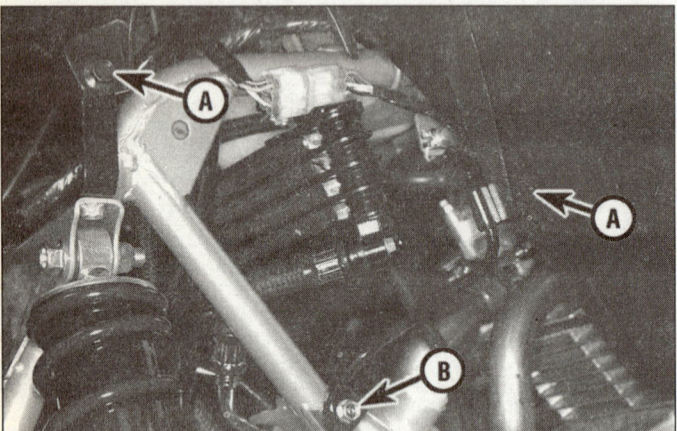

4.13 Here are the TRX400EX front fender bolts (A) and fender stay bolts (B) (left side shown)

4.14 Remove the TRX400EX side cover Phillips screw (arrow) . . .

TRX400EX models

Refer to illustrations 4.10a, 4.10b, 4.13 and 4.14

Note: *It's unnecessary to detach the front fender unit from the side covers in order to remove the fenders. Leave the side covers attached to the fenders.*

8   Remove the seat and rear fenders (see Section 2).
9   Remove the fuel filler cap and breather hose.
10   Release and remove the four trim clips from the upper cover **(see illustrations)**. Pop up the center pin to release each clip, then

5.4 . . . pry up the trim clip's center pin and pull out the clip (arrow) . . .

remove the clip.
11   Slide the upper cover to the rear to disengage its four tabs from their corresponding slits in the front fender. Remove the upper cover.
12   Install the fuel filler cap and breather hose.
13   Remove the four fender bolts **(see illustration)**.
14   Remove the screw from each side cover **(see illustration)** to detach the side covers from the frame (it's not necessary to detach the side covers from the front fenders).
15   To remove the front fenders, spread the side covers apart, slide the fender unit forward, unplug the electrical connectors and lift the fender unit off the vehicle.
16   Installation is the reverse of removal.

5   Side cover (TRX400EX models) - removal and installation

Refer to illustrations 5.4 and 5.5

Note: *This procedure applies to both the left and the right side covers.*
1   Remove the seat and rear fenders (see Section 2).
2   Remove the screw from the side cover **(see illustration 4.14)**.
3   Remove the trim clip from the upper cover **(see illustration 4.10a)**.
4   Remove the side cover trim clip from the front fender **(see illustration)**.
5   Raise the front fender bodywork slightly and disengage the three tabs on the lower edge of the fender from their corresponding slots in the upper edge of the side cover by sliding the side cover forward slightly **(see illustration)**.
6   Installation is the reverse of removal.

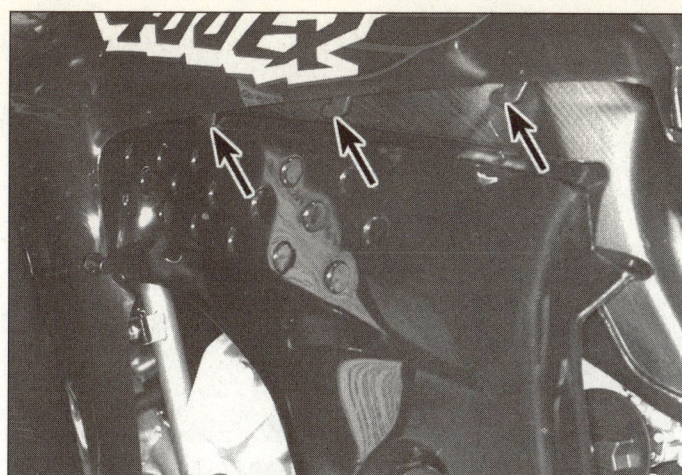

5.5 . . . slide the side cover forward and disengage the fender tabs (arrows) from their slots in the side cover

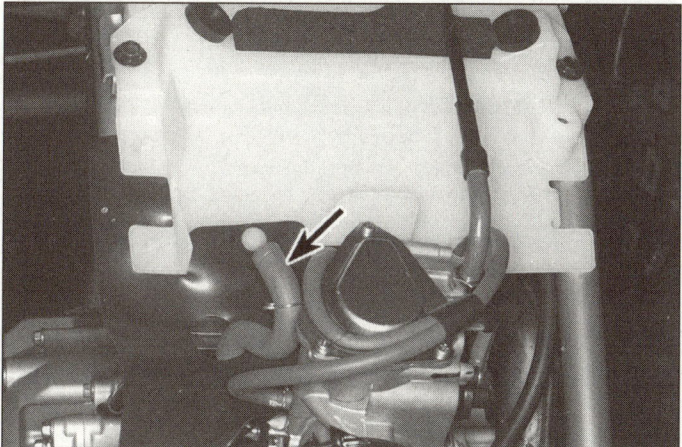

6.6 . . . pull the fuel hose out of the rubber protector (arrow)

6 Heat protector (TRX400EX models) - removal and installation

Refer to illustrations 6.3 and 6.6
1   Remove the front fender assembly (see Section 4).
2   Remove the fuel tank (see Chapter 3).

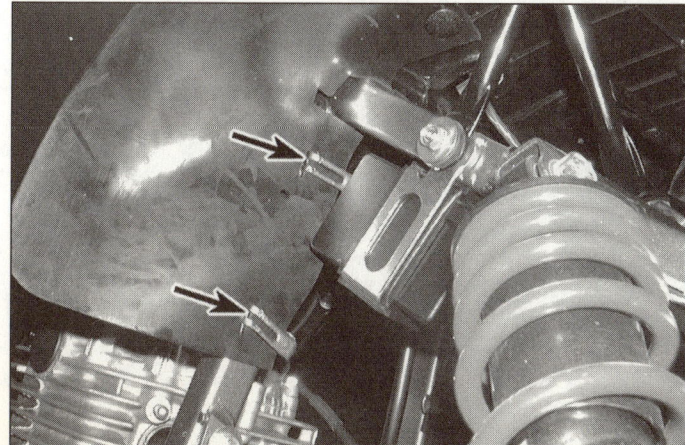

7.1 Bend open the hooks (arrows) to detach each inner fender (TRX300EX)

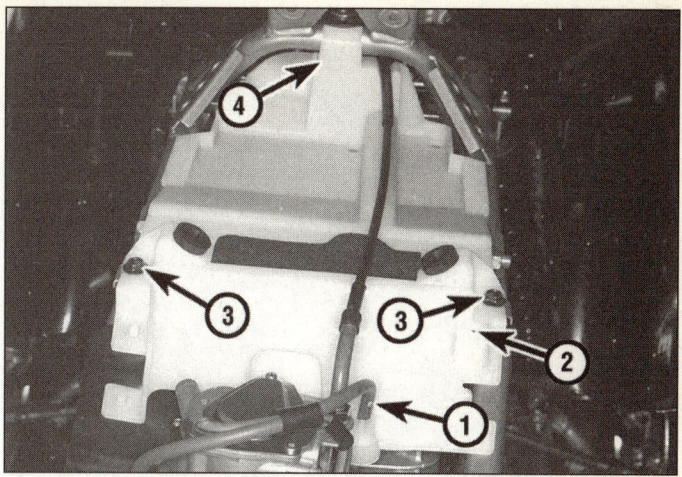

6.3 Heat protector details (TRX400EX)

1	Carburetor vent hose	3	Bolts
2	Trim clip (if equipped)	4	Front hook

3   Pull the carburetor air vent tube from the tube holder in the heat protector **(see illustration)**.
4   Disengage the clutch cable from the cable guide at the front of the heat protector.
5   Release and remove the trim clip (if equipped) from the heat protector **(see illustration 6.3)**. Pop up the center pin to release the clip, then remove the clip.
6   Pull the fuel hose through the hole in the rubber protector **(see illustration)**.
7   Remove the two heat protector retaining bolts **(see illustration 6.3)**.
8   Disengage the tabs at the rear corners of the heat protector and remove the heat protector.
9   Installation is the reverse of removal.

7 Inner fender and air guide (TRX300EX models) – removal and installation

Refer to illustrations 7.1, 7.2 and 7.3
1   Bend open the two retaining hooks **(see illustration)** from each inner fender and detach the inner fender from the frame.
2   Remove the two clips from the rear air guide **(see illustration)** and remove the rear air guide. Remove the two air guide bracket nuts, visible through the holes in the front air guide.

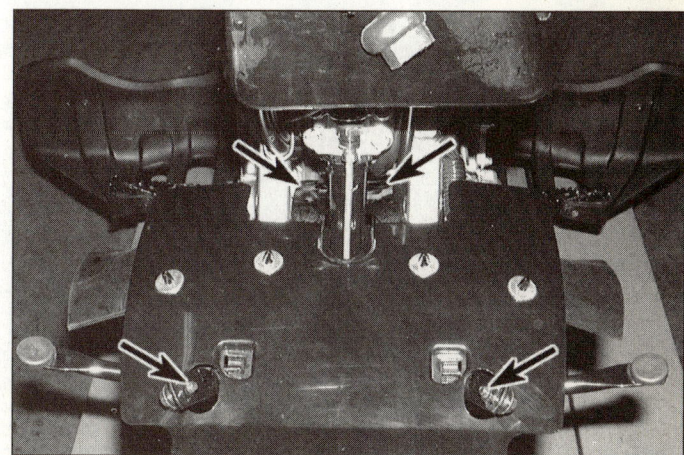

7.2 Remove the clips and nuts (arrows) and remove the rear air guide (TRX300EX)

7

7.3   Disengage the brackets on the underside of the front air guide from the pins (circled) on the air guide support bracket (TRX300EX)

8.1   Footpeg bracket bolts (left arrows) and mudguard mounting bolt (right arrow) (TRX300EX)

3   Disengage the front air guide from the air guide brackets **(see illustration)** and remove the front air guide and the inner fenders as a single assembly.
4   If you're replacing the inner fender(s) or the front air guide, remove the clips and separate the inner fenders from the front air guide.
5   Installation is the reverse of removal.

8   Footpeg/mudguard - removal and installation

TRX300EX models

Refer to illustrations 8.1, 8.2 and 8.3

1   To remove the footpeg, the mudguard and the mudguard stay as a single assembly, remove the footpeg mounting bolts and the mudguard mounting bolt **(see illustration)** and remove the footpeg/mudguard/stay assembly.
2   To remove only the mudguard, remove the four mudguard mounting screws **(see illustration)**.
3   To remove the mudguard stay, remove the mudguard (see Step 2), remove the stay bolts **(see illustration)** and disengage the stay from the outer end of the footpeg.
4   To remove the footpeg, remove the mudguard and the mudguard stay (see Steps 2 and 3), and then remove the footpeg mounting bolts **(see illustration 8.1)**.
5   Installation is the reverse of removal. Tighten all fasteners securely.

TRX400EX models

Refer to illustrations 8.6 and 8.7

6   To remove the mudguard, remove the five mudguard mounting screws **(see illustration)**.
7   To remove the footpeg, remove the stay bolt from the underside of the footpeg and remove the two footpeg mounting bolts **(see illustration)**.
8   Installation is the reverse of removal. Tighten all fasteners securely.

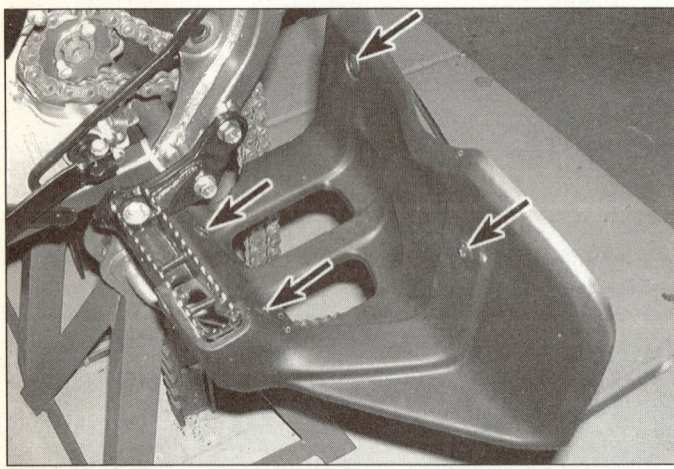

8.2   Mudguard mounting screws (arrows) (TRX300EX)

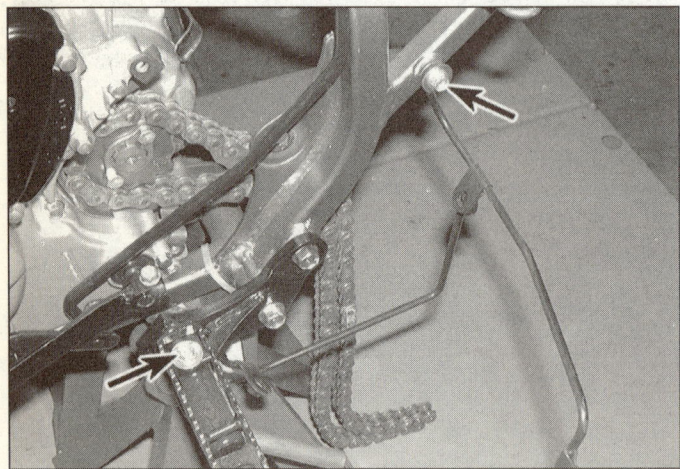

8.3   Mudguard stay bolts (arrows) (TRX300EX)

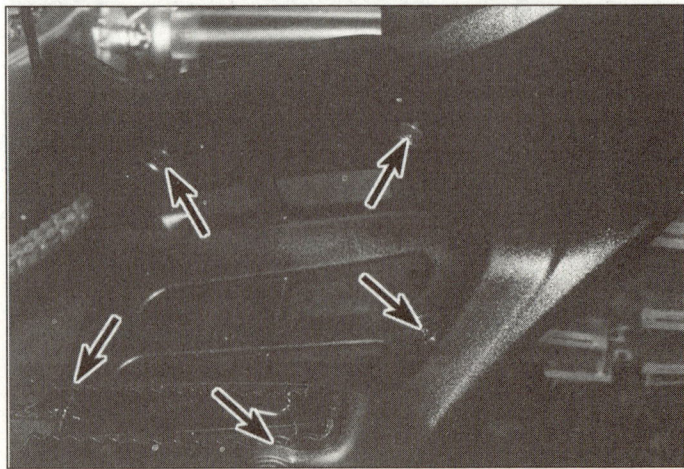

8.6   Mudguard mounting screws (arrows) (TRX400EX)

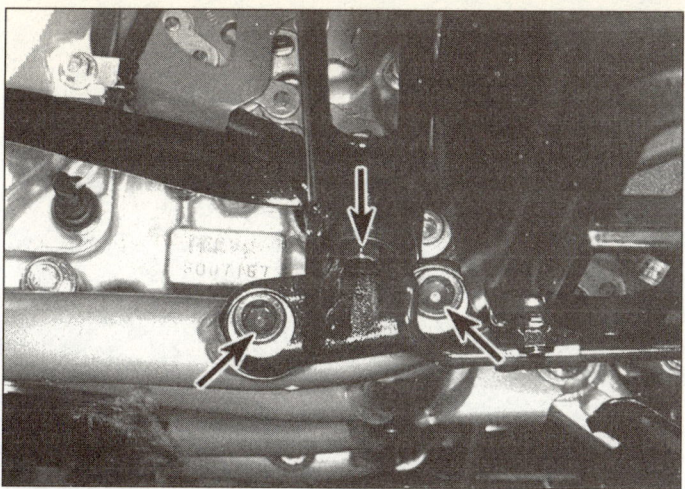

8.7 Footpeg stay bolt (upper arrow) and footpeg mounting bolts (lower arrows) (TRX400EX)

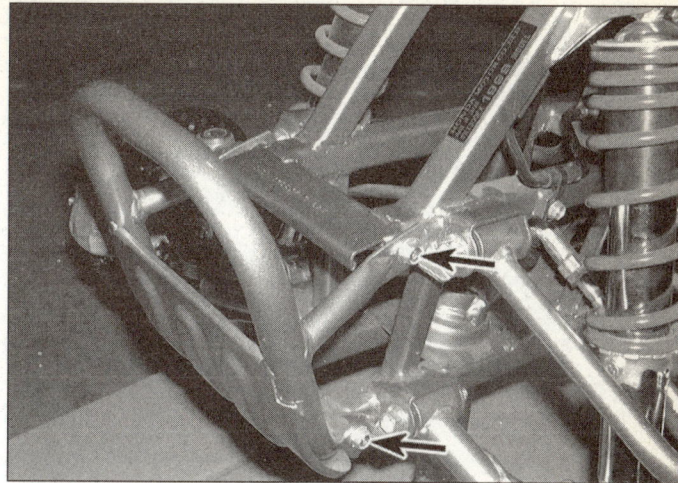

9.1 The front bumper is secured by two bolts on each side (arrows) (TRX300EX shown) . . .

9   Front bumper - removal and installation

Refer to illustration 9.1

1   Remove the four bumper mounting bolts (see illustration) and remove the bumper.
2   Installation is the reverse of removal. Tighten the bumper bolts securely.

10   Skid plate - removal and installation

Refer to illustrations 10.3a and 10.3b

1   Remove the fuel tank and drain the carburetor float chamber (see Chapter 3).
2   Carefully lay the machine on its side.
3   Remove the skid plate bolts and remove the skid plate from the swingarm (see illustrations).
4   Installation is the reverse of removal. Tighten all fasteners securely.

11   Frame - general information, inspection and repair

1   All models use a double-cradle frame made of steel tubing.
2   The frame shouldn't require attention unless accident damage has occurred. In most cases, frame replacement is the only satisfactory remedy for such damage. A few frame specialists have the jigs and other equipment necessary for straightening the frame to the required standard of accuracy, but even then there is no simple way of assessing to what extent the frame may have been over-stressed.
3   After the machine has accumulated a lot of miles, the frame should be examined closely for signs of cracking or splitting at the welded joints. Corrosion can also cause weakness at these joints. Loose engine mount bolts can cause elongation of the bolt holes and can fracture the engine mounting points. Minor damage can often be repaired by welding, depending on the nature and extent of the damage.
Remember that a frame that is out of alignment will cause handling problems. If misalignment is suspected as the result of an accident, it will be necessary to strip the machine completely so the frame can be thoroughly checked.

10.3a The skid plate is secured by four bolts underneath (arrows) (TRX300EX shown) . . .

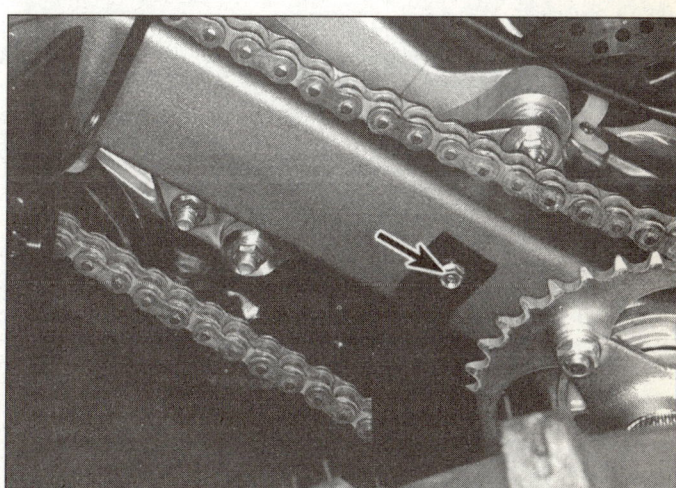

10.3b . . . and one bolt on the left side of the swingarm (TRX400EX shown)

7

Notes

Chapter 8
Electrical system

Contents

Specifications

Battery

Type	Maintenance free
Standing voltage (engine off)..	13.0 to 13.2 volts

Bulbs

Headlight(s)
TRX300EX	60/55 watts
TRX400EX	30/30 watts each
Tail light (TRX300EX and TRX400EX)	5 watts

Indicator light(s)
TRX300EX (Neutral and Reverse)...	1.7 watts
TRX400EX (Neutral)	3.4 watts

Charging system

Regulated charging voltage (engine running)	14.0 to 15.5 volts at 5000 rpm
Current leakage limit..	0.1 mA

Stator (charging) coil resistance
TRX300EX	0.2 to 0.6 ohm at 20 degrees C (68 degrees F)
TRX400EX	0.1 to 1.0 ohm at 20 degrees C (68 degrees F)

Starter motor

Brush length	
Standard..	12.5 mm (0.49 inch)
Minimum..	8.5 mm (0.33 inch)
Commutator diameter	Not specified
Fuse rating	15 amps

Torque specifications

Left crankcase cover bolts	
TRX300EX	10 Nm (84 in-lbs)
TRX400EX	Not specified
Stator coil Allen bolts (TRX300EX)...	10 Nm (84 in-lbs)
Stator coil bolts (TRX400EX)..	Not specified
Starter reduction gear cover bolts (TRX300EX)	10 Nm (84 in-lbs)
Idle gear cover bolts (TRX400EX)	Not specified
Rotor bolt	
TRX300EX	110 Nm (80 ft-lbs)
TRX400EX	127 Nm (94 ft-lbs)
Neutral and Reverse switches (TRX300EX)............................	13 Nm (108 in-lbs)*
Neutral switch (TRX400EX)..	13 Nm (108 in-lbs)*
Starter clutch-to-flywheel Torx bolts (TRX300EX)	16 Nm (144 in-lbs)*
Flywheel-to-starter clutch Torx bolts (TRX400EX).................	30 Nm (22 ft-lbs)*

Apply non-permanent thread locking agent to the threads.

1   General information

The machines covered by this manual are equipped with a 12-volt electrical system. The components include a three-phase permanent magnet alternator and a regulator/rectifier unit. The regulator/rectifier unit maintains the charging system output within the specified range to prevent overcharging and converts the AC (alternating current) output of the alternator to DC (direct current) to power the lights and other components and to charge the battery.

An electric starter mounted to the engine case behind the cylinder is standard equipment. The starting system includes the starter motor, the battery, the starter relay and the various wires and switches. If the engine kill switch and the main key switch are both in the On position, the circuit relay allows the starter motor to operate only if the transmission is in Neutral. **Note:** *Keep in mind that electrical parts, once purchased, can't be returned. To avoid unnecessary expense, make very sure the faulty component has been positively identified before buying a replacement part.*

2   Electrical troubleshooting

A typical electrical circuit consists of an electrical component, the switches, relays, etc. related to that component and the wiring and connectors that hook the component to both the battery and the frame. To aid in locating a problem in any electrical circuit, wiring diagrams are included at the end of this Chapter.

Before tackling any troublesome electrical circuit, first study the appropriate diagrams thoroughly to get a complete picture of what makes up that individual circuit. Trouble spots can often be identified by noting whether components related to the circuit are operating correctly. If several components or circuits fail at one time, chances are the fault lies in the fuse or ground connection, as several circuits often are routed through the same fuse and ground connections.

Electrical problems often stem from simple causes, such as loose or corroded connections or a blown fuse. Prior to any electrical troubleshooting, always visually check the condition of the fuse, wires and connections in the problem circuit.

If testing instruments are going to be utilized, use the diagrams to plan where you will make the necessary connections in order to accurately pinpoint the trouble spot.

The basic tools needed for electrical troubleshooting include a test light or voltmeter, a continuity tester (which includes a bulb, battery and set of test leads) and a jumper wire, preferably with a circuit breaker incorporated, which can be used to bypass electrical components. Specific checks described later in this Chapter may also require an ammeter or ohmmeter.

Voltage checks should be performed if a circuit is not functioning properly. Connect one lead of a test light or voltmeter to either the negative battery terminal or a known good ground. Connect the other lead to a connector in the circuit being tested, preferably nearest to the battery or fuse. If the bulb lights, voltage is reaching that point, which means the part of the circuit between that connector and the battery is problem-free. Continue checking the remainder of the circuit in the same manner. When you reach a point where no voltage is present, the problem lies between there and the last good test point. Most of the time the problem is due to a loose connection. Since these vehicles are designed for off-road use, the problem may also be dirt, water or corrosion in a connector. Keep in mind that some circuits only receive voltage when the ignition key is in the On position.

One method of finding short circuits is to remove the fuse and connect a test light or voltmeter in its place to the fuse terminals. There should be no load in the circuit. Move the wiring harness from side-to-side while watching the test light. If the bulb lights, there is a short to ground somewhere in that area, probably where insulation has rubbed off a wire. The same test can be performed on other components in the circuit, including the switch.

A ground check should be done to see if a component is grounded properly. Disconnect the battery and connect one lead of a self-powered test light (such as a continuity tester) to a known good ground. Connect the other lead to the wire or ground connection being tested. If the bulb lights, the ground is good. If the bulb does not light, the ground is not good.

A continuity check is performed to see if a circuit, section of circuit or individual component is capable of passing electricity through it. Disconnect the battery and connect one lead of a self-powered test light (such as a continuity tester) to one end of the circuit being tested and the other lead to the other end of the circuit. If the bulb lights, there is continuity, which means the circuit is passing electricity through it properly. Switches can be checked in the same way.

Remember that all electrical circuits are designed to conduct electricity from the battery, through the wires, switches, relays, etc. to the electrical component (light bulb, motor, etc.). From there it is directed to the frame (ground) where it is passed back to the battery. Electrical problems are basically an interruption in the flow of electricity from the battery or back to it.

3.2 To remove the battery, disconnect the cables (negative first), then unbolt the hold-down strap and lift out the battery (TRX400EX shown)

3.8 Check battery standing voltage with a voltmeter connected between the terminals

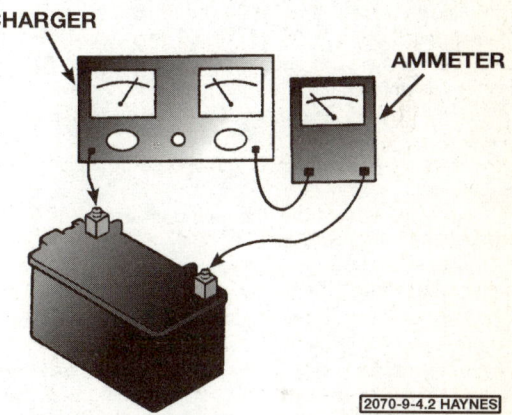

4.2 If the charger doesn't have an ammeter built in, connect one in series as shown; DO NOT connect the ammeter between the battery terminals or it will be ruined

3   Battery - inspection and maintenance

Refer to illustrations 3.2 and 3.8

1   Most battery damage is caused by heat, vibration, and/or low electrolyte levels, so keep the battery securely mounted, and make sure the charging system is functioning properly. The battery used on these vehicles is a maintenance free (sealed) type and therefore does-n't require the addition of water. However, the following checks should still be regularly performed. **Warning:** *Always disconnect the negative cable first and connect it last to prevent sparks which could the battery to explode.*

2   To remove the battery, disconnect the cables, negative cable first, then remove the hold-down bolt **(see illustration)** and lift out the battery. If necessary, remove the battery box.

3   Inspect the battery terminals and cables for tightness and corrosion. If corrosion is evident, disconnect the cables from the battery, disconnecting the negative (-) terminal first, and clean the terminals and cable ends with a wire brush or knife and emery cloth. Reconnect the cables, connecting the negative cable last, and apply a thin coat of petroleum jelly to the cables to slow further corrosion.

4   The battery and box must be kept clean to prevent current leak-age, which can discharge the battery over a period of time (especially when it sits unused). Wash the outside of the battery and box with a solution of baking soda and water. Rinse the battery and box thor-oughly, then dry them.

5   Look for cracks in the battery case and the battery box and replace the battery and/or box if any are found. If acid has spilled onto the battery frame, neutralize it with a baking soda and water solution, and then touch up any damaged paint. If the frame is not repairable, replace it.

6   Make sure the battery vent tube is directed away from the frame and is not kinked or pinched (there is no vent tube on sealed original equipment batteries, but some conventional aftermarket batteries are vented).

7   If the vehicle sits unused for long periods of time, disconnect the cables from the battery terminals. Refer to Section 4 and charge the battery approximately once every month.

8   To assess the condition of the battery, measure the voltage at the battery terminals **(see illustration)**. Connect the voltmeter positive probe to the battery positive terminal and the negative probe to the negative terminal. When fully charged, the battery should have a standing voltage of about 13 volts. If the voltage falls below 12.3 volts, remove and charge the battery (see Section 4).

4   Battery - charging

Refer to illustration 4.2

1   If the machine sits idle for extended periods or if the charging sys-tem malfunctions, the battery can be charged from an external source.

2   The battery should be charged at no more than the rate printed on the charging rate and time label fixed to the battery. Honda recom-mends a special battery tester and charger that are unlikely to be avail-able to the vehicle owner. To measure the charging rate, connect an ammeter is series with a battery charger **(see illustration)**.

3   When charging the battery, always remove it from the machine **(see illustration 3.2a or 3.2b)**.

4   Disconnect the battery cables (negative cable first), then connect a voltmeter between the battery terminals and measure the voltage.

5   If terminal voltage is within the range listed in this Chapter's Specifications, the battery is fully charged. If it's lower, recharge the battery.

6   A quick charge can be used in an emergency, provided the maxi-mum charge rate and time printed on the battery are not exceeded (exceeding the maximum rate or time may buckle the battery plates, rendering it useless). A quick charge should always be followed as soon as possible by a charge at the standard rate and time.

7   Hook up the battery charger leads (positive lead to battery posi-tive terminal, negative lead to battery negative terminal), then, and only

8

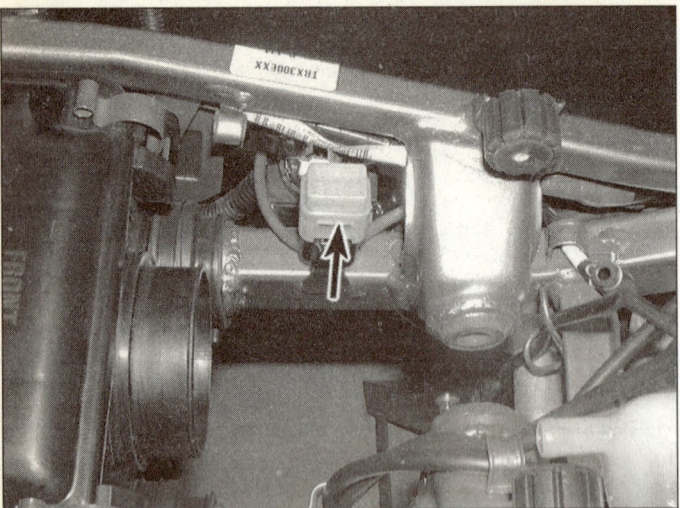

5.1a On TRX300EX models, the fuse holder (arrow) is located under the seat, on the left frame rail

5.1b On TRX400EX models, the fuse holder (arrow) is located on the right side of the machine, in front of the battery

then, plug in the battery charger. **Warning:** *The hydrogen gas escaping from a charging battery is explosive, so keep open flames and sparks well away from the area. Also, the electrolyte is extremely corrosive and will damage anything it comes in contact with.*

8   Allow the battery to charge for the specified time. If the battery overheats or gases excessively, the charging rate is too high. Either disconnect the charger or lower the charging rate to prevent damage to the battery.

9   After the specified time, unplug the charger first, then disconnect the leads from the battery.

10   If the recharged battery discharges rapidly when left disconnected, it's likely that an internal short caused by physical damage or sulfation has occurred. A new battery will be required. A sound battery will tend to lose its charge at 1% per day.

11   When the battery is fully charged, unplug the charger first, then disconnect the leads from the battery. Wipe off the outside of the battery case and install the battery in the vehicle.

5   Fuse - check and replacement

Refer to illustrations 5.1a, 5.1b, 5.3a and 5.3b

1   All models have a 15-amp fuse, which is located under the seat

on TRX300EX models, and in front of the battery on TRX400EX models **(see illustrations)**.

2   To inspect or replace the fuse, remove the seat and rear fenders (see Chapter 7).

3   Unlock and remove the cover **(see illustration)** and pull out the fuse. A blown fuse is easily identified by a break in the element **(see illustration)**.

4   If a fuse blows, be sure to check the wiring harnesses very carefully for evidence of a short circuit. Look for bare wires and chafed, melted or burned insulation. If a fuse is replaced before the cause is located, the new fuse will blow immediately.

5   Never, under any circumstances, use a higher rated fuse or bridge the fuse terminals, as damage to the electrical system could result.

6   Occasionally a fuse will blow or cause an open circuit for no obvious reason. Corrosion of the fuse ends and fuse holder terminals may occur and cause poor fuse contact. If this happens, remove the corrosion with a wire brush or emery paper, then spray the fuse end and terminals with electrical contact cleaner.

7   Install the seat and rear fenders (see Chapter 7).

6   Lighting system - check

1   The battery provides power for operation of the headlights, tail light, brake light and instrument cluster lights. If none of the lights operate, always check battery voltage before proceeding. Low battery voltage indicates a faulty battery, low battery electrolyte level or a defective charging system. Refer to Chapter 1 and Section 3 of this

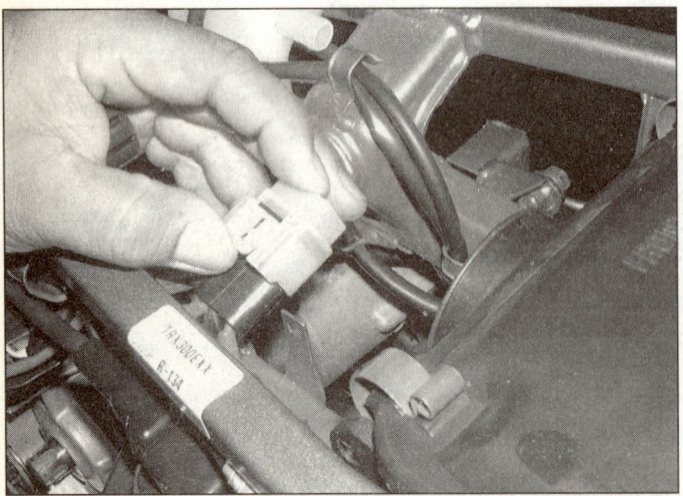

5.3a To inspect or replace the fuse, unlock and remove the plastic cover

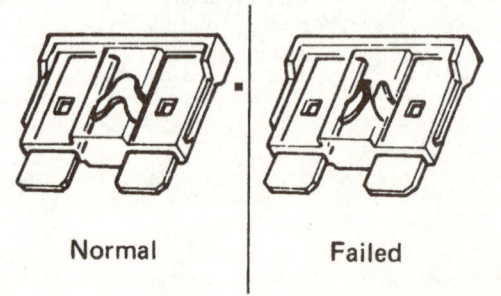

Normal   Failed

5.3b A blown fuse can be identified by a broken element; be sure to replace a blown fuse with one of the same amperage rating

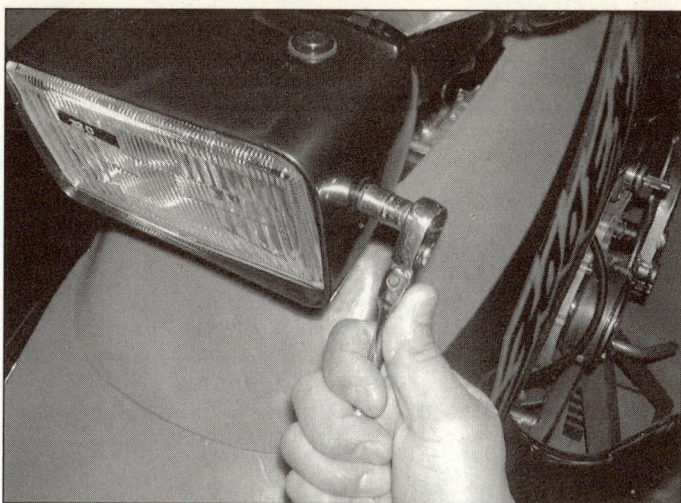

7.1 To replace the headlight bulb on a TRX300EX, remove the two headlight bolts . . .

Chapter for battery checks and Sections 24 through 27 for charging system tests. Also, check the condition of the fuses and replace any blown fuses with new ones.

Headlights

2   If a headlight bulb is out with the lighting switch in the On position, check the main fuse with the key On (see Section 5).
3   If the fuse is okay, unplug the electrical connector for the headlight bulb (see Section 7) and use a jumper wire to connect the bulb directly to the battery terminals as follows:
 a) Green wire terminal to battery negative terminal
 b) White wire terminal to battery positive terminal
4   If the light comes on, there's an open, short or bad connection in the wiring or in one of the switches in the circuit. Refer to Sections 13 and 14 for the switch testing procedures, and also the wiring diagrams at the end of this Chapter.
5   If either filament (high or low beam) of the bulb doesn't light, replace the bulb (see Section 7).

Taillight

6   If the taillight fails to work, check the bulb and the bulb terminals first, then check for battery voltage at the power wire in the taillight.
7   If voltage is present, check the ground circuit for an open or poor connection.
8   If no voltage is indicated, check the wiring between the taillight and the lighting switch, then check the switch.

Neutral indicator light

9   If the neutral light fails to operate when the transmission is in Neutral, check the main fuse (see Section 5) and the bulb (see Section 11).
10   If the bulb and fuse are in good condition, check for battery voltage at the wire attached to the neutral switch on the right side of the engine (TRX300EX models) or the left side (TRX400EX models).
11   If battery voltage is present, check the neutral switch (see Section 16).
12   If no voltage is indicated, check the wiring to the bulb, to the switch and between the switch and the bulb for open circuits and poor connections.

Reverse indicator light (TRX300EX models)

13   If the reverse indicator light doesn't come on when the transmission is in Reverse, check the main fuse and the bulb (see Section 11).
14   If the bulb and fuse are in good condition, check for battery voltage at the wire attached to the reverse switch on the right side of the engine.
15   If battery voltage is present, check the reverse switch (see Section 16).
16   If no voltage is indicated, check the wiring to the bulb, to the switch and between the switch and the bulb for open circuits and poor connections.

7   Headlight bulb - replacement

Warning: *If the headlight has just burned out, give it time to cool off before changing the bulb, to avoid burning your fingers.*

TRX300EX models
Refer to illustrations 7.1, 7.2, 7.3, 7.4a and 7.4b
1   Remove one bolt from each side and remove the headlight housing **(see illustration)**.
2   Unplug the electrical connector from the headlight housing **(see illustration)**.
3   Remove the rubber dust cover from the headlight housing **(see illustration)**.
4   Flip open the wire bulb socket retainer, twist the bulb socket counterclockwise and remove it from the headlight housing **(see illustrations)**.

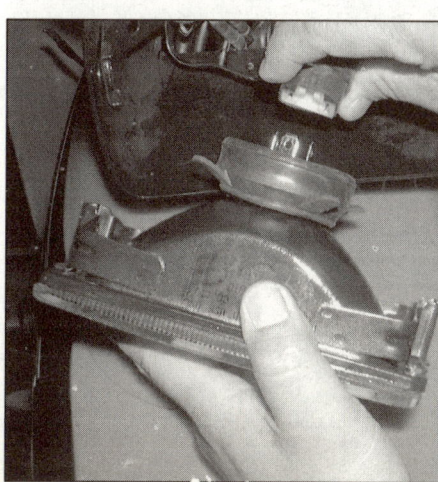

7.2 . . . remove the headlight assembly from the headlight housing, disconnect the electrical connector . . .

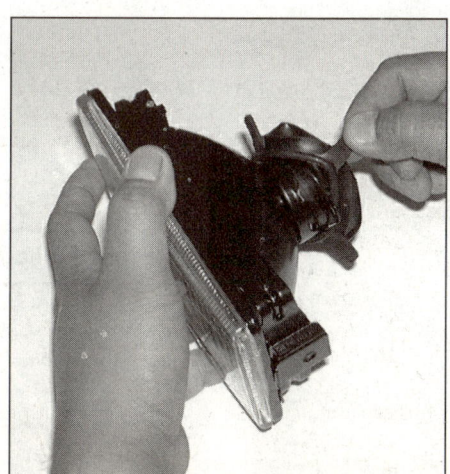

7.3 . . . pull off the rubber dust cover . . .

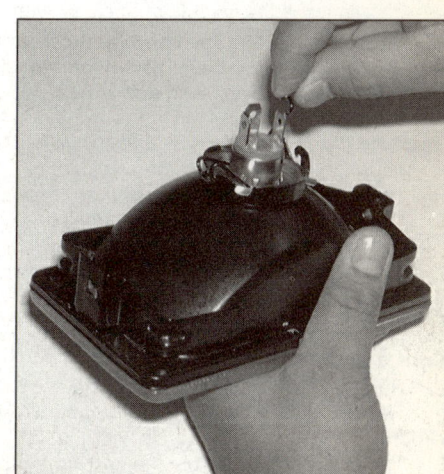

7.4a . . . unlock the wire bulb socket retainer . . .

8

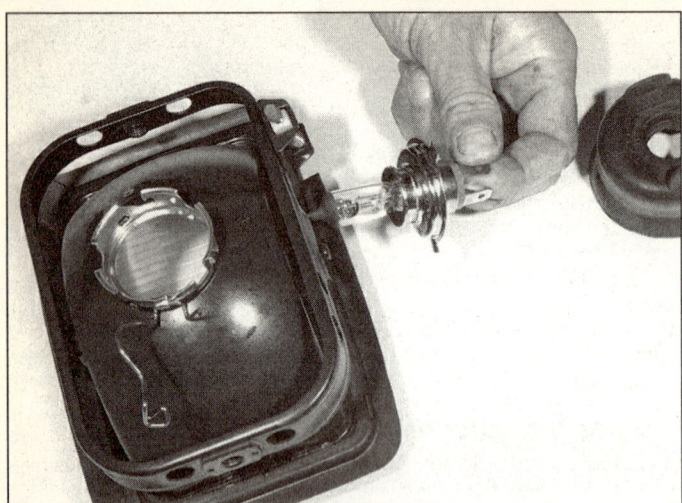

7.4b . . . and remove the bulb and socket

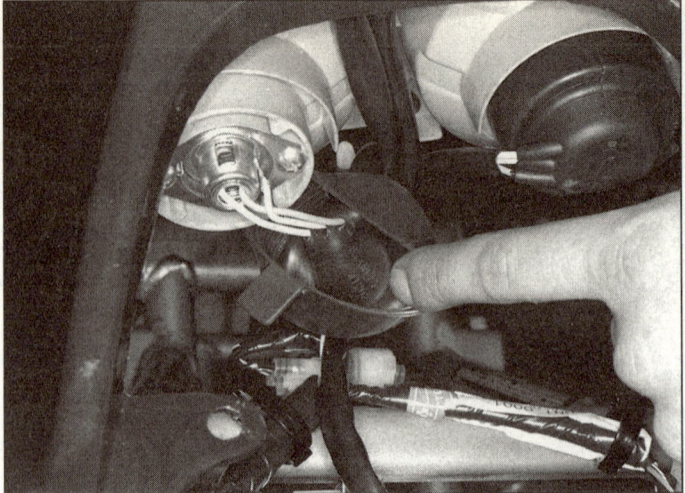

7.8 To replace a TRX400EX headlight bulb, remove the rubber
dust cover . . .

7.9 . . . push the bulb in, turn it counterclockwise and pull it out

8.2 Remove the TRX300EX headlight housing bracket bolts
(arrows), pull out the wiring harness and remove the housing . . .

5   To remove the bulb from the bulb socket, pull it straight out.

6   Installation is the reverse of removal, with the following additions:

a) *Be sure not to touch the new bulb with your fingers - oil from your skin will cause the bulb to overheat and fail prematurely. If you do touch the bulb, wipe it off with a clean rag dampened with rubbing alcohol.*

b) *Align the tab on the metal bulb flange with the slot in the headlight housing.*

c) *There are no headlight adjustment screws, but the headlight (not the housing) must be adjusted vertically before tightening the headlight bolts. Before performing the adjustment, make sure the fuel tank is at least half full, and have an assistant sit on the seat.*

TRX400EX models

Refer to illustrations 7.8 and 7.9

7   Remove the front fenders (see Chapter 7).

8   Remove the rubber dust cover from the headlight case **(see illustration)**.

9   To remove the bulb from the headlight case, twist the bulb socket counterclockwise and push it in at the same time. Pull the bulb out without touching the glass **(see illustration)**.

10   Installation is the reverse of the removal procedure, with the following additions:

a) *Be sure not to touch the bulb with your fingers - oil from your skin will cause the bulb to overheat and fail prematurely. If you do touch the bulb, wipe it off with a clean rag dampened with rubbing alcohol.*

b) *Align the tab on the metal bulb flange with the slot in the headlight case.*

c) *Make sure that the dust cover is installed with the TOP marking facing up.*

---

8   Headlight housing – removal and installation

TRX300EX models

Refer to illustrations 8.2 and 8.4

1   Remove the headlight from the headlight housing (see Section 7).

2   Remove the headlight housing bracket bolts **(see illustration)** and washers.

3   Remove the headlight housing and bracket and pull the wiring harness out of the housing.

4   Remove the bracket bolt collars from the housing bracket **(see illustration)**.

5   Separate the housing bracket from the housing.

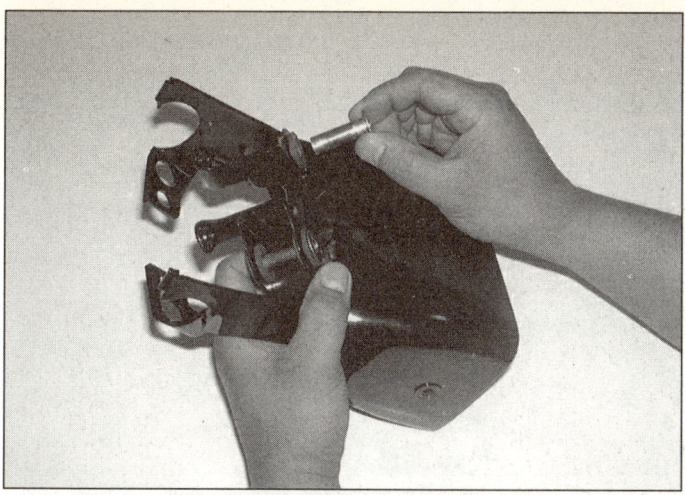

8.4 . . . and remove the bolt collars

8.8 On the TRX400EX, disengage the wiring harness from the clip and unbolt the headlight housing

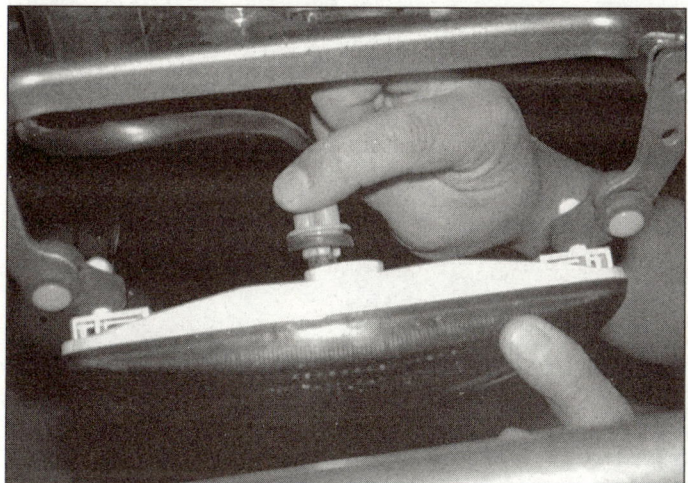

9.1 Push the bulb socket in, turn it counterclockwise and pull it out of the housing . . .

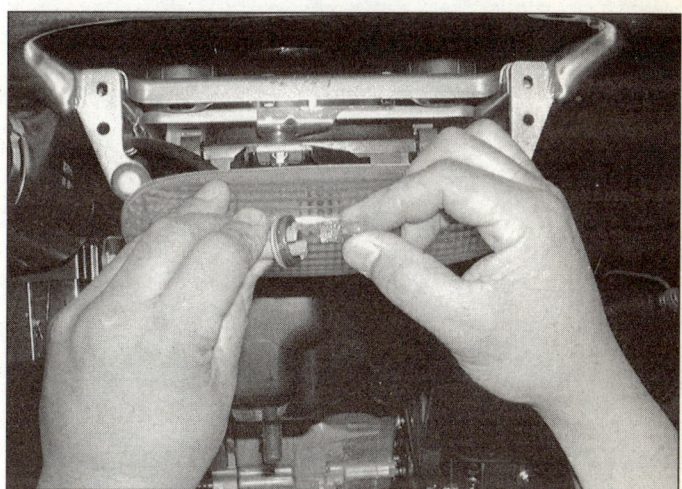

9.2 . . . and pull the bulb out of the socket

6   Installation is the reverse of removal. Be sure to adjust the headlight (see Step 6 in Section 7) before tightening the headlight bolts.

TRX400EX models

Refer to illustration 8.8

7   Remove the rubber dust covers and the headlight bulbs from the headlight case (see Section 7).
8   Disengage the wiring harness from the clip (**see illustration**).
9   Remove the two headlight housing mounting bolts from the front fender bracket (**see illustration 8.8**).
10   Remove the headlight case.
11   Installation is the reverse of removal, with the following additions:
  a) *When installing the headlight housing, make sure that the tab on top of the headlight case is aligned with the slit in the fender.*
  b) *Before tightening the headlight housing mounting bolts, make sure that the bolt centers are aligned with the index marks on the fender bracket.*

9   Tail light bulb - replacement

Refer to illustrations 9.1, 9.2, 9.4 and 9.5

1   To remove the tail light bulb socket from the tail light housing, push it in and turn it counterclockwise, then pull it out (**see illustration**).

2   Pull the bulb out of the socket (**see illustration**).
3   Check the socket terminals for corrosion and clean them if necessary.
4   If you need to replace the tail light bulb *socket*, follow the wires forward to their electrical connectors and disconnect them (**see illustration**).

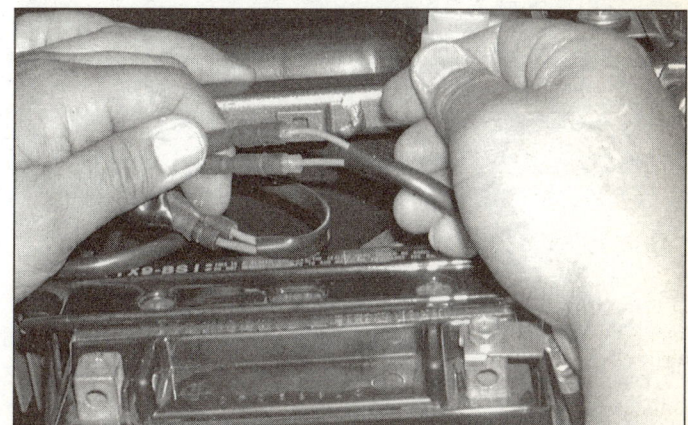

9.4 Disconnect the electrical connectors if you're removing the tail light socket

8

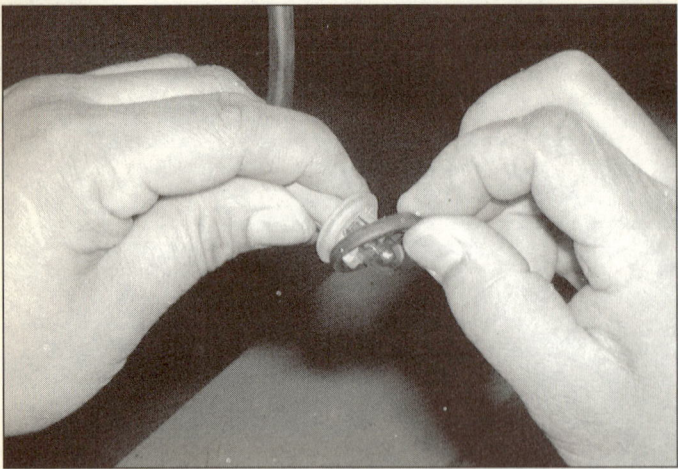

9.5 If the rubber gasket for the tail light bulb socket is in bad shape, replace it

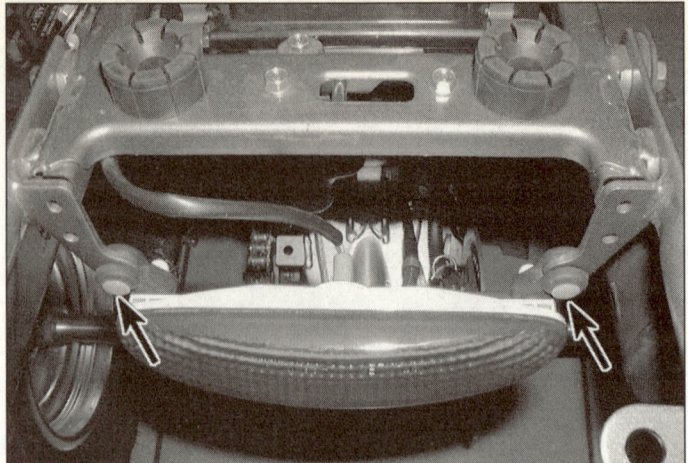

10.2 To remove the taillight housing, disengage the rubber grommets (arrows) from the brackets

11.2a Pull the indicator lens out of the socket in the headlight housing (TRX300EX) . . .

11.2b . . . or the front fender (right arrow) (TRX400EX); (left arrow indicates ignition switch)

5   Make sure the socket's rubber gasket is in good condition. If it isn't, replace it **(see illustration)**.
6   Installation is the reverse of removal. Make sure that the arrows on the bulb socket and on the tail light housing are aligned.

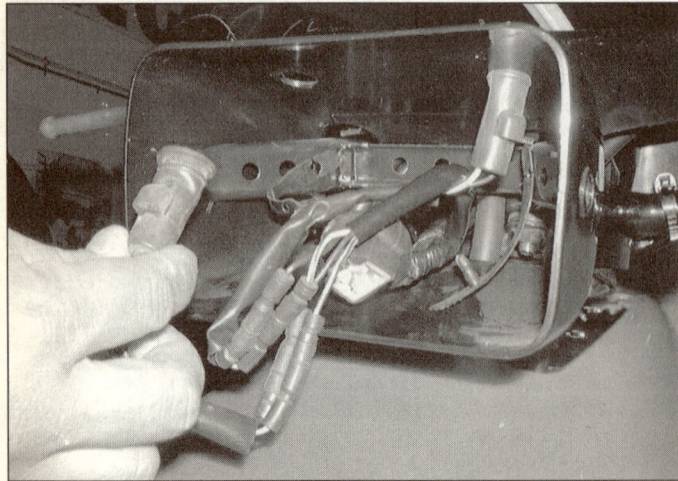

11.3 . . . then pull the socket out of the headlight housing (TRX300EX) or front fender (TRX400EX) . . .

10   Tail light housing - removal and installation

Refer to illustration 10.2
1   Remove the tail light bulb socket (see Section 9).
2   Disengage the taillight housing rubber grommets **(see illustration)** from the frame brackets.
3   Installation is the reverse of removal.

11   Indicator bulbs - replacement

Refer to illustrations 11.2a, 11.2b, 11.3, 11.4 and 11.6
1   On TRX300EX models, remove the headlight (see Section 7).
2   Remove the indicator lens from the bulb socket **(see illustrations)**.
3   Pull the bulb socket out of the headlight housing (TRX300EX) or the front fender (TRX400EX) **(see illustration)**.
4   Pull the bulb out of the socket **(see illustration)**. **Note:** *Indicator bulbs are deeply recessed in their sockets. If you have trouble pulling a bulb out of the socket, push a piece of rubber or soft plastic tubing down over the bulb, then wiggle it back and forth a little and pull on it simultaneously to work the bulb out of the socket. You can use the same method to install the new bulb.*
5   If the socket contacts are dirty or corroded, they should be scraped clean and sprayed with electrical contact cleaner before new

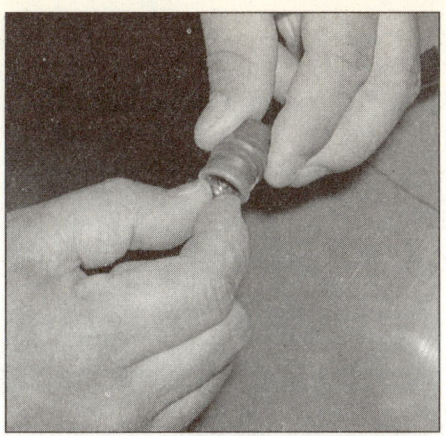

11.4 . . . and pull the bulb out of the socket (use a piece of rubber tubing if necessary)

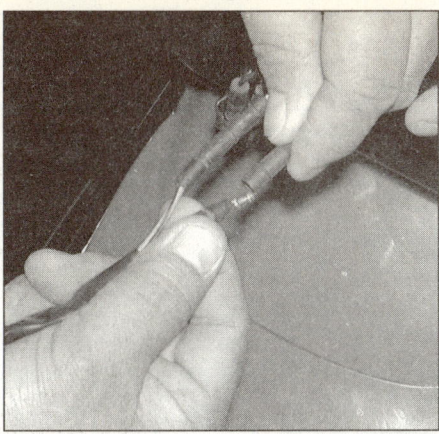

11.6 To replace an indicator bulb socket, disconnect the electrical connectors

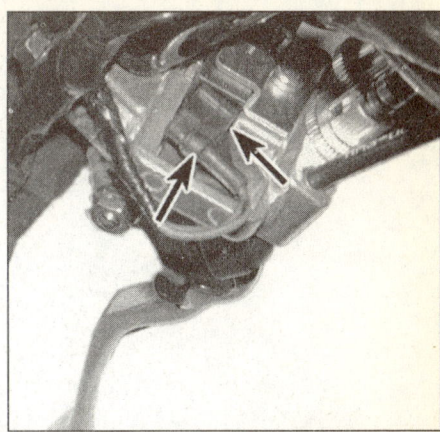

12.2a Disconnect the clutch switch electrical connectors (arrows) . . .

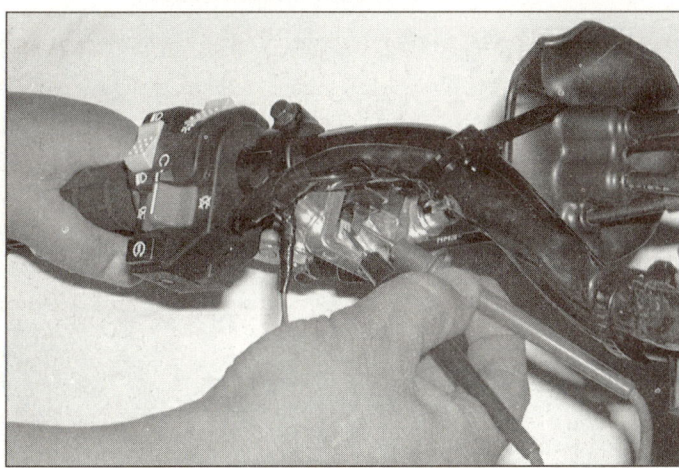

12.2b . . . and check continuity between the terminals with the lever applied, then released

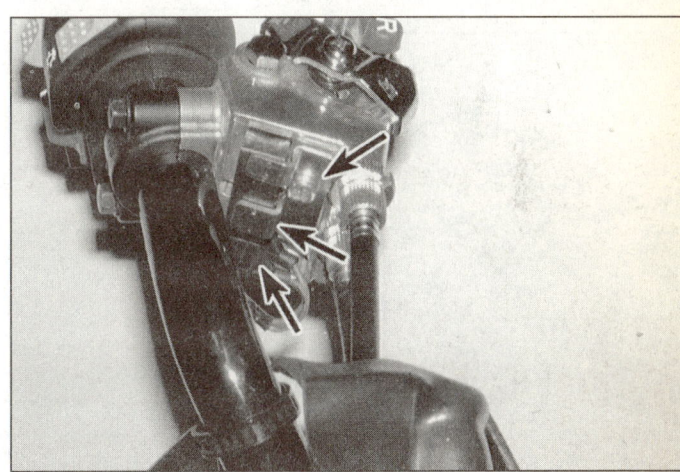

12.4 Disconnect the clutch switch electrical connectors (lower arrows), remove the retaining screw (upper arrow) . . .

bulbs are installed.

6   If you need to replace a socket, trace the leads back to the wire harness and unplug the connectors **(see illustration)**.

7   Carefully push the new bulb into position, using the same tubing that was used for removal, if necessary, then push the socket into the headlight housing or front fender. Make sure that the headlight housing or front fender bodywork is correctly seated in the groove in the socket, with the circumference of the socket's upper end protruding uniformly from the headlight housing or front fender.

12 Clutch switch – check and replacement

Check

Refer to illustrations 12.2a and 12.2b

1   Normally, the engine is started in Neutral. It cannot be started in any other gear. However, if the engine stalls in gear, the clutch switch, which is located on the clutch lever bracket, allows the engine to be started in gear, when the clutch lever is pulled in. But if the starter motor can crank the engine even when the clutch lever is not pulled in, the clutch switch is probably defective. Check the clutch switch as follows.

2   Disconnect the electrical connector from the clutch switch **(see illustration)** and check the continuity between the terminals **(see illustration)**. When the clutch lever is pulled in, there should be continuity between the terminals; when the clutch lever is released, there should be no continuity.

3   If the clutch switch doesn't operate as described, replace it.

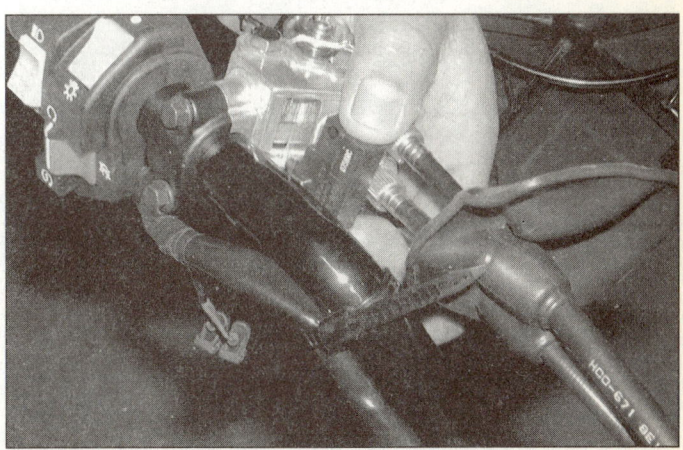

12.5 . . . and pull the switch out of the clutch lever bracket

Replacement

Refer to illustrations 12.4 and 12.5

4   Unplug the switch electrical connector and remove the switch retaining screw **(see illustration)**.

5   Remove the clutch switch **(see illustration)**.

6   Installation is the reverse of removal. Tighten the switch retaining screw securely.

8

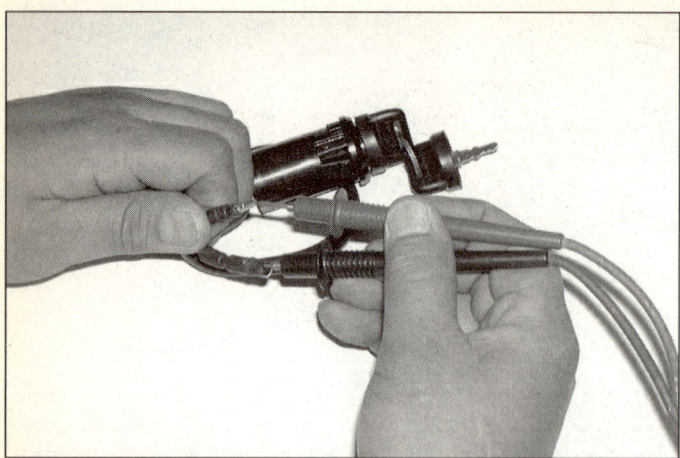

13.2 Check continuity between the ignition switch terminals with the switch in each position

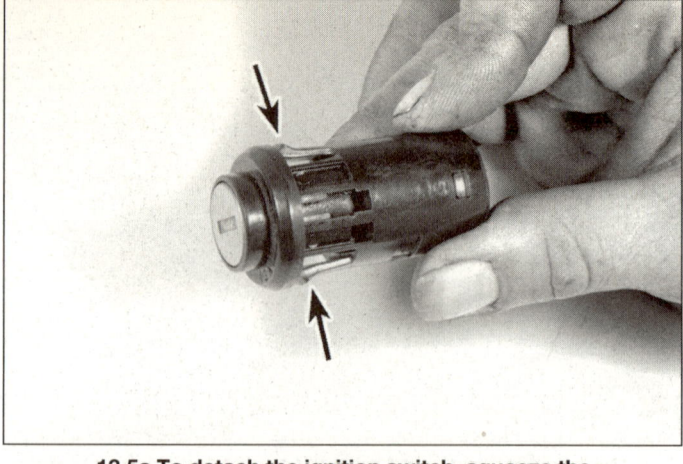

13.5a To detach the ignition switch, squeeze the prongs (arrows) . . .

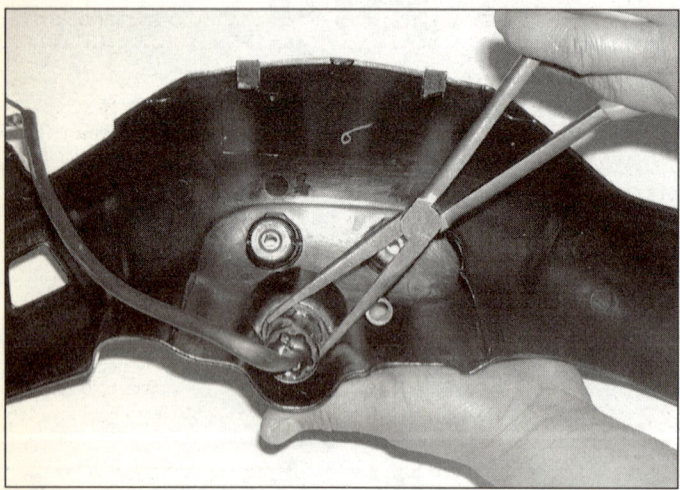

13.5b . . . with a pair of needle-nose pliers and pull out the switch

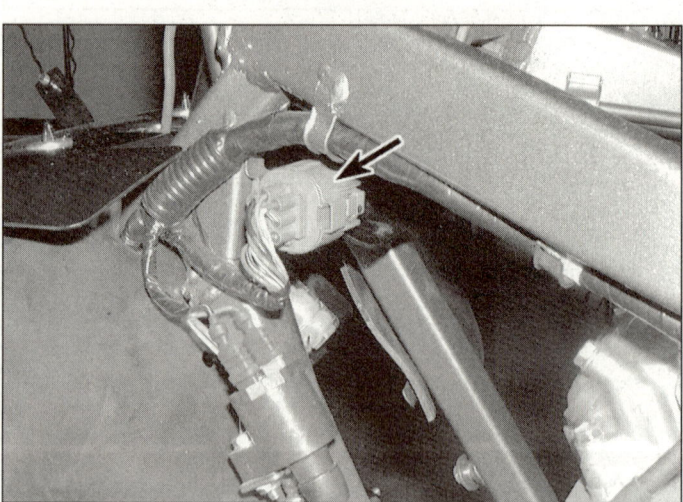

14.3 The TRX300EX handlebar switch connector (arrow) is on the left side of the machine, above the ignition coil

13   Ignition main (key) switch - check and replacement

Check

Refer to illustration 13.2

1   On TRX300EX models, detach the handlebar cover from the handlebar; on TRX400EX models, remove the front fender (see Chapter 7). Trace the ignition switch leads from the ignition switch to the connectors and unplug the connectors. On TRX300EX models, there are just two ignition switch connectors. On TRX400EX models, there are two connectors: a single-pin connector and a white three-pin connector, both of which are located at the front crossmember, to the left of the ignition control module.

2   Using an ohmmeter, check the continuity between the switch terminals **(see illustration)** at each switch position indicated in the continuity tables which are included with the wiring diagrams at the end of this Chapter.

3   If the switch fails any test, replace it.

Replacement

Refer to illustrations 13.5a and 13.5b

4   Remove the handlebar cover (TRX300EX models) or the front fender (TRX400EX models) and unplug the switch electrical connectors.

5   The ignition switch is secured to the handlebar cover (TRX300EX models) or to the front fender (TRX400EX models) by two plastic

prongs **(see illustration)**. Squeeze the prongs **(see illustration)** and pull the switch out of the handlebar cover or front fender.

6   Installation is the reverse of the removal procedure.

14   Handlebar switches - check

Refer to illustrations 14.3 and 14.4

1   The handlebar switches are generally reliable and trouble-free. Most problems are caused by dirty or corroded contacts, but internal parts occasionally wear or break. If a switch breaks, it must be replaced; individual parts are not available.

2   If a switch malfunctions, check it for continuity with an ohmmeter or a continuity test light. Always disconnect the battery negative cable, to prevent the possibility of a short circuit, before testing a switch.

3   Trace the wiring harness from the faulty switch to the electrical connector. On TRX300EX models, the handlebar switch connector is located on the left side of the machine, right above the coil **(see illustration)**. On TRX400EX models, the handlebar switch connector is located at the left end of the oil cooler **(see illustration 21.10 in Chapter 2B)**.

4   Using the ohmmeter or test light, check for continuity between the terminals of the switch connector **(see illustration)** with the switch in each position. Refer to the continuity tables contained in the wiring diagrams at the end of this Chapter.

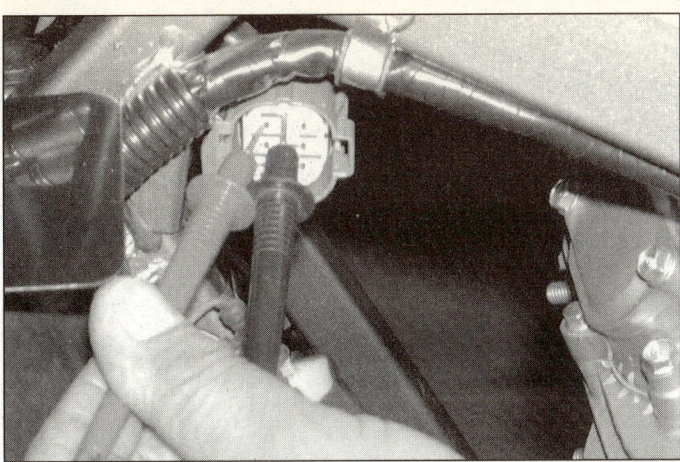

14.4 Check continuity between the switch terminals with the switch in each position

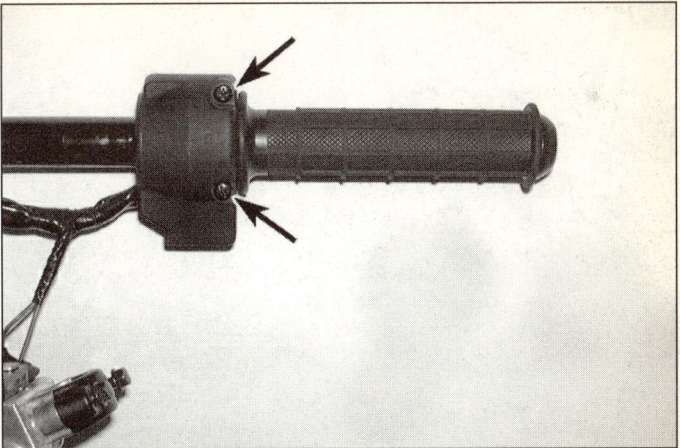

15.1a Remove the screws (arrows) and separate the halves of the handlebar switch housing

15 Handlebar switch housing - removal and installation

Refer to illustrations 15.1a and 15.1b

1   The handlebar switches are housed in a two-piece housing that clamps around the handlebar. To remove a switch for cleaning or inspection, remove the clamp screws and pull the switch halves away from the handlebar **(see illustrations)**.

2   To completely remove a switch housing, unplug the electrical connector **(TRX300EX models, see illustration 14.3; TRX400EX models, see illustration 21.10 in Chapter 2B)** and disengage the harness from all cable ties and harness clips.

3   When installing the switches, make sure the wiring harness is correctly routed to avoid pinching or stretching the wires.

15.1b The handlebar switches are inside the housing

16 Neutral and reverse switches - check and replacement

Neutral and reverse switches (TRX300EX models)

Refer to illustrations 16.1a, 16.1b and 16.5

1   Remove the switch cover from the right side of the engine and disconnect the wiring connector from the reverse and neutral switches **(see illustrations)**. **Warning:** *Label the reverse and neutral switch*

5   If the continuity check indicates a problem, remove and disassemble the switch (see Section 15). If a switch is damaged or broken, replace it. If there's no obvious damage, spray the switch contacts with electrical contact cleaner and scrape the contacts clean with crocus cloth, or a small wire brush or penknife. Retest the switch to verify that it now functions correctly.

16.1a The TRX300EX reverse and neutral switches are under this cover . . .

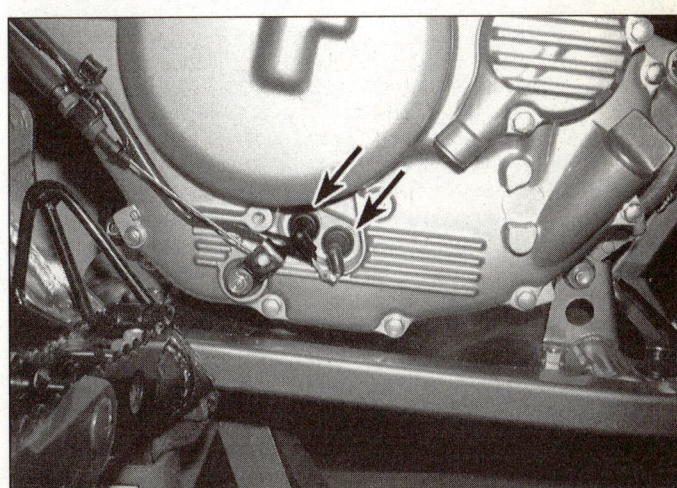

16.1b . . . the neutral switch is on the left, the reverse switch on the right

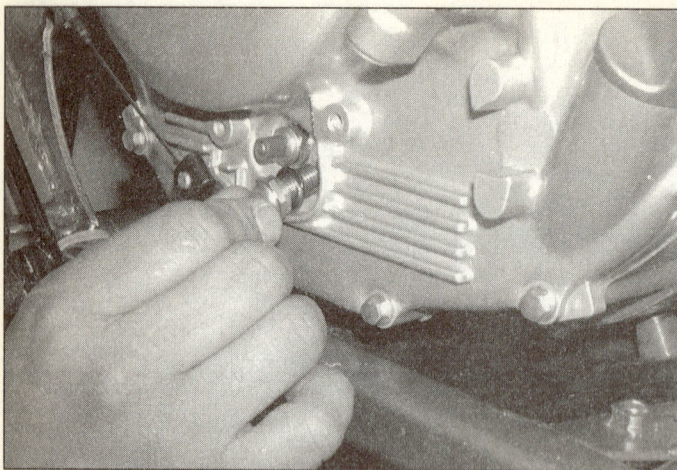

16.5 Unscrew the switches; use new sealing washers when you install them

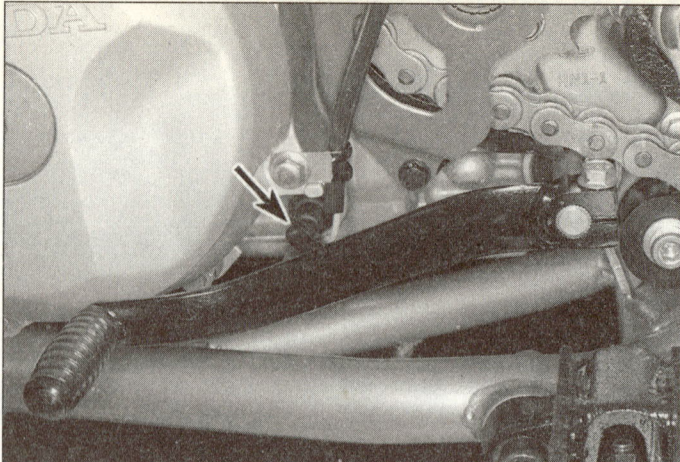

16.11 The TRX400EX neutral switch (arrow) is on the left side of the engine, below the sprocket cover

wires so there's no chance of mixing them up when you reconnect them. If this happens, the neutral light will come on when the transmission is actually in reverse. This may cause the vehicle to move backward when you don't expect it.

2   Connect one lead of an ohmmeter to a good ground and the other lead to the terminal post on the switch being tested.

3   When the transmission is in neutral, the ohmmeter should read 0 ohms between the neutral switch and ground - in any other position, the meter should read infinite resistance.

4   When the transmission is in reverse, the ohmmeter should read 0 ohms between the reverse switch and ground - in any other position, the meter should read infinite resistance.

5   If either switch fails the first test, unscrew the switch **(see illustration)**, discard the old sealing washer and connect the ohmmeter between the terminal post on the switch and the switch body. It should read 0 ohms when the switch is pressed in and infinite resistance when it's released.

6   If the switch doesn't check out as described, replace it.

7   If the switch fails the first (on-vehicle) test, but functions correctly when it's removed, remove the right crankcase cover and inspect the reverse or neutral switch rotor mechanism (see Section 19 in Chapter 2A).

8   Install the switch with a new sealing washer and tighten it to the torque listed in this Chapter's Specifications.

9   Reconnect the switch wire (note the Warning in Step 1).

10   The remainder of installation is the reverse of removal.

Neutral switch (TRX400EX models)

Refer to illustration 16.11

11   The neutral switch **(see illustration)** is located on the left side of the engine. To test it, refer to Steps 2, 3 and 5 above.

12   If the switch doesn't check out as described, replace it. Disconnect the electrical connector and unscrew the switch from the engine. If the wiring harness retainer is in the way, bend it out of the way or unbolt it from the engine,

13   Install the switch with a new sealing washer and tighten it to the torque listed in this Chapter's Specifications.

14   Reconnect the switch wire. Bend back or reinstall the harness retainer if necessary.

17   Reverse inhibitor unit (TRX300EX models) - check and replacement

Check

Refer to illustrations 17.1 and 17.2

1   The reverse inhibitor unit **(see illustration)**, which is located on the left side of the frame, above the left front shock absorber, prevents the engine from being started with the transmission in Reverse, unless the clutch lever is pulled in or the transmission is in Neutral. If the inhibitor switch allows the engine to be started in Reverse, or doesn't allow it to be started when the clutch lever is pulled in or the transmission is in Reverse, test it as follows.

2   Disconnect the three electrical connectors from the inhibitor **(see**

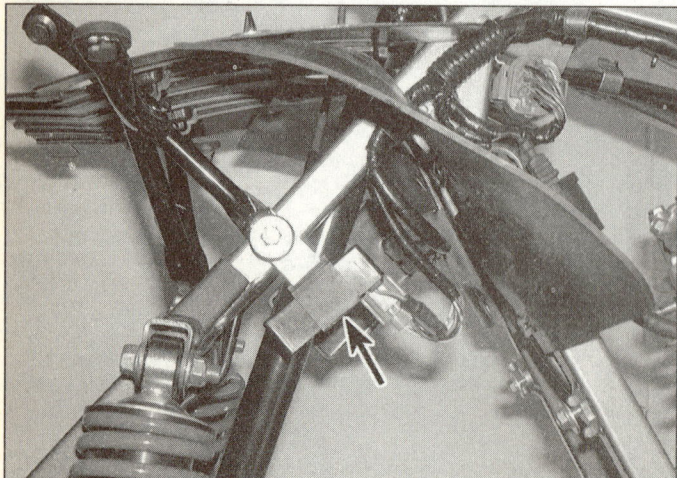

17.1 The TRX300EX inhibitor unit (arrow) is on the left side of the frame, above the left front shock absorber

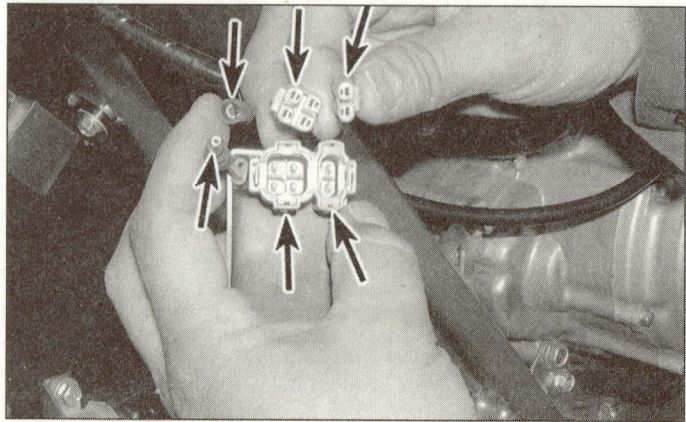

17.2 Here are the inhibitor unit connectors

18.2a The TRX300EX starter relay is on the left side; pull back the cover to expose the terminals

18.2b The TRX400EX starter relay is on the right side; pull back the cover to expose the terminals

illustration). The connectors include: a single-pin connector (light green/red wire to the neutral indicator light); two-pin connector (green/red wire from clutch switch; yellow/green wire to starter relay); and a four-pin connector (gray/black wire to reverse indicator light; black wire from neutral and reverse indicators, light green wire to neutral switch, and gray wire to reverse switch)

3   Disconnect the connector at the other end of the yellow/green wire between the inhibitor unit and the starter relay (it's in the large rubber boot above the battery). Verify that this yellow/green wire has continuity (this is the signal wire that closes or opens the ground circuit for the starter relay).

4   If the yellow/green wire does not have continuity, repair it before proceeding.

5   If the yellow/green wire has continuity, make the following three checks on the harness side (not the inhibitor unit side) of the connectors you unplugged in Step 2.

6   First, pull in the clutch lever and verify that there's continuity between the terminal for the green/red wire (in the two-pin connector) and ground. If there isn't continuity, check and, if necessary, replace the clutch switch (see Section 12), and then repeat this test.

7   Second, put the transmission in Neutral and verify that there's continuity between the terminal for the light green wire (in the green four-pin connector) and ground. If there isn't continuity, check and, if necessary, replace the Neutral switch (see Section 16), and then repeat this test.

8   Third, put the transmission in Reverse and verify that there is NO continuity between the terminal for the green wire (in the green four-pin connector) and ground. If there isn't continuity, check and, if necessary, replace the Reverse switch (see Section 16), and then repeat this test.

9   If there is continuity in the wire between the inhibitor unit and the starter relay, and the clutch, Neutral and Reverse switches are functioning correctly, but the inhibitor unit is not functioning correctly, replace it.

Replacement

10   Unplug the electrical connectors **(see illustration 17.2)** and slip the inhibitor unit out of its rubber mount.

11   Installation is the reverse of removal.

18 Starter relay - check and replacement

Check

Refer to illustrations 18.2a, 18.2b and 18.5

1   Remove the seat and rear fenders (see Chapter 7).

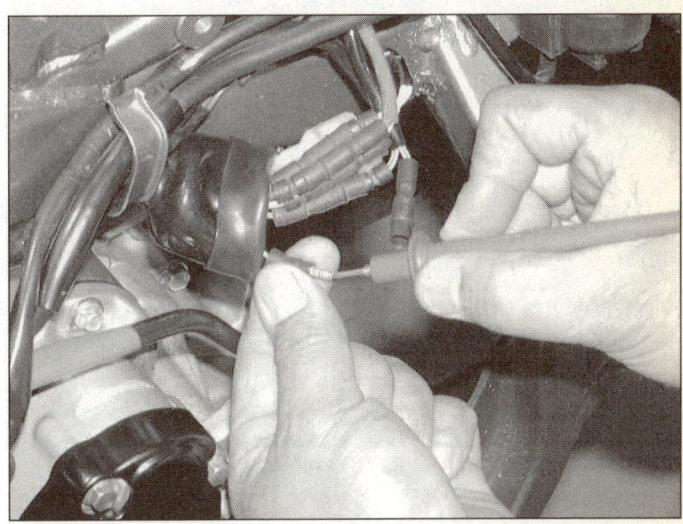

18.5 Check for battery voltage at the starter relay connector

2   Turn the ignition switch to ON, push the starter button and verify that the starter relay switch clicks **(see illustrations)**.

3   If the relay clicks, it's okay.

4   If it doesn't click, inspect the relay terminals. Make sure they're clean and tight. Clean the terminals with a small wire brush and contact cleaner, if necessary. Try the "click" test again.

5   If the relay still doesn't click, verify that there's voltage to the relay when the ignition switch is ON and the starter button is pushed. Locate the single-pin connector for the yellow/red wire (inside the large rubber boot above the battery) and disconnect it (this is the wire that connects the starter button to the starter relay). Touch the positive voltmeter probe to the starter-button side of the connector **(see illustration)** and the negative lead to ground. With the ignition switch turned to ON, push the starter button. There should be battery voltage.

6   If there's voltage to the starter relay, replace the relay.

7   If there's no voltage to the relay, troubleshoot the wire harness between the starter button and the relay (see wiring diagrams at the end of this Chapter).

8

Replacement

Refer to illustration 18.8

8   Disconnect the cables from the relay terminals **(see illustration**

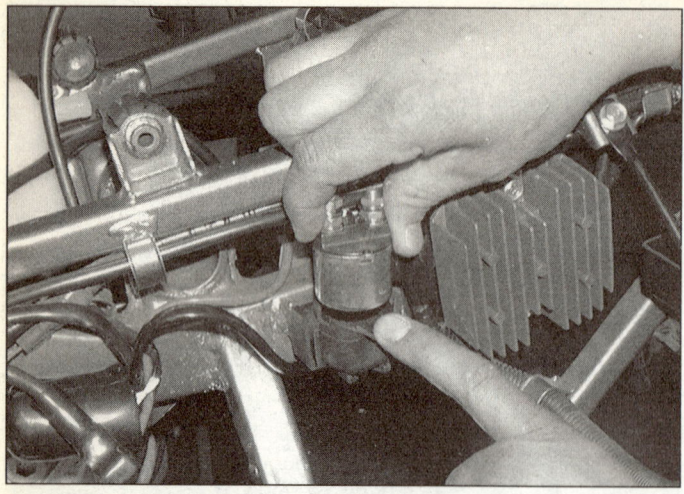

18.8 Pull the starter relay straight up out of its rubber mount

20.4 Pull up the cover and remove the starter cable nut

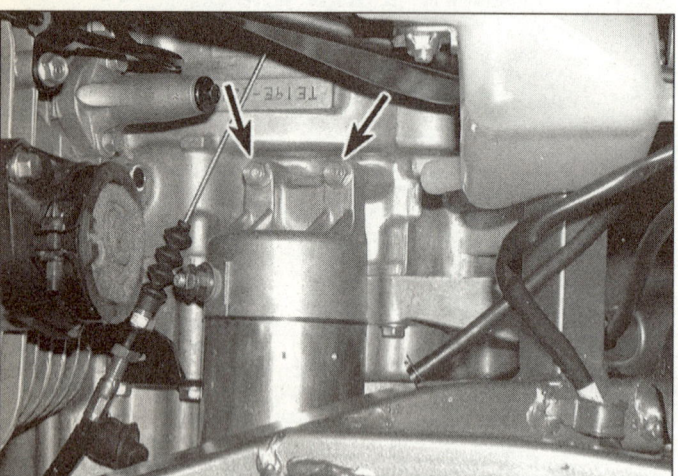

20.5a Starter motor mounting bolts (TRX300EX)

20.5b Starter motor mounting bolts (TRX400EX); the rear bolt attaches the brake lock cable guide to the crankcase

18.2a or 18.2b) and pull the relay out of its rubber mount **(see illustration)**.

9   Inspect the rubber mount. If it's cracked, torn or deteriorated, pull it off the metal bracket and install a new rubber mount.

10   Installation is the reverse of removal. Reconnect the negative battery cable after all the other electrical connections are made.

19 Clutch diode (TRX400EX models) - check and replacement

1   Remove the sear/rear fender assembly (see Chapter 7).

2   Disconnect the starter relay switch connector (yellow/green wire) and the neutral switch connector (light green wire). Both connectors are located inside the large rubber boot located right above the battery **(see illustration 4.5 in Chapter 4)**.

3   The clutch diode is located inside the wrapped portion of the wiring harness located immediately above the rubber boot **(see illustration 4.5 in Chapter 4)**. The clutch diode is connected to the light green and yellow/green wires.

4   Connect an ohmmeter between the diode terminals, then switch the ohmmeter leads so they're connected to the opposite terminals. There should be continuity (little or no resistance) in one direction and infinite resistance in the other direction. If the diode doesn't function as described, replace it.

20 Starter motor - removal and installation

Removal

Refer to illustrations 20.4, 20.5a, 20.5b and 20.7

1   Disconnect the cable from the negative terminal of the battery.

2   On TRX300EX models, remove the starter gear cover bolts, remove the starter gear cover and remove the cover dowel pins and gasket **(see illustrations 22.1a and 22.1b)**.

3   On TRX300EX models, remove the starter reduction gear and shaft **(see illustration 22.2)**.

4   Pull back the rubber boot and remove the nut retaining the starter cable to the starter **(see illustration)**.

5   Remove the two starter mounting bolts **(see illustration)**. On TRX400EX models, note the brake lock cable guide is attached to the crankcase by the rear starter bolt **(see illustration)**.

6   Lift the outer end of the starter up a little bit and slide the starter out of the engine case. **Caution:** *Don't drop or strike the starter or its magnets may be demagnetized, which will ruin it.*

7   Inspect the O-ring on the end of the starter and replace it if necessary **(see illustration)**.

8   Remove any corrosion or dirt from the mounting lugs on the starter and the mounting points on the crankcase.

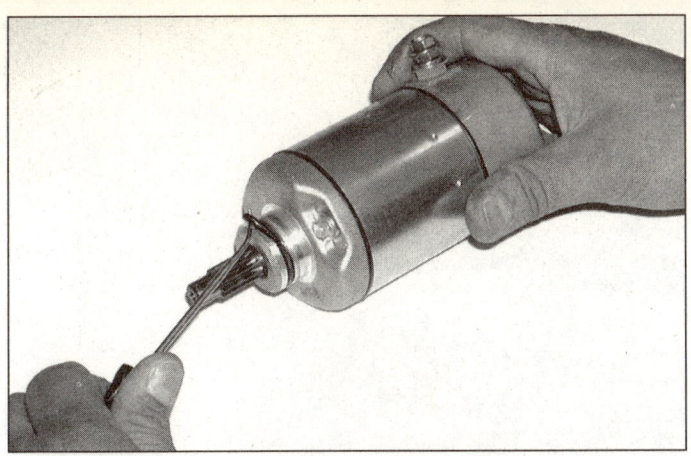

20.7 Inspect the starter motor O-ring; if it's damaged, replace it

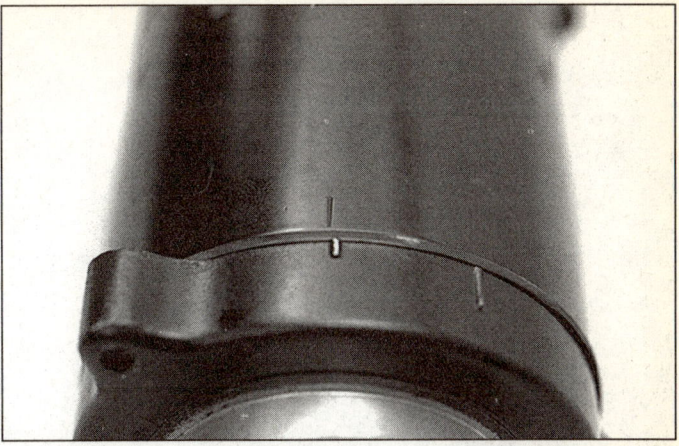

21.2 Make alignment marks between the housing and end covers before disassembly

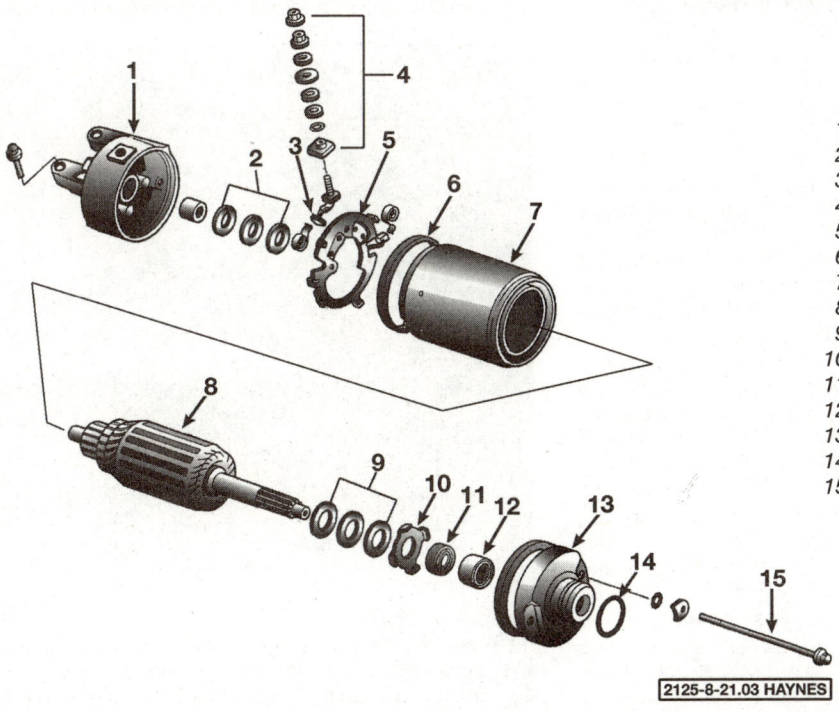

21.3 Starter - exploded view

1 Rear end cover
2 Shims, washer and dust seal
3 Brush
4 Terminal bolt components
5 Brush plate
6 O-ring
7 Starter housing
8 Armature
9 Shims
10 Lockwasher
11 Dust seal
12 Needle roller bearing
13 Front end cover
14 O-ring
15 Through-bolt

2125-8-21.03 HAYNES

Installation

9 Before installing the starter motor, apply a little engine oil to the O-ring. Tighten the starter motor bolts securely.
10 Installation is otherwise the reverse of removal, with the following additions:

a) On TRX400EX models, don't forget to reattach the brake lock cable guide to the crankcase with the rear starter bolt.
b) On TRX300EX models, be sure to tighten the starter gear cover bolts to the torque listed in this Chapter's Specifications.

21 Starter motor - disassembly, inspection and reassembly

1 Remove the starter motor (see Section 20).

Disassembly

Refer to illustrations 21.2 and 21.3

2 Make alignment marks between the housing and covers **(see illustration)**.

3 Unscrew the two long bolts, then remove the cover with its O-ring from the motor. Remove the shim(s) from the armature, noting their correct locations **(see illustration)**.
4 Remove the front cover with its O-ring from the motor. Remove the toothed washer from the cover and slide off the insulating washer and shim(s) from the front end of the armature, noting their locations.
5 Withdraw the armature from the housing.

Inspection

Refer to illustrations 21.7, 21.8, 21.9, 21.11a and 21.11b

Note: Check carefully which components are available as replacements before starting overhaul procedures.

6 Connect an ohmmeter between the terminal bolt and the insulated brush holder (or the indigo colored wire). There should be continuity (little or no resistance). When the ohmmeter is connected between the rear cover and the insulated brush holder (or the indigo colored wire), there should be no continuity (infinite resistance). Replace the brush holder if it doesn't test as described.
7 Unscrew the nut from the terminal bolt and remove the plain washer, insulating washers and rubber ring, noting carefully how

8

21.7 Unscrew the nut and remove the washers from the terminal bolt, noting their correct installed order

21.8 Lift the brush springs and slide the brushes out of their holders

they're installed **(see illustration)**. Withdraw the terminal bolt and brush assembly from the housing and recover the insulator.

8   Lift the brush springs and slide the brushes out of their holders **(see illustration)**.

9   The parts of the starter motor that most likely will require attention are the brushes. If one brush must be replaced, replace both of them. The brushes are replaced together with the terminal bolt and the brush plate. Brushes must be replaced if they are worn excessively, cracked, chipped, or otherwise damaged. Measure the length of the brushes and compare the results to the brush length listed in this Chapter's Specifications **(see illustration)**. If either of the brushes is worn beyond the specified limits, replace them both.

10   Inspect the commutator for scoring, scratches and discoloration. The commutator can be cleaned and polished with fine emery paper, but do not use sandpaper and do not remove copper from the commutator. After cleaning, clean out the grooves and wipe away any residue with a cloth soaked in an electrical system cleaner or denatured alcohol.

11   Using an ohmmeter or a continuity test light, check for continuity between the commutator bars **(see illustration)**. Continuity should exist between each bar and all of the others. Also, check for continuity between the commutator bars and the armature shaft **(see illustration)**. There should be no continuity between the commutator and the shaft. If the checks indicate otherwise, the armature is defective.

12   Check the dust seal in the front cover for wear or damage. Check the needle roller bearing in the front cover for roughness, looseness or

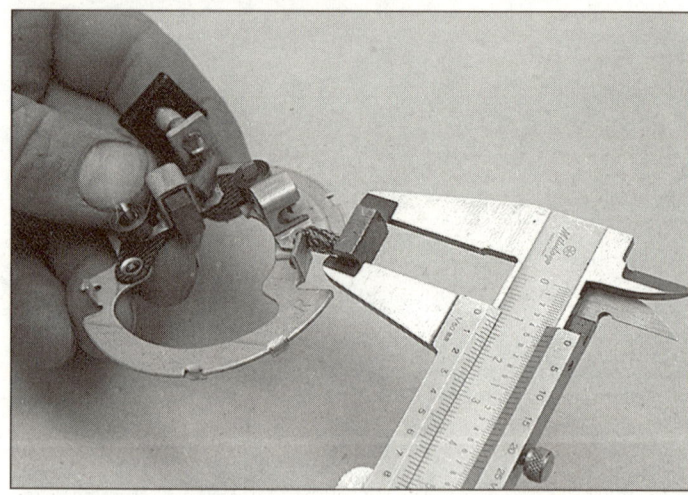

21.9 Measure the brush length and replace them if they're worn

loss of lubricant. Check with a motorcycle shop or Honda dealer to see if the bearing can be replaced separately; if this isn't possible, replace the starter motor.

13   Inspect the bushing in the rear cover. Replace the starter motor if

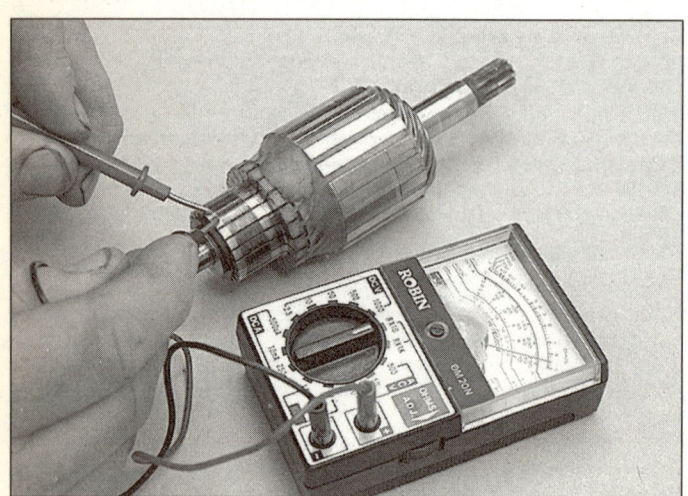

21.11a Check for continuity between the commutator bars . . .

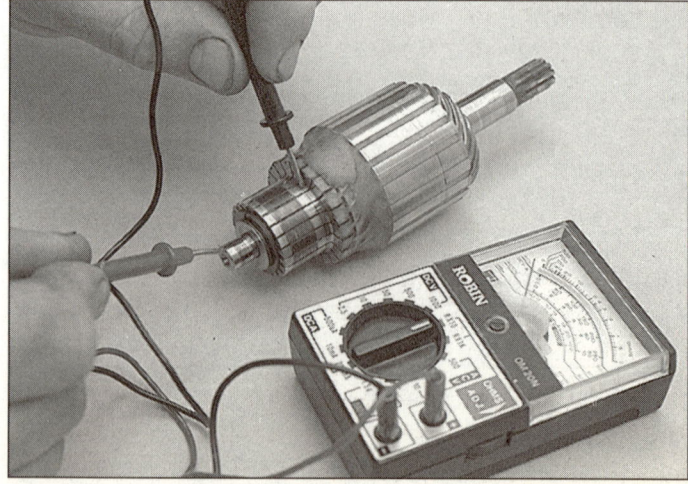

21.11b . . . and for no continuity between each commutator bar and the armature shaft

21.16 Install the brush plate assembly and terminal bolt

21.17 Install the terminal bolt nut and tighten it securely

21.19a Install the shims on the armature, then insert the armature . . .

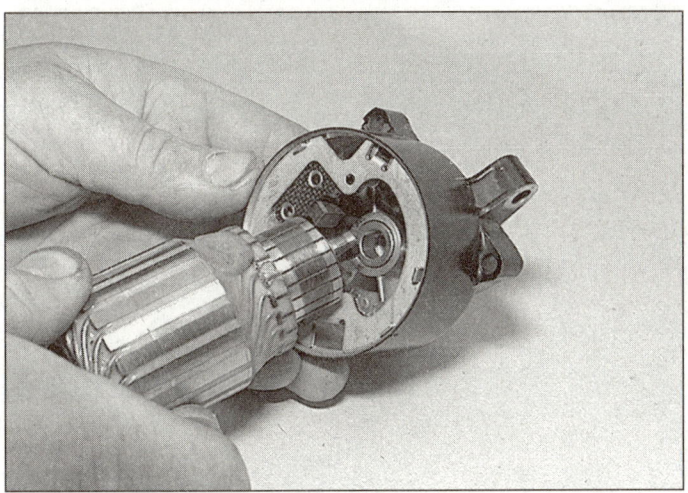

21.19b . . . and locate the brushes on the commutator

21.20 Fit the lockwasher to the front cover so its teeth engage the cover ribs

the bushing is worn or damaged.

14   Check the starter pinion for worn, chipped or broken teeth. If the gear is damaged or worn, replace the starter motor. Inspect the insulating washers for signs of damage and replace if necessary.

Reassembly

Refer to illustrations 21.16, 21.17, 21.19a, 21.19b, 21.20, 21.21, 21.22 and 21.23

15   Lift the brush springs and slide the brushes back into position in their holders.

16   Fit the insulator to the housing and install the brush plate. Insert the terminal bolt through the brush plate and housing **(see illustration)**.

17   Slide the rubber ring and small insulating washer onto the bolt, followed by the large insulating washers and the plain washer. Fit the nut to the terminal bolt and tighten it securely **(see illustration)**.

18   Locate the brush assembly in the housing, making sure its tab is correctly located in the housing slot.

19   Fit the shims to the armature shaft **(see illustration)** and insert the armature in the housing, locating the brushes to the commutator bars. Check that each brush is securely pressed against the commutator by its spring and is free to move easily in its holder **(see illustration)**.

20   Fit the lockwasher to the front cover so that its teeth are correctly located with the cover ribs **(see illustration)**. Apply a smear of grease to the cover dust seal lip.

21   Slide the shim(s) onto the front end of the armature shaft, then fit the insulating washer **(see illustration)**. Fit the sealing ring to the housing and carefully slide the front cover into position, aligning the marks made on removal.

21.21 Install the shims and washer on the armature, making sure they're in the correct order

8

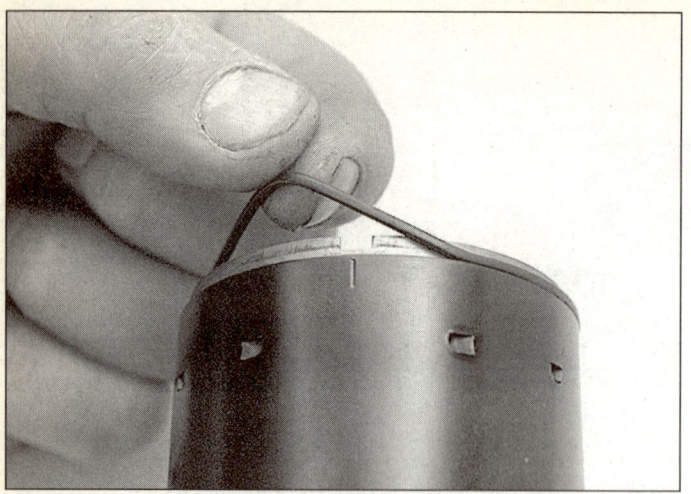

21.22 Fit the O-ring . . .

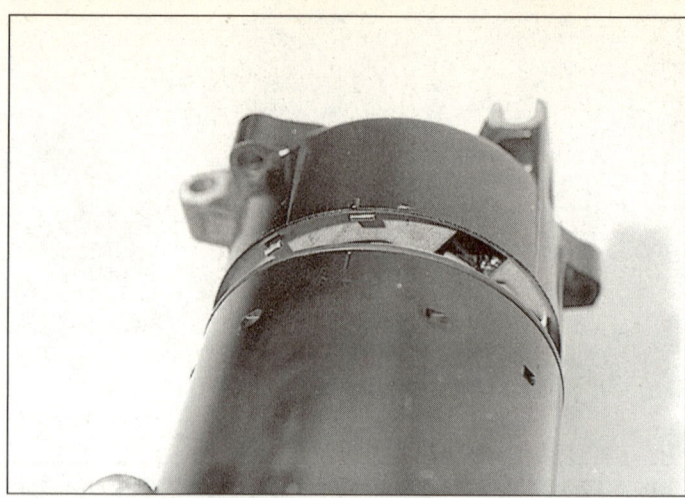

21.23 . . . and install the rear cover, aligning its groove with the brush plate outer tab

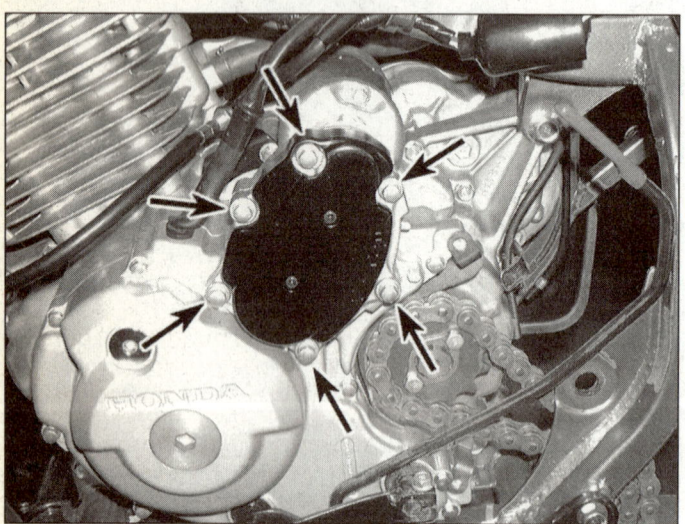

22.1a Remove the TRX300EX starter gear cover bolts (arrows), the cover and gasket . . .

22.1b . . . the dowels (arrows) and gears may come off with the cover or stay in the engine

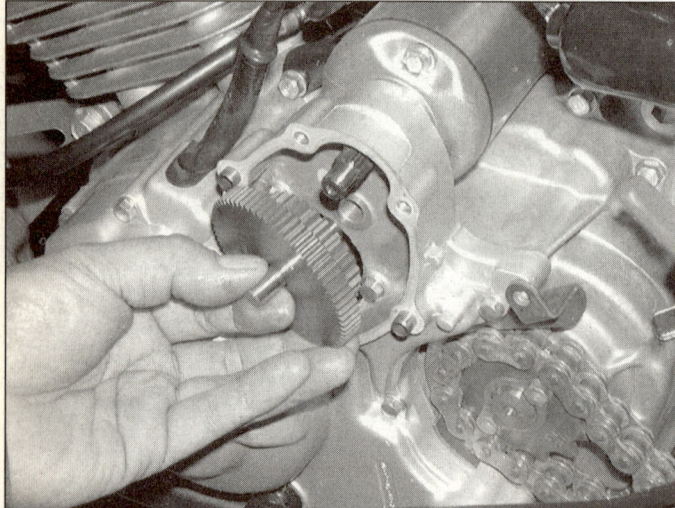

22.2 Remove the starter reduction gear and shaft

22   Ensure the brush plate inner tab is correctly located in the housing slot and fit the O-ring to the housing **(see illustration)**.

23   Align the rear cover groove with the brush plate outer tab and install the cover **(see illustration)**.

24   Check that the marks made on disassembly are correctly aligned, then fit the long bolts and tighten them securely **(see illustration 21.3)**.

25   Install the starter as described in Section 20.

---

22   Starter reduction gears - removal, inspection, bearing replacement and installation

TRX300EX models

Removal

Refer to illustrations 22.1a, 22.1b, 22.2, 22.3a, 22.3b and 22.3c

1   Remove the reduction gear cover, dowel pins and gasket **(see illustrations)**. Sometimes the smaller reduction gears come off with the cover, sometimes they don't. If they didn't come off with the cover, remove them now.

2   Remove the big starter reduction gear and the reduction gear shaft **(see illustration)**.

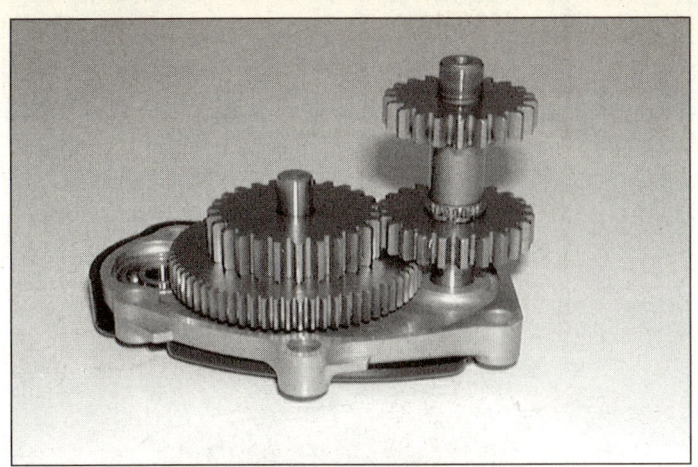

22.3a The assembled reduction gears look like this

22.3b Pull out the shaft and snap-ring . . .

22.3c . . . then pull out the remaining gears and washer

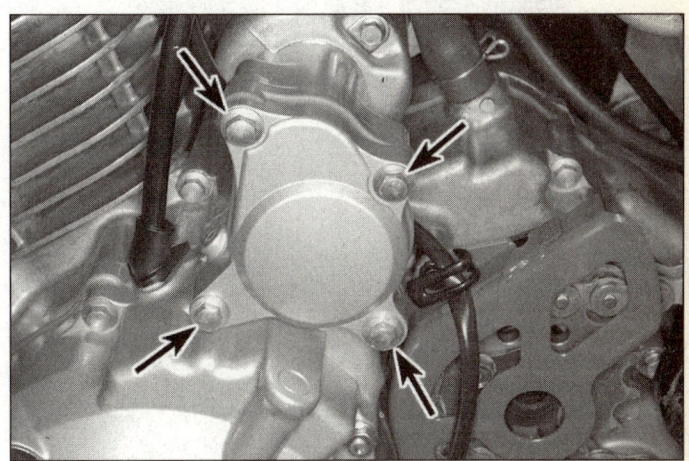

22.16 Remove the TRX400EX starter idle gear cover bolts
(arrows), then remove the cover, gasket and dowels

3   If the gears came off with the cover, remove them now **(see illustrations)**.

Inspection

4   Check the gears for worn or broken teeth. Check the shafts and the friction surface on the gears for wear or damage and replace any parts that show defects.

5   Turn the bearing with a finger. If it's rough, loose or noisy, replace it.

Bearing replacement

6   Remove the plastic protector from the aluminum reduction gear cover.

7   Heat the cover in an oven, then hold it with the bearing facing down and tap around the bearing with a soft faced mallet until the bearing drops out.

8   Tap a new bearing in with a socket or bearing driver that bears against the bearing outer race.

9   Install the plastic protector on the reduction gear cover.

Installation

10   Remove all old gasket and sealant from the cover and crankcase.

11   Install the big starter reduction gear and shaft **(see illustration 22.2)**.

12   Install the smaller reduction gears on the cover **(see illustration 22.3a)**.

13   Install a new gasket and make sure the dowels are in position **(see illustration 22.1c)**.

14   Install the cover and the gears **(see illustration 22.1b)**.

15   Install the cover bolts and tighten them evenly in a criss-cross pattern to the torque listed in this Chapter's Specifications.

TRX400EX models

Removal

Refer to illustration 22.16

16   Remove the four starter idle gear cover bolts **(see illustration)** and remove the cover, gasket and dowel pins.

17   Slide the spacer off the starter idle gear shaft, then pull out the shaft and remove the gear from the left crankcase cover.

18   Remove the seat and rear fenders (see Chapter 7).

19   Disconnect the alternator connectors, which are located on the left side of the machine, in the rubber boot above the battery **(see illustration 4.5 in Chapter 4)**.

20   Remove the gearshift pedal from the gearshift spindle (see Section 20 in Chapter 2B).

21   Remove the left crankcase cover (see Section 26).

22   Remove the starter reduction gear and shaft out of the gear (it's above and to the right of the alternator rotor), and lift the gear away from the engine.

Inspection

23   Inspect the gears for worn and broken teeth.

24   Inspect the shafts and the friction surface on the gears for wear and damage.

25   Replace all worn or damaged parts.

23.2 While holding the rotor, the gear should turn smoothly clockwise, but not at all counterclockwise

23.3 Lift the driven gear off the rotor

23.4a Remove the Torx bolts (TRX300EX shown; the TRX400EX bolts are removed from the flywheel side) . . .

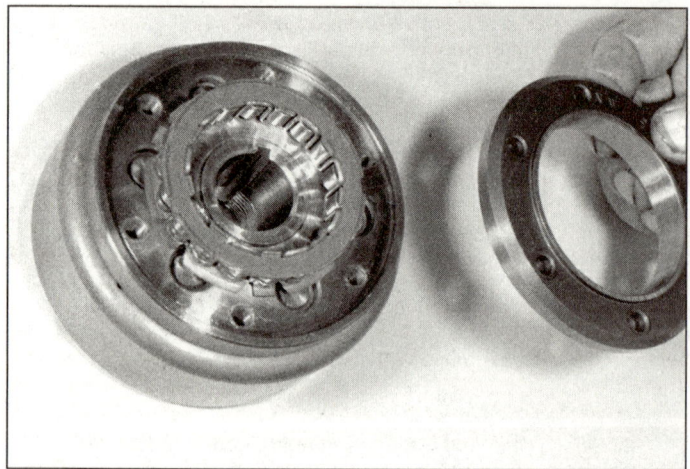

23.4b . . . then remove the outer clutch and the roller assembly

Installation

26   Remove all old gasket and sealant from the cover and crankcase.
27   Position the starter reduction gear over its shaft hole in the crankcase, then slip the shaft through the gear and into the hole.
28   Install the left crankcase cover (see Section 26).
29   Install the gearshift pedal (see Section 20 in Chapter 2B).
30   Plug in the alternator connectors.
31   Install the seat and rear fenders (see Chapter 7).
32   Install the idle gear shaft, then install the starter idle gear and spacer on the shaft.
33   Install the dowel pins, a new starter idle cover gasket and the starter idle cover. Install the cover bolts and tighten them securely.

23 Starter clutch - removal, inspection and installation

Refer to illustrations 23.2, 23.3, 23.4a and 23.4b
1   Remove the left crankcase cover and rotor (see Section 26). The starter clutch is mounted on the back of the alternator rotor.
2   Hold the alternator rotor with one hand and try to rotate the starter driven gear with the other hand **(see illustration)**. It should rotate clockwise smoothly, but not rotate counterclockwise (anti-clockwise) at all.
3   If the gear rotates both ways or neither way, or if its movement is rough, remove it from the alternator rotor **(see illustration)**.
4   Remove the Torx bolts and lift off the outer clutch **(see illustra-**

tions). Inspect the outer clutch and one-way clutch for wear and damage and replace them if any problems are found.
5   Installation is the reverse of the removal steps. Before you tighten the Torx bolts, turn the starter driven gear (see Step 2). It should rotate clockwise only; if it rotates counterclockwise (anti-clockwise), the one-way clutch is installed backwards. Apply non-permanent thread locking agent to the threads of the Torx bolts and tighten them to the torque listed in this Chapter's Specifications.

24 Charging system testing - general information and precautions

1   If the performance of the charging system is suspect, the system as a whole should be checked first, followed by testing of the individual components (the alternator and the regulator/rectifier). **Note:** *Before beginning the checks, make sure the battery is fully charged and that all system connections are clean and tight.*
2   Checking the output of the charging system and the performance of the various components within the charging system requires the use of a voltmeter, ammeter and ohmmeter or the equivalent multimeter.
3   When making the checks, follow the procedures carefully to prevent incorrect connections or short circuits, as irreparable damage to electrical system components may result if short circuits occur.
4   If the necessary test equipment is not available, it is recommended that charging system tests be left to a dealer service department or a reputable motorcycle repair shop.

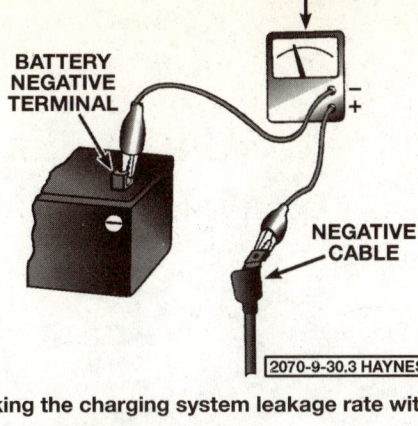

25.3 Checking the charging system leakage rate with an ammeter

25 Charging system - leakage and output test

1 If a charging system problem is suspected, perform the following checks. Start by removing the seat and rear fenders (see Chapter 7).

Leakage test

Refer to illustration 25.3

2 Turn the ignition switch Off and disconnect the cable from the battery negative terminal.

3 Set the multimeter to the mA (milliamperes) function and connect its negative probe to the battery negative terminal, and the positive probe to the disconnected negative cable **(see illustration)**. Compare the reading to the leakage limit listed in this Chapter's Specifications.

4 If the indicated leakage exceeds the specified leakage limit, there is probably a short circuit in the wiring. Thoroughly check the wiring between the various components (see the wiring diagrams at the end of the book).

5 If the indicated leakage is less than the specified leakage limit, the leakage rate is satisfactory. Disconnect the meter and connect the negative cable to the battery, tightening it securely. Check the alternator output as described below.

Output test

6 Start the engine and let it warm up to normal operating temperature.

7 With the engine idling, attach the positive (red) voltmeter lead to the positive (+) battery terminal and the negative (black) lead to the battery negative (-) terminal. The voltmeter selector switch (if equipped) must be in the 0-20 DC volt range.

8 Slowly increase the engine speed to 5000 rpm and compare the voltmeter reading to the value listed in this Chapter's Specifications.

9 If the output is as specified, the alternator is functioning properly.

10 Low voltage output may be the result of damaged windings in the alternator stator coils or wiring problems. Make sure all electrical connections are clean and tight, then refer to the following Sections to check the alternator stator coils and the regulator/rectifier.

11 High voltage output (above the specified range) indicates a defective voltage regulator/rectifier. Refer to Section 27 for regulator testing and replacement procedures.

26 Alternator stator coils and rotor - check and replacement

Stator (charging) coil resistance check

Refer to illustration 26.2

1 Locate and disconnect the alternator coil connector on the left side of the vehicle, in the rubber boot above the battery. On TRX300EX

26.2 Check stator coil resistance at the alternator connector (TRX300EX shown)

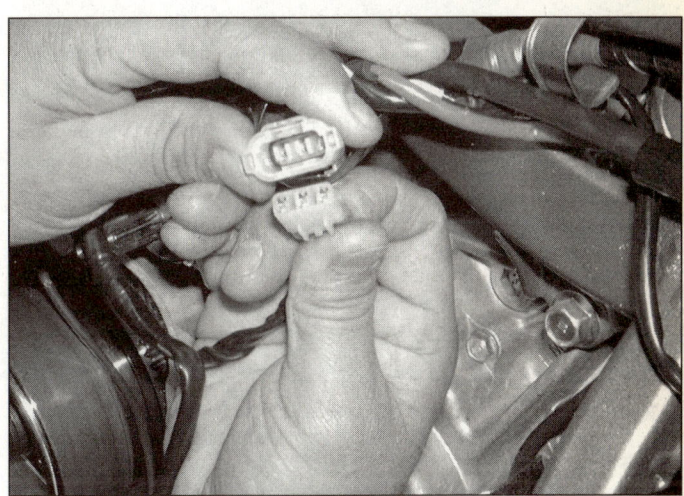

26.5 On TRX300EX models, the alternator/ignition pulse generator connector is a green three-pin connector

models, look for a three-pin round white connector with three yellow wires. On TRX400EX models, look for a four-pin connector with a blue/yellow wire, a green wire, a pink wire and a yellow wire.

2 Using an ohmmeter, check stator coil resistance. On TRX300EX models, measure the resistance between the connector terminals for the three yellow wires **(see illustration)**, on the alternator side of the connector. On TRX400EX models, measure the resistance between the connector terminals for the pink wire and the yellow wire, on the alternator side of the connector. Compare your measurements to the stator coil resistance listed in this Chapter's Specifications. If the readings are not within this range, replace the stator coil.

3 Connect the ohmmeter between a good ground on the vehicle and each of the alternator connector terminals. The meter should indicate infinite resistance between each terminal and ground. If there's continuity between any of the terminals and ground, replace the stator coil.

Replacement

TRX300EX models

Stator coil

Refer to illustrations 26.5, 26.10, 26.11, 26.12 and 26.13

4 Drain the engine oil (see Chapter 1).

5 Disconnect the alternator connector **(see illustration 26.2)** and the green three-pin connector for the ignition pulse generator **(see illustration)**.

8

26.10 Left crankcase cover bolts (TRX300EX); use a new copper washer on the circled bolt

26.11 The dowel pins (arrows) may stay in the crankcase or cover

6   Remove the drive sprocket cover, the drive sprocket and the final drive chain (see Chapter 5).

7   Remove the gearshift pedal (see Section 22 in Chapter 2A).

8   Remove the reverse shift rod (see Section 19 in Chapter 2A).

9   Remove the starter reduction gear cover and the starter reduction gears (see Section 22).

10   Remove the left crankcase cover bolts **(see illustration)**. The rotor magnets may create some resistance, so pull firmly if necessary. Don't force the cover off, though, and don't pry against the gasket surfaces; if the cover won't come off, make sure all fasteners have been removed.

11   Remove the crankcase cover gasket and dowel pins **(see illustration)**.

12   To detach the stator coils from the cover, remove the wiring harness bolt and the three Allen bolts **(see illustration)**.

13   Installation is the reverse of the removal steps, with the following additions:

   a) *Tighten the stator coil Allen bolts to the torque listed in this Chapter's Specifications.*

   b) *Remove all old gasket material from the left crankcase cover and from the gasket surface of the crankcase. Use a new gasket on the left crankcase cover and smear a film of sealant across the wiring harness grommet* **(see illustration)**.

   c) *Make sure the cover dowels are in position.*

   d) *Be sure to install a new sealing washer on the circled cover bolt* **(see illustration 26.10)**.

   e) *Tighten the cover bolts evenly, in a criss-cross pattern, to the torque listed in this Chapter's Specifications.*

Rotor

Refer to illustrations 26.15a, 26.15b, 26.16a, 26.16b, 26.17, 26.18a and 26.18b

Note: *To remove the alternator rotor, a special Honda puller (part no. 07733-0020001 or 07933-3950000) or an aftermarket equivalent will be required. Don't try to remove the rotor without the right puller, because it will be damaged unless it's removed correctly. Pullers are readily available from motorcycle dealers and aftermarket tool suppliers.*

14   Remove the left crankcase cover (Steps 4 through 11).

15   To loosen the rotor bolt, lock the rotor in place with a strap wrench **(see illustration)**. If you don't have one and the engine is in the frame, the rotor can be locked by placing the transmission in gear and holding the rear brake on. Unscrew the rotor bolt and remove the washer **(see illustration)**.

16   Thread an alternator puller into the center of the rotor and use it to remove the rotor **(see illustration)**. If the rotor doesn't come off easily, tap sharply on the end of the puller to release the rotor's grip on the

26.12 Remove the wiring harness bolt and stator coil Allen bolts (arrows)

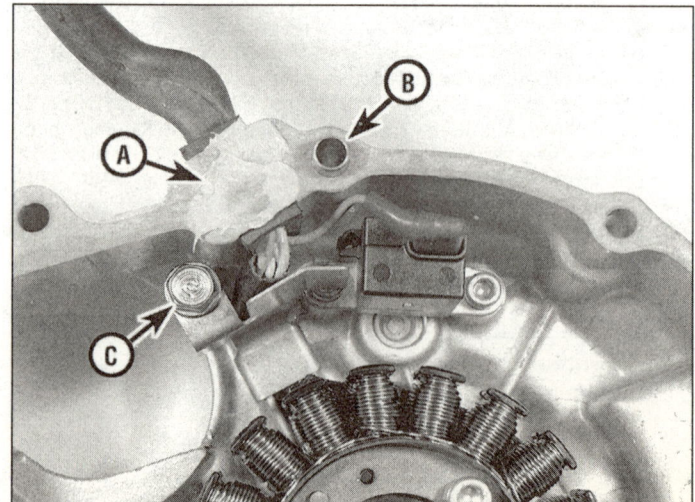

26.13 Smear sealant across the grommet (A); this dowel (B) stayed in the cover; install the wiring harness bolt (C)

26.15a Use a strap wrench to hold the rotor so that you can break loose the rotor bolt

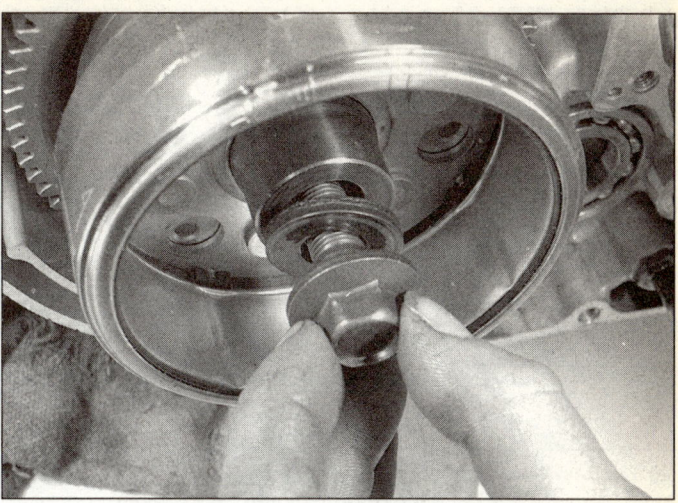

26.15b Remove the rotor bolt and washer

26.16a Remove the rotor with a puller designed for the purpose

26.16b Pull the rotor off together with the starter clutch and driven gear

tapered crankshaft end. Pull off the rotor off, together with the starter clutch and starter driven gear **(see illustration)**.

17   Remove the needle roller bearing and washer from the crankshaft **(see illustration)**.

18   If the Woodruff key **(see illustration)** is not secure in its slot, pull it out and set it aside for safekeeping. A convenient method is to stick the Woodruff key to the magnets inside the rotor **(see illustration)**, but

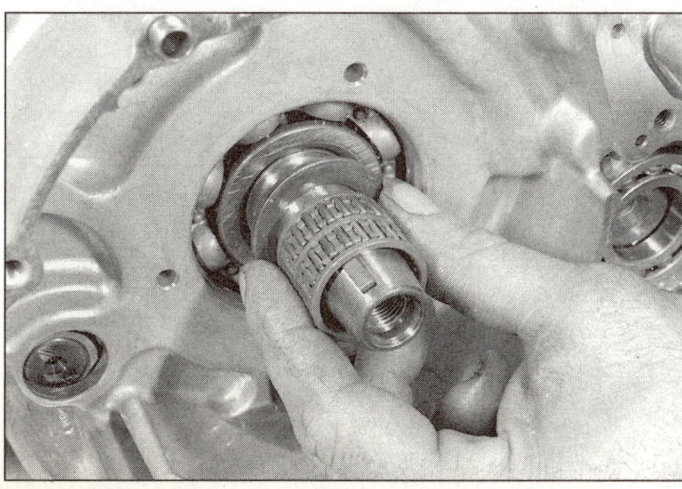

26.17 Remove the needle roller bearing and washer

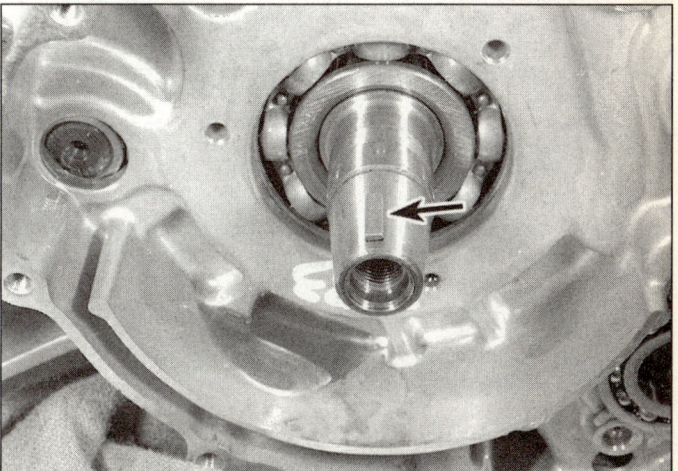

26.18a Make sure the Woodruff key (arrow) is securely in its slot before installing the rotor

8

26.18b Make sure there's nothing (like this Woodruff key) stuck to the rotor magnets

26.28 The TRX400EX left crankcase cover bolts (arrows) and nut (circled)

make SURE you don't forget it's there! The rotor and stator coils will be seriously damaged if the engine is started with anything stuck to the magnets.

19   Degrease the center of the rotor and the end of the crankshaft.

20   Make sure the Woodruff key is positioned securely in its slot **(see illustration 26.18a)**.

21   Align the rotor slot with the Woodruff key. Place the rotor, together with the starter clutch and starter driven gear, on the crankshaft.

22   Install the washer and rotor bolt. Immobilize the rotor from turning with one of the methods described in Step 15 and tighten the bolt to the torque listed in this Chapter's Specifications.

23   Take a look to make sure there isn't anything stuck to the inside of the rotor **(see illustration 26.18b)**.

24   Install the left crankcase cover (see Steps 4 through 11).

TRX400EX models

Stator coil and ignition pulse generator assembly

Refer to illustrations 26.28 and 26.30

25   Remove the starter idle gear cover and remove the spacer, starter idle gear and shaft (see Section 22).

26   Disconnect the alternator connectors, which are located on the left side of the machine, in the rubber boot above the battery **(see illustration 4.5 in Chapter 4)**.

27   Remove the gearshift pedal from the gearshift spindle (see Section 20 in Chapter 2B).

28   Remove the seven left crankcase cover bolts and the single nut **(see illustration)**. Remove the dowel pins and the old gasket.

29   Remove the starter reduction gear and shaft (see Section 22).

30   Remove the stator coil bolts, the ignition pulse generator bolts and the wire harness clamp bolt **(see illustration)** and remove the stator coil and ignition pulse generator assembly from the left crankcase cover.

31   Installation is the reverse of the removal steps, with the following additions:

    a)   *Tighten the stator coil bolts securely.*

    b)   *Remove all old gasket material from the left crankcase cover and from the gasket surface of the crankcase. Use a new gasket on the left crankcase cover and smear a film of sealant across the wiring harness grommet.*

    c)   *Make sure the cover dowels are in position.*

    d)   *Be sure to install the nut at the circled position* **(see illustration 26.28)**.

    e)   *Tighten the cover bolts gradually and evenly, in a criss-cross pattern (there is Not specified for these bolts).*

Rotor

32   Remove the left crankcase cover (see Steps 25 through 29).

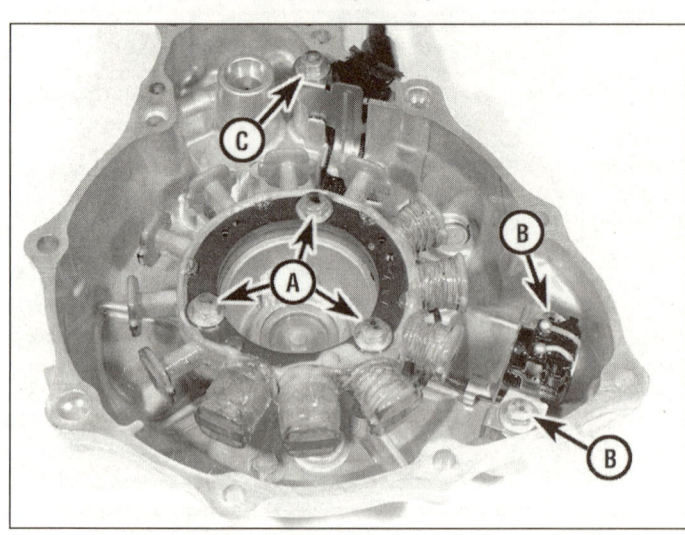

26.30 Stator coil and pulse generator (TRX400EX)

    a)   *Stator coil bolts*

    b)   *Ignition pulse generator bolts*

    c)   *Harness retainer bolt*

33   Remove the rotor (see Steps 15 through 23).

34   Installation is the reverse of removal.

---

27   Voltage regulator/rectifier - check and replacement

Check

Refer to illustrations 27.1, 27.2a, 27.2b and 27.3

1   The regulator/rectifier is mounted on the left side of the vehicle frame in front of the battery **(see illustration)**.

2   On TRX300EX models, follow the wiring leads from the regulator/rectifier to the connectors and unplug the connectors **(see illustration)**. On TRX400EX models, unplug the six-pin connector from the regulator/rectifier **(see illustration)**. Inspect the connector(s) for corrosion, dirt and a loose fit. The connector(s) must be in good shape to test the regulator/rectifier.

3   Verify that there is battery voltage to the regulator/rectifier. On TRX300EX models, connect a voltmeter between the single-pin connector terminals for the red wire and the green/black wire **(see illustra-**

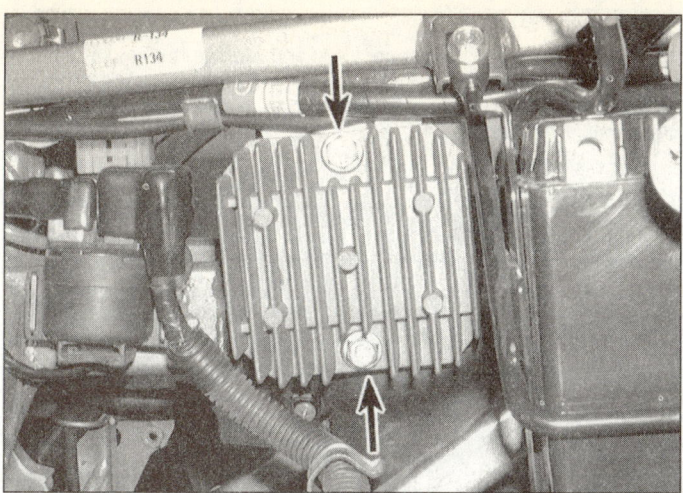

27.1 Remove the bolts (arrows) to detach the TRX300EX regulator/rectifier from the left side of the frame

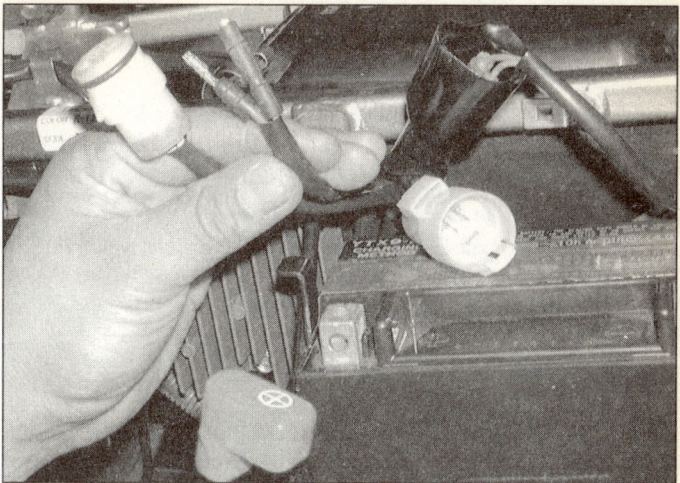

27.2a The TRX300EX regulator/rectifier has a pair of single-pin connectors and a round white three-pin connector

27.2b Disconnect the six-pin connector and remove the bolts (arrows) to detach the TRX400EX regulator/rectifier

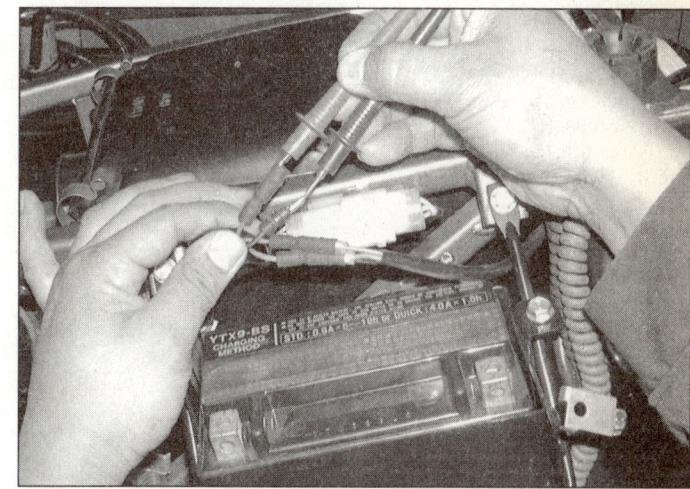

27.3 Check the TRX300EX regulator/rectifier for battery voltage at the harness side of the red wire and green/black wire connectors

tion), on the harness side of the connectors; it should indicate battery voltage. On TRX400EX models, connect the voltmeter leads between the connector terminal for the red wire, on the harness side of the connector, and ground; it should indicate battery voltage.

4　Verify that there is continuity on the ground side of the regulator/rectifier. On TRX300EX models, connect an ohmmeter to the connector terminal for the green/black wire, on the harness side of the connector, and ground; there should be continuity. On TRX400EX models, connect an ohmmeter to the connector terminal for the green wire, on the harness side of the connector, and ground; there should be continuity.

5　If there is no battery voltage to the regulator/rectifier, or continuity to ground, inspect the wiring harness for breaks or poor connections, make any necessary repairs and repeat the above tests.

6　If the wiring and connectors are in good shape, check the resistance between the stator (charging) coils and verify that none of the stator coils are shorted to ground (see Section 26).

7　On TRX400EX models, check the voltage feedback line. Measure the voltage between the terminal for the black wire and ground. There should be battery voltage with the ignition switch turned to ON, but no voltage with the ignition switch turned to OFF.

8　If everything else checks out, the regulator/rectifier is probably defective, by a process of elimination. No further testing of the regulator/rectifier on TRX400EX models is possible at home; have it tested by a Honda dealer.

9　The internal resistance of the regulator/rectifier on TRX300EX models can be checked at home, but only if you have one of the two multimeters specified by Honda (Kowa digital 07411-0020000 or Sanwa analog 07308-0020001). According to Honda, other multimeters will provide incorrect readings. If you don't have one of these two multimeters, have the regulator/rectifier tested by a Honda dealer.

10　If you have one of the specified multimeters, make sure that the ohmmeter battery is in good condition; an old battery can cause inaccurate readings. And keep your fingers off the ohmmeter probes during the following tests; touching the probes will also produce inaccurate readings. If you're using a Sanwa multimeter, set the resistance range to "k-ohms." If you're using the Kowa multimeter, multiply all readings by 100. All of the following checks are conducted at the two connectors for the regulator/rectifier, on the regulator/rectifier side of the connectors.

11　Connect the negative ohmmeter probe to the red/white wire terminal and connect the positive probe to the following connectors in turn and note the readings:

a)　*To each of the terminals for the three yellow wires: infinite resistance*

b)　*To the terminal for the green wire: infinite resistance*

12　Connect the ohmmeter negative probe to one yellow wire terminal, then connect the positive probe to each of the following terminals

8

and note the readings:

    a) *To the terminal for the red/white wire: 500 to 10 k-ohms*
    b) *To each of the two other terminals for yellow wires: infinite resistance*
    c) *To the terminal for the green wire: infinite resistance*

13   Repeat the test in Steps 12 for each of the other two yellow wire terminals (negative probe to second yellow wire terminal and positive probe to a, b and c; negative probe to third yellow wire, positive probe to a, b and c).

14   Connect the ohmmeter negative probe to the green wire terminal, then connect the positive probe to each of the following terminals and note the readings:

    a) *To the terminal for the red/white wire: 700 to 15 k-ohms*
    b) *To each of the yellow wire terminals: 500 to 10 k-ohms*

15   If the TRX300EX regulator/rectifier doesn't check out as described, replace it.

Replacement

16   Unplug the electrical connector(s) and remove the regulator/rectifier retaining bolts **(TRX300EX models, see illustrations 27.1 and 27.2a; TRX400EX models, see illustration 27.2b)**. Remove the regulator/rectifier.

17   Installation is the reverse of removal.

28   Wiring diagrams

Prior to troubleshooting a circuit, check the fuses to make sure they're in good condition. Make sure the battery is fully charged and check the cable connections.

When checking a circuit, make sure all connectors are clean, with no broken or loose terminals or wires. When unplugging a connector, don't pull on the wires - pull only on the connector housings themselves.

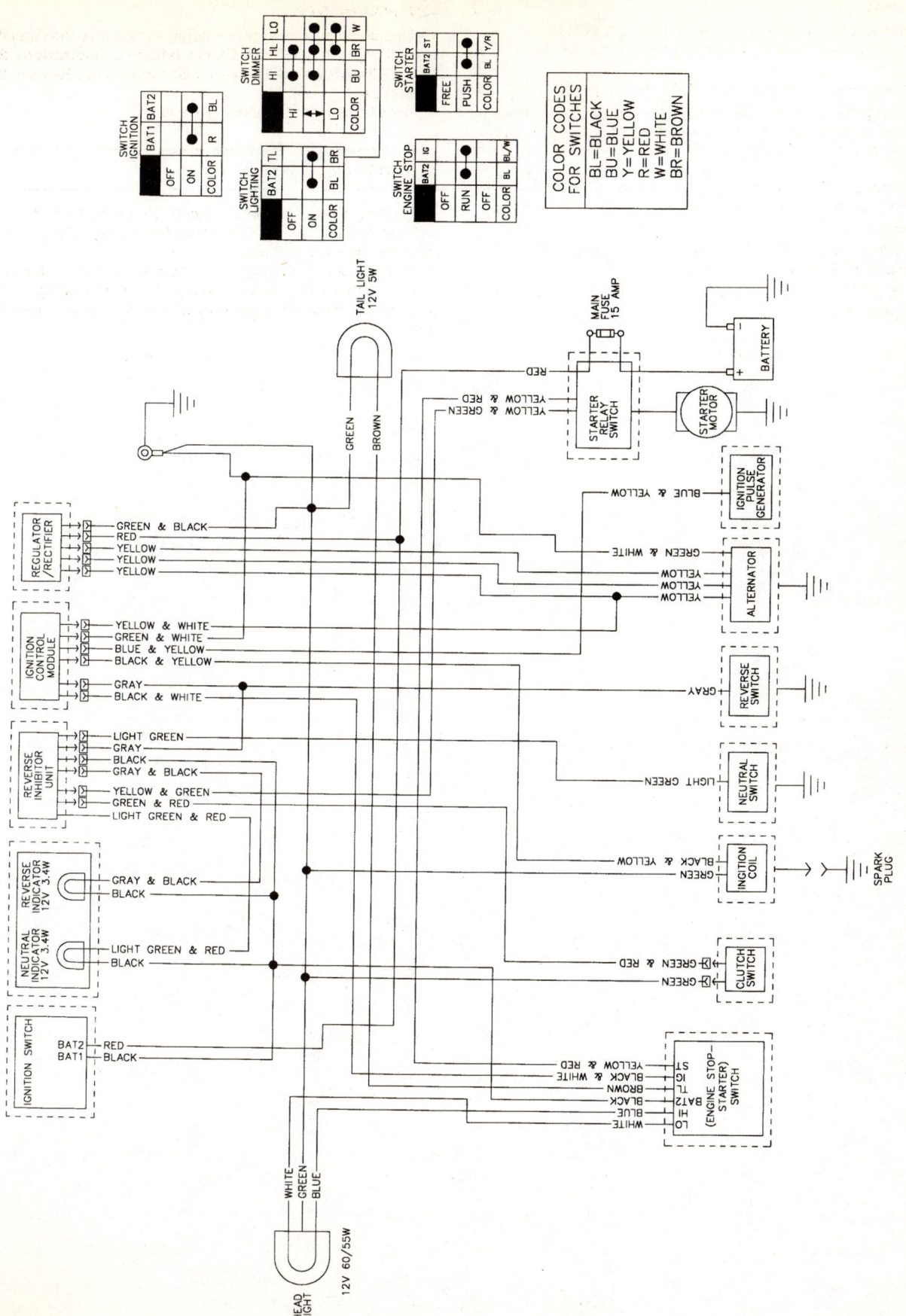

Wiring diagram - TRX300EX

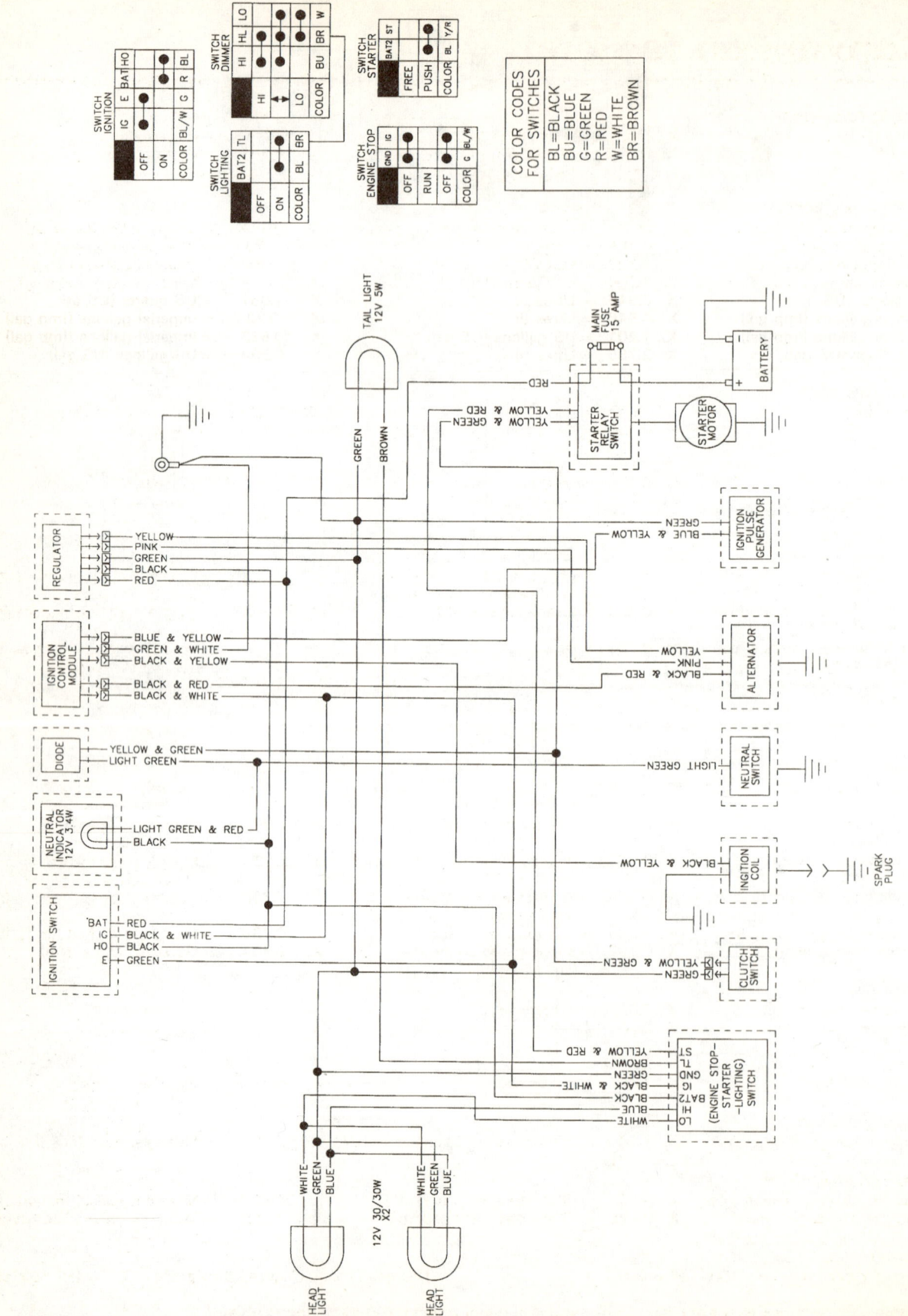

Wiring diagram - TRX400EX

Conversion factors

Length (distance)
Inches (in)	X	25.4	= Millimetres (mm)	X	0.0394	= Inches (in)
Feet (ft)	X	0.305	= Metres (m)	X	3.281	= Feet (ft)
Miles	X	1.609	= Kilometres (km)	X	0.621	= Miles

Volume (capacity)
Cubic inches (cu in; in^3)	X	16.387	= Cubic centimetres (cc; cm^3)	X	0.061	= Cubic inches (cu in; in^3)
Imperial pints (Imp pt)	X	0.568	= Litres (l)	X	1.76	= Imperial pints (Imp pt)
Imperial quarts (Imp qt)	X	1.137	= Litres (l)	X	0.88	= Imperial quarts (Imp qt)
Imperial quarts (Imp qt)	X	1.201	= US quarts (US qt)	X	0.833	= Imperial quarts (Imp qt)
US quarts (US qt)	X	0.946	= Litres (l)	X	1.057	= US quarts (US qt)
Imperial gallons (Imp gal)	X	4.546	= Litres (l)	X	0.22	= Imperial gallons (Imp gal)
Imperial gallons (Imp gal)	X	1.201	= US gallons (US gal)	X	0.833	= Imperial gallons (Imp gal)
US gallons (US gal)	X	3.785	= Litres (l)	X	0.264	= US gallons (US gal)

Mass (weight)
Ounces (oz)	X	28.35	= Grams (g)	X	0.035	= Ounces (oz)
Pounds (lb)	X	0.454	= Kilograms (kg)	X	2.205	= Pounds (lb)

Force
Ounces-force (ozf; oz)	X	0.278	= Newtons (N)	X	3.6	= Ounces-force (ozf; oz)
Pounds-force (lbf; lb)	X	4.448	= Newtons (N)	X	0.225	= Pounds-force (lbf; lb)
Newtons (N)	X	0.1	= Kilograms-force (kgf; kg)	X	9.81	= Newtons (N)

Pressure
Pounds-force per square inch (psi; lbf/in^2; lb/in^2)	X	0.070	= Kilograms-force per square centimetre (kgf/cm^2; kg/cm^2)	X	14.223	= Pounds-force per square inch (psi; lbf/in^2; lb/in^2)
Pounds-force per square inch (psi; lbf/in^2; lb/in^2)	X	0.068	= Atmospheres (atm)	X	14.696	= Pounds-force per square inch (psi; lbf/in^2; lb/in^2)
Pounds-force per square inch (psi; lbf/in^2; lb/in^2)	X	0.069	= Bars	X	14.5	= Pounds-force per square inch (psi; lbf/in^2; lb/in^2)
Pounds-force per square inch (psi; lbf/in^2; lb/in^2)	X	6.895	= Kilopascals (kPa)	X	0.145	= Pounds-force per square inch (psi; lbf/in^2; lb/in^2)
Kilopascals (kPa)	X	0.01	= Kilograms-force per square centimetre (kgf/cm^2; kg/cm^2)	X	98.1	= Kilopascals (kPa)

Torque (moment of force)
Pounds-force inches (lbf in; lb in)	X	1.152	= Kilograms-force centimetre (kgf cm; kg cm)	X	0.868	= Pounds-force inches (lbf in; lb in)
Pounds-force inches (lbf in; lb in)	X	0.113	= Newton metres (Nm)	X	8.85	= Pounds-force inches (lbf in; lb in)
Pounds-force inches (lbf in; lb in)	X	0.083	= Pounds-force feet (lbf ft; lb ft)	X	12	= Pounds-force inches (lbf in; lb in)
Pounds-force feet (lbf ft; lb ft)	X	0.138	= Kilograms-force metres (kgf m; kg m)	X	7.233	= Pounds-force feet (lbf ft; lb ft)
Pounds-force feet (lbf ft; lb ft)	X	1.356	= Newton metres (Nm)	X	0.738	= Pounds-force feet (lbf ft; lb ft)
Newton metres (Nm)	X	0.102	= Kilograms-force metres (kgf m; kg m)	X	9.804	= Newton metres (Nm)

Vacuum
Inches mercury (in. Hg)	X	3.377	= Kilopascals (kPa)	X	0.2961	= Inches mercury
Inches mercury (in. Hg)	X	25.4	= Millimeters mercury (mm Hg)	X	0.0394	= Inches mercury

Power
Horsepower (hp)	X	745.7	= Watts (W)	X	0.0013	= Horsepower (hp)

Velocity (speed)
Miles per hour (miles/hr; mph)	X	1.609	= Kilometres per hour (km/hr; kph)	X	0.621	= Miles per hour (miles/hr; mph)

Fuel consumption*
Miles per gallon, Imperial (mpg)	X	0.354	= Kilometres per litre (km/l)	X	2.825	= Miles per gallon, Imperial (mpg)
Miles per gallon, US (mpg)	X	0.425	= Kilometres per litre (km/l)	X	2.352	= Miles per gallon, US (mpg)

Temperature
Degrees Fahrenheit = (°C x 1.8) + 32   Degrees Celsius (Degrees Centigrade; °C) = (°F - 32) x 0.56

8

It is common practice to convert from miles per gallon (mpg) to litres/100 kilometres (l/100km), where mpg (Imperial) x l/100 km = 282 and mpg (US) x l/100 km = 235

Notes

Index